Inclusive STEM
Transforming Disciplinary Writing Instruction for a Socially Just Future

Across the Disciplines Books

Series Editor: Michael A. Pemberton
Associate Editor: Kathryn M. Northcut

The Across the Disciplines Books series is closely tied to published themed issues of the online, open-access, peer-reviewed journal *Across the Disciplines*. In keeping with the editorial mission of *Across the Disciplines*, books in the series are devoted to language, learning, academic writing, and writing pedagogy in all their intellectual, political, social, and technological complexity.

The WAC Clearinghouse and University Press of Colorado are collaborating so that these books will be widely available through free digital distribution and low-cost print editions. The publishers and the series editors are committed to the principle that knowledge should freely circulate and have embraced the use of technology to support open access to scholarly work.

Other Books in This Series

Jonathan Hall and Bruce Horner (Eds.), *Toward a Transnational University: WAC/WID Across Borders of Language, Nation, and Discipline* (2023)

Marilee Brooks-Gillies, Elena G. Garcia, Soo Hyon Kim, Katie Manthey, and Trixie G. Smith (Eds.), *Graduate Writing Across the Disciplines: Identifying, Teaching, and Supporting* (2020)

Steven J. Corbett, Jennifer Lin LeMesurier, Teagan E. Decker, and Betsy Cooper (Eds.). *Writing In and About the Performing and Visual Arts: Creating, Performing, and Teaching* (2019).

Alice S. Horning, Deborah-Lee Gollnitz, and Cynthia R. Haller (Eds.). *What is College Reading?* (2017)

Frankie Condon and Vershawn Ashanti Young (Eds.), *Performing Antiracist Pedagogy in Rhetoric, Writing, and Communication* (2017)

Inclusive STEM
Transforming Disciplinary Writing Instruction for a Socially Just Future

Edited by Heather M. Falconer and LaKeisha McClary

The WAC Clearinghouse
wac.colostate.edu
Fort Collins, Colorado

University Press of Colorado
upcolorado.com
Denver, Colorado

The WAC Clearinghouse, Fort Collins, Colorado 80523

University Press of Colorado, Denver, Colorado 80203

Copyright © 2024 by Heather M. Falconer and LaKeisha McClary. This work is licensed under a Creative Commons Attribution-NonCommercial-NoDerivatives 4.0 International License.

ISBN: 978-1-64215-236-4 (PDF) | 978-1-64215-237-1 (ePub) | 978-1-64642-687-4 (pbk.)

DOI: 10.37514/ATD-B.2024.2364

Library of Congress Cataloging-in-Publication Data

Names: Falconer, Heather M., 1974– editor | McClary, LaKeisha, 1979– editor

Title: Inclusive STEM : transforming disciplinary writing instruction for a socially just future / edited by Heather M. Falconer and LaKeisah McClary.

Description: Fort Collins, Colorado : The WAC Clearinghouse, University Press of Colorado | Series: Across the disciplines books | Includes bibliographical references.

Identifiers: LCCN 2024045609 (print) | LCCN 2024045610 (ebook) | ISBN 9781646426874 paperback | ISBN 9781642152364 adobe pdf | ISBN 9781642152371 epub

Subjects: LCSH: Technical writing—Study and teaching—Moral and ethical aspects | Minorities in engineering | Minorities in science | Discrimination in science | Discrimination in education | Social justice

Classification: LCC T11 .I475 2025 (print) | LCC T11 (ebook)

LC record available at https://lccn.loc.gov/2024045609

LC ebook record available at https://lccn.loc.gov/2024045610

Copyeditor: Samantha Maloney
Book Design: Mike Palmquist
Cover Art and Design: Raw Pixel Image 13225061. Licensed.
Series Editor: Michael A. Pemberton
Series Associate Editor: Kathryn M. Northcut

The WAC Clearinghouse supports teachers of writing across the disciplines. Hosted by Colorado State University, it brings together scholarly journals and book series as well as resources for teachers who use writing in their courses. This book is available in digital formats for free download at wac.colostate.edu.

Founded in 1965, the University Press of Colorado is a nonprofit cooperative publishing enterprise supported, in part, by Adams State University, Colorado State University, Fort Lewis College, Metropolitan State University of Denver, University of Alaska Fairbanks, University of Colorado, University of Denver, University of Northern Colorado, University of Wyoming, Utah State University, and Western Colorado University. For more information, visit upcolorado.com.

Citation Information: Falconer, Heather M. & LaKeisha McClary. (2024). Inclusive STEM: Transforming Disciplinary Writing Instruction for a Socially Just Future. The WAC Clearinghouse; University Press of Colorado. https://doi.org/10.37514/ATD-B.2024.2364

Land Acknowledgment. The Colorado State University Land Acknowledgment can be found at https://landacknowledgment.colostate.edu.

Contents

3   Introduction
         Heather M. Falconer

17   Section 1. Disrupting the Status Quo

19   Student Vignette
         Dhatri Badri

21   Student Vignette
         Riya Sharma

23   STEM Writing as Disruption: Views from First Year Writing
         Jameta Nicole Barlow and Kylie E. Quave

41   The Inclusive Potential of Teaching the History of (White Mainstream) English as the International Language of Science
         Elizabeth Blomstedt

59   "Science has always been about asking questions": Critical Science Literacy in STEM Writing
         Megan Callow and Holly Shelton

83   Integrating Social Justice Data and Scaffolded Writing with Universal Design Principles Into Introductory Statistics
         Laura Kyser Callis

101   A Curriculum Exploring Arab and Muslim Science: Opening Space for Other Epistemologies of Science
         Alicia Bitler and Ebtissam Oraby

123   Creating Assignments that Put Programmatic Inclusion and Diversity Work into Practice
         Justiss Wilder Burry, Carolyn Gubala, Jessica Griffith, Tanya Zarlengo, and Lisa Melonçon

147   Section 2. Challenging Orientations to Instruction and Assessment

149   Student Vignette
Madison Brown

151   Student Vignette
Madeline Dougherty

153   Promoting Inclusion Through Participation in and Construction of Engineering Judgments
Rachel C. Riedner, Royce A. Francis, and Marie C. Paretti

173   Engineering an Inclusive Integrated Writing Course
Jennifer C. Mallette

199   Putting Science in Black and White: Intensive Technical Writing Through Non-disposable Assignments as a Path for Decolonizing STEM
Sally B. Seraphin

223   Exploring Ungrading in a Biochemistry Laboratory Course
Jennifer Newell-Caito

247   A Call to Action for More Inclusive STEM
Janelle M. Johnson, Kimberlee Bourelle, Adrian Clifton, Mary Coleman, Parker Edingfield, Amanda Myers, Madeline Onstott, Joseph Schneiderwind, and Katie Weaver

263   Teaching Neuroethics in a Time of Crisis: Lessons in Liberatory Pedagogy
Ann E. Fink

285   Conclusion: Lessons from the Front Lines
LaKeisha McClary and Heather M. Falconer

309   Contributors

Inclusive STEM
Transforming Disciplinary Writing Instruction for a Socially Just Future

Introduction

Heather M. Falconer
University of Maine

In early 2021, LaKeisha McClary (co-editor of this collection and Assistant Professor of Chemistry at The George Washington University) and I found ourselves in a conversation about writing instruction in science, technology, engineering, and mathematics (STEM) disciplines, as well as the persistent, pernicious inequities that continue in these spaces. It was a casual conversation in the midst of a global pandemic where everyone seemed to be "pivoting" left, right, and center. Across academia, instructors were still sorting out how to move their large, in-person lectures into online modalities while simultaneously deciding if and how the social unrest being experienced nationally in the US[1] should find its way into classroom conversations. In our conversation, LaKeisha and I shared some of our own approaches toward creating inclusive spaces in our classrooms and the challenges we faced in doing so. It also involved a significant airing of grievances about our conditions of operation (e.g., institutional barriers, resistant faculty, resistant students), but much of it focused on steps toward improvement. What would a socially just future in STEM look like? What role might writing play in that future, and how could inclusive instruction be enacted in STEM spaces? How can we help STEM instructors be more equitable in their writing assessments and explicit in their instruction? What can be done, short of blowing the whole system up and starting all over again?

That conversation was the impetus for this book. We set out to highlight the ways in which this work can be done both in writing *and* disciplinary courses, providing a firsthand look at the types of interdisciplinary conversations we would love to see more of on campuses across the US. We also aimed for a bottom-up approach, one where the underlying assumption from day one was that equity should simply be part of the new normal. Making our classroom spaces accessible and welcoming to *all students* is just how operating in the 21st century should be. Part of that equity and accessibility is making explicit the ways in which the writing and meaning-making we do in our disciplines is unique and specific to our fields, and as such the *teaching* of those practices falls on anyone who is invested in language education and writing in STEM spaces. Hence, this book is both for those whose primary academic home is STEM as well as those who are focused on writing instruction.

---

1   I am referring, here, to the 2020 Presidential election, the storming of the U.S. Capitol Building, the very public murders of Black Americans at the hands of the police, and the marches around gender and LGBTQIA+ rights. All of these are tensions that continue to persist.

To be clear, this book is not arguing for the teaching of STEM through an interdisciplinary lens. We are not attempting to bridge the divide, for example, between STEM and the arts and humanities (i.e., a STEAM approach) or STEM and the public. There are already excellent collections taking up this work (see Kao & Kiernan, 2022 and Yu & Northcut, 2017, respectively, as recent examples). Similarly, we knew that another text arguing for the importance of disciplinary writing instruction itself was unnecessary. Effective communication skills have been recognized for some time as a critical aspect of being a STEM practitioner. Research has shown that explicitly teaching the ways in which language and forms of writing (i.e., genres) are representative of the various procedural and communicative tasks scientists and engineers regularly perform has positive impacts not only on persistence but on the development of disciplinary identity and agency—particularly for those from historically marginalized groups within those fields (Falconer, 2019a, 2019b; Hyland, 2012; Paretti et al., 2019; Poe et al., 2010). Accreditation boards and national STEM organizations have also recognized the necessary role of communication instruction in higher education: ABET Criterion 3 (2022–23) identifies the need for students in engineering, as well as applied and natural science programs, to develop an ability to communicate effectively with a range of audiences; the 2011 American Association for the Advancement of Science (AAAS) Vision and Change report identified the "ability to communicate and collaborate with other disciplines" (p. 15) as a core competency for biology undergraduates; the National Research Council (2012) has explicitly called out the need for further research into the ways educational conditions and strategies like writing across the curriculum (WAC) can "limit or promote metacognition" (p. 175) and have an impact on retention and persistence in STEM disciplines.

Compiling another text drawing attention to inequities in STEM was also not our goal. Concerns of equity, retention, and persistence for minoritized groups in STEM have been a topic of discussion for a considerable amount of time, with initiatives supported through the U.S. government (e.g., the National Science Foundation, President Obama's "STEM for All") as well as programs designed to offer high-impact practices like undergraduate research experiences. Scholars such as Ebony O. McGee (2020a, 2020b) have well-documented structural racism in U.S. STEM higher education and its impact on retention and persistence of Black students and scholars, particularly as it relates to performativity expectations (McGee & Martin, 2011). McGee and William H. Robinson (2020) have published compelling research into the ways in which inequity (both structural as well as social—i.e., microaggressions) impacts racial minorities in STEM, offering suggestions for remediation. Both Kathi N. Miner and colleagues (2018) and Mary Blair-Loy and Erin A. Cech (2022) have similarly examined the ways in which STEM inequity is structured as it relates to historically marginalized communities, highlighting the fact that epistemological and cultural beliefs perpetuate unequal and unfair outcomes. Remediating discrepancies must begin with a shift in lens in how the problems are viewed—from the individual's challenge to the group's responsibility. These conversations have been circulating for long enough that in a 2017 letter published in the journal *Science,* Amanda J. Zelmer and Aleksandra Sherman noted that "the failure of long-standing efforts to effect substantial change [in STEM diversity] reflects a deeper issue: the widespread cultural belief that science is neutral, objective, and apolitical" (p. 312–313). In their explicit call for STEM instructors to use culturally relevant teaching practices and materials in their classrooms to dismantle barriers, the authors asserted that "the idea that science is separate from social and cultural issues is flawed and alienates women and underrepresented minorities" (p. 313).

Yet, we found ourselves wondering to what degree instructors feel comfortable *doing* this work. How does engaging with these questions of ontology and epistemology force educators to confront what Mark Skopec and co-authors (2021) refer to as "epistemic fragility:" "an effortful reinstatement of an epistemic status quo, as a reaction against introducing ideas, narratives and research associated with decolonizing the higher education curriculum" (p. 3)? And what about resistances to writing instruction? Despite significant research related to writing and writing instruction in STEM and the recognized need for direct instruction, gaps continue to persist between WAC scholarship and its implementation in STEM education. Reynolds et al. (2012) have attributed this siloing of knowledge to a "lack of awareness of the research on the effectiveness of [WAC pedagogy], since most published findings are in journals not regularly read by STEM faculty and the majority of studies use methods unfamiliar to most scientists" (p. 18). More recently, research into STEM faculty beliefs related to writing illustrated reluctance due to understandings of what constitutes writing in their courses (Bathgate et al., 2019; Hora et al., 2019; Lund & Stains, 2015), whether writing is of benefit to students within these contexts (Thompson et al., 2021), and whether writing is even part of the knowledge-making process in their field (Gere et al., 2019; Moon et al., 2018). The humanistic aspect of writing—that it is a process of thinking and creating knowledge, not just a skill to document information, and is rooted in culture—often gets lost.

These are heavy challenges, to be sure. They don't have easy solutions, and they don't fall onto STEM instructors alone to resolve. Those who work with STEM students in writing courses and initiatives also bear some of this burden. From a writing studies perspective, we have known for some time that writing plays an important role in how knowledge is constructed and disseminated in STEM disciplines. For decades, scholars have examined the role of stases and topoi in scholarly arguments (e.g., Fahnestock & Secor, 1988; Wolfe et al., 2014), the ways in which language shapes how scientific knowledge is constructed and communicated (e.g., Bazerman, 2000, 1981; Myers, 1990, 1985), and various approaches to the incorporation of writing into STEM disciplinary spaces (e.g., Finkenstaedt-Quinn et al., 2021; Gallagher et al., 2020; Gere et al., 2019; Venters et al., 2018). In short, we have a good idea of how STEM researchers write and how those practices reify particular ways of knowing and doing. In writing studies, we also have a rich body of scholarship related to inequity (e.g., Condon & Young, 2016; Inoue, 2019; Poe et al., 2018), though that has not quite yet merged with the scholarship related to STEM from writing in the disciplines—and neither seems to have effectively crossed the disciplinary divide to reach STEM practitioners directly. Topics of writing in STEM journals, particularly as they relate to Discipline-Based Education Research (DBER), tend to focus on the use of inquiry-based writing in laboratories to improve students' critical thinking skills and knowledge acquisition (e.g., Badenhorst et al., 2020; Jeon et al., 2021; Larsen & Gärdebo, 2017), not explicitly to increase access.

It is with all of these questions and challenges in mind that we began cultivating the chapters that appear in this collection, as well as the vignettes that offer important insights into the lived experiences of students in STEM. We sought contributions that moved beyond typical disciplinary writing and content instruction and instead focused on work that was intentionally, sometimes subtly, disrupting the assumptions of STEM writing, communication, and knowledge-making. In our call for submissions, we asked contributors to think critically about how we create a sense of belonging for students from groups that have historically been kept out of these disciplines, how faculty can consciously create space for student voices to be heard, and specifically how we can do this with an eye toward discursive practices of STEM disciplines. Contributors were asked to offer us specific cases—classroom- or research-based contexts—that described their intents and goals, the interventions they enacted, how students responded, and the unexpected elements that presented themselves. We asked contributors to be self-reflective in ways that were transparent and showed the ugly bits; to share the lessons they learned and the errors they made.

In selecting chapters for this collection, we intentionally chose contributions that worked to disrupt the status quo, challenge assumptions, and embrace inclusive writing pedagogies. To be sure, these are not quick-fix solutions to appease the diversity, equity, and inclusion committees on campus, nor are they a one-off to allow instructors to check a box and feel that they have done their part. Rather, these chapters serve as entry points; they are the *beginning* of a conversation and set of practices that we hope educators and scholars will take up, expand on, and incorporate into programs so that, together, we can materialize a vision of a socially just future in STEM. We aim to create, as Rebecca Walton, Kristen R. Moore, and Natasha N. Jones (2019) have argued, spaces that "value ways of learning and knowing beyond [our] own and challenge complicity in oppressive intellectual practices" (p. 95).

Once chapters were accepted, we also circulated a request for vignettes from STEM students (either current or former) who had experiences that invoked a sense of belonging in their fields. This request, which was circulated via our authors as well as through social networks, resulted in short reflections about what helped make these students feel welcome in STEM spaces. These vignettes are included so that readers can see the power of microinclusions—subtle practices that tell our students that they are valued, that their perspective matters, and that they belong. The vignettes illustrate the ways that small changes and the creation of space can have lasting impacts on students from historically marginalized groups in STEM.

With this book, our goal is to create a "way in" for instructors in a wide range of disciplines to incorporate inclusive practices into STEM spaces—whether that is in a disciplinary writing classroom, teacher preparation program, traditional classroom, or undergraduate research. We seek to inspire, while also providing useful resources that can immediately be incorporated into existing courses and programs. This collection aims to show how meaningful change does not need to be drastic or involve tension or massive curricular reform. Simply modifying an assignment or replacing an assessment practice can create microinclusion opportunities. While we cannot change the system as a whole all at once, we *can* make efforts in the places we control (our classrooms and laboratories) to help counteract the negative messages students encounter elsewhere. Small efforts by individuals lead to larger, collective change.

Our Guiding Principles

As faculty who actively engage in interdisciplinary work, we began this project with certain assumptions about what instructors need—assumptions based specifically on U.S. educational contexts. We recognize that many of the inequities we experience in the US regarding STEM education are present in other countries, but we also recognize that different contexts and systems require different solutions and that some of our assumptions may not apply. We offer our assumptions here so that readers outside of the US can determine what applies and what does not, and those within can see how we are oriented.

Despite coming from very different fields (Heather from writing studies; LaKeisha from chemistry), we recognize some important considerations that impact this work:

- Faculty in STEM rarely have access to courses in pedagogy and, outside of WAC programs, typically do not receive instruction on how to teach disciplinary writing.
- Faculty in writing programs may have a firm grasp of writing pedagogy but not the disciplinary orientations or discourse knowledge to effectively teach STEM writing.
- There is often tension surrounding who has the authority—who is *allowed*—to teach disciplinary writing (the people who do it versus the people who study it).

- Equity and inclusion work is new to most instructors, and though interest is often there, a "way in" can be very hard to find.
- Balancing the course content for a traditional STEM class with writing instruction *and* inclusion work is a big lift.

In this collection, readers will find detailed information about the practices our authors have tested within their classroom spaces, as well as the relevant resources (reading lists, assignment prompts, etc.) that were used to effectively conduct the course. Importantly, we have asked authors to speak to the challenges they experienced in teaching the material, what they might change, and other frictions encountered or anticipated for the future. Our goal with these inclusions is to highlight the often-messy, imperfect ways in which inclusion *and* writing work gets done. We wish to destigmatize who is able to do this work, as well as offer some guidance in avoiding pitfalls. This collection is about action, not only theoretical orientations. We wish to offer actionable steps faculty can enact to make their STEM writing spaces more inclusive and challenge assumptions about disciplinary writing. We want readers to read a chapter, be inspired and empowered to modify the materials to fit their local context and try something new. That isn't to say that conscious, careful consideration of students and disciplinarity are not at the forefront. Rather, these considerations are already built into the chapters so that readers start at a place of accessibility and positive action.

At the same time that we strive for accessibility and positivity, we don't shy away from the hard truths. As Ann Fink notes in her chapter (this collection), "Practitioners must decide how and when they will resist oppressive practices around them, knowing that this also, inevitably, involves risk." Throughout this book, readers will encounter theoretical orientations and frameworks from a wide variety of disciplines. Some of these may be familiar (such as feminism or colonialism); others may be new or have connotations from the public sphere that need to be disentangled from political rhetoric (critical race theory, for example, or linguistic justice). Our authors present the scholarly definitions of these terms, as they were introduced in their original formulation, to help readers separate evidence-based frameworks from speculation or misinterpretation. In the end, though, the agency is with the reader as to whether these approaches work within their specific institutional contexts and needs, as well as if they feel prepared to enact these evidence-based theories effectively.

The chapters in this collection are organized around the themes of disruption to epistemic beliefs and challenge to traditional pedagogical practice. We believe these themes will resonate with instructors broadly rather than arranging chapters by disciplinary area. This is because the authors have worked hard to present their approaches in ways that transcend disciplinary boundaries. An instructor who works with engineering students, for example, can learn as much from chapters that discuss technical communication and mathematics as they do from those chapters that focus on engineering contexts. Likewise, instructors who primarily teach writing will be able to find and use concepts and assignments presented in distinctly disciplinary courses. We believe that the interdisciplinarity of this collection is one of its strengths. For readers who are interested in specific topics or disciplines, however, we offer a matrix at the end of this introduction that identifies common elements addressed in the chapters. This allows for more of a "pick-your-own" journey through the collection.

In *Section 1. Disrupting the Status Quo*, contributing authors share stories of building critical awareness of inequity throughout the curriculum. Jameta Barlow and Kylie Quave open the section with an exploration of what this work might look like within the context of a first-year writing course. They offer us ways to use decolonial, Black Feminist, and queer theoretical frameworks both as a way to teach writing and communicate scientific information about the world. This is followed by Blomstedt, who advocates for STEM writing instruction to begin with teaching students the history of how English became "the language of science" (this collection). Responding to calls from those in writing studies to resist linguistic imperialism and white language supremacy in our teaching (Canagarajah, 1996; Baker-Bell, 2020) and instead teach writing from a translingual approach, the series of lessons described by Blomstedt teaches students the precise means by which English became and has remained "dominant" in STEM writing.

Megan Callow and Holly Shelton continue this theme of challenging historical accounts of STEM knowledge with a discussion of a novel partnership between writing scholars and STEM faculty at the University of Washington. In this chapter, the authors describe how they designed and implemented a Critical Science Literacy course to help students think critically about the nature of science through the analysis and production of texts and about the ways that scientific knowledge shifts as it traverses platforms and audiences. The course emphasizes an understanding of the nature of science as contingent, contested, and situated; engages a diversity of ways of knowing and doing in science across cultures and nations; and traces the genealogies of ideas in circulation as information moves through pipelines and networks.

Laura Callis expands on this topic of knowledge-in-circulation with a discussion of the roles mathematics and statistics have historically played as tools of oppression, as well as how they can be leveraged to highlight and address injustice. Her chapter describes two assignments used in introductory statistics courses at a neurodiverse college that welcomes learners with a range of educational backgrounds. The assignments use real data about social justice topics and low-stakes, scaffolded writing prompts to support students in working through the statistical inquiry process, developing conceptual understanding and technological fluency, and improving their precision of language both mathematically and contextually.

Callis uses this chapter to show how using data sets that address injustice can be a solution for statistics faculty who feel the tensions of covering an ambitious syllabus, developing students' conceptual understanding, and recruiting interest in the quantitative fields.

This is followed by a chapter by Alicia Bitler and Ebtissam Oraby that discusses a course meant to destabilize and challenge the prevailing view of science as Western, male, and white. The authors of this chapter created a course that allows students to explore a non-Western epistemology of science and think of science as diverse and inclusive. Throughout the course, students explore Muslim and Arab science history and culture as part of a globally shared human heritage to open a space for other ways of thinking about and doing science. Muslim and Arab scholars have contributed to science in meaningful and often unacknowledged ways, founding disciplines like chemistry, algebra, modern surgery, and optics, shaping science as we know it. The course highlights the achievements and ways of knowing in science of prominent Arab and Muslim scientists.

The section concludes with a chapter by Justiss Burry, Carolyn Gubala, Jessica Griffith, Tanya Zarlengo, and Lisa Melonçon, who take up similar considerations of justice and ask: "What happens in a large [Technical and Professional Communication] program when it creates a programmatic inclusion vision and then sets out to enact it?" (this collection) In this chapter, the authors discuss the answer to this question as a way to address this collection's emphasis on actionable steps faculty can enact to make their STEM writing spaces more inclusive and challenge assumptions about disciplinary writing.

Section 2. Challenging Orientations to Instruction and Assessments moves from an exploration of ontology and epistemology into one of application. Contributors in this section present ways to enact elements of disruption into considerations of genre and disciplinary practice, while also asking STEM educators to turn the lens back onto themselves and what they value. In the opening chapter to this section, Rachel Riedner, Royce Francis, and Marie Paretti ask questions of common classroom practices by looking specifically at the intersection of writing and identity in engineering in the context of engineering judgment. Their goal is to consider how one might design assignments and create group work practices that help students to actively position themselves as engineers. This chapter discusses the theoretical framework and praxis implications from an instrumental case study that explores how writing in the disciplines (WID) assignments do and do not support students' engineering identities in an existing capstone course.

Continuing with a consideration of writing in engineering, Jennifer Mallette's chapter examines the situated learning and integrated approaches that facilitated one engineering communication course's success, with a focus on the ways the course was planned and designed and the approaches built into that design that were aimed at supporting student success, particularly in a year where more students struggled because of remote classes and various pandemic challenges. The first part of the chapter examines the impacts of designing a course in collaboration with the College of Engineering and the specific department, implementing a backward planning approach that also incorporated inclusive excellence pedagogical strategies and equitable assessment. The second part of the chapter explores the course's preliminary impact on student learning, given the course's built-in flexibility and use of contract grading in an online environment. The chapter concludes with key takeaways for designing a course with inclusion and equity as a core value, as well as approaches to implement in a course to support student success.

Sally B. Seraphin continues the theme of supporting student success by presenting a framework for creating relevant, meaningful writing assignments that leave space for students to perform at their best and grow in their learning. Her "non-disposable assignments" engage students at a variety of tiers of engagement in a manner that leads to sharing of resources and materials in a multitude of ways. These assignments, Seraphin argues, provide entry points for students to thrive—to capitalize on their skills and knowledge in a way that moves beyond completing activities for assessment and toward having an impact in the world.

Similarly pushing traditional notions of writing instruction and assessment, Jennifer Newell-Caito discusses what it looks like to incorporate "ungrading" (Kohn & Blum, 2020) strategies into an upper-level analytical biochemistry course. Newell-Caito explains how her use of flexible deadlines, authentic assessment, contract grading, and process letters support student learning and aids in building metacognition for students.

Continuing with the theme of meaningful writing, Janelle Johnson et al. present a strategy for engaging students with questions of their own positionality within STEM education. Explicitly focusing on disability, the authors (which include participants in the course) present an assignment sequence that asks students to choose an educational inequity they are passionate about and combine a synthesis of the policy context with a sharp focus on a particular community. They learn to create a series of concrete actions they can take to address the inequity, and those actions are captured in a public service announcement. The project concludes with an exposition where students publicly share their call to action.

In the final chapter of this section, Ann Fink offers educators an example of how to enact liberatory pedagogy in STEM content courses. By focusing on a course in neuroethics, Fink discusses the way she has built an inclusive curriculum that disrupts traditional ways of thinking about neuroethics as well as the pedagogical approaches used to make the classroom more equitable.

The collection concludes with a discussion of the kinds of questions we anticipate readers will have regarding the practical realities of implementing these practices. We offer individual perspectives from our respective fields as well as additional resources for those who wish to continue their social justice journey.

Our Call to Action

As noted earlier, this collection is not about convincing STEM instructors that writing is important; nor is it about convincing anyone that diversity and inclusion are paramount concerns for their fields. Rather, this book is for the educator who wishes to do something about it. As editors, we recognize that there are important conversations missing from this collection and equity conversations more broadly. Though we actively sought scholars and educators doing work specifically around neurodiversity in STEM, for example, our outreach yielded very little response (Johnson et al. being the exception). Similarly, finding scholars exploring the residual effect of this work—tracking what sticks and what fades away—proved elusive. We sought contributions from scholars doing work and practices that pushed the boundaries of what typically gets addressed in equity and inclusion (looking toward disability and socioeconomics, for example, or experimenting with language use) but encountered similar challenges.

To be sure, there are individuals doing this work, and we have made strong efforts to provide direction to that scholarship in the Conclusion. Finding unpublished work related to STEM writing (and not education broadly), however, turned out to be more difficult than expected—particularly in interdisciplinary spaces and international contexts. What that highlighted, though, was that inclusive writing instruction in STEM spaces is an area of scholarly and practical interest, and with many lines of inquiry still left to be explored and amplified. We are hopeful that the approaches presented in this collection will empower educators to start (or continue) equity and inclusion work in their STEM-relevant classrooms and inspire researchers to consider new lines of inquiry aimed at the long-term impacts of this work and how it transfers to spaces outside of the classroom. To that end, the collection's Conclusion provides some reflections by each of us (including a final, powerful call to action from LaKeisha), a series of questions and considerations for educators and resources to continue learning and contributing.

References

ABET (2018). Criteria for accrediting engineering programs. http://www.abet.org/.
American Association for the Advancement of Science (2011). Vision and change in undergraduate biology education: A call to action. https://visionandchange.org/wp-content/uploads/2013/11/aaas-VISchange-web1113.pdf.
Badenhorst, C. M., Moloney, C. & Rosales, J. (2020). New literacies for engineering students: Critical reflective-writing practice. *The Canadian Journal for the Scholarship of Teaching and Learning, 11*(1), 1–20. https://doi.org/10.5206/cjsotl-rcacea.2020.1.10805.

Baker-Bell, A. (2020). *Linguistic justice: Black language, literacy, identity, and pedagogy.* Routledge.

Bathgate, M. E., Aragón, O. R., Cavanagh, A. J., Waterhouse, J. K., Frederick, J. & Graham, M. J. (2019). Perceived supports and evidence-based teaching in college STEM. *International Journal of STEM Education, 6*(11). https://doi.org/10.1186/s40594-019-0166-3.

Bazerman, C. (2000). *Shaping written knowledge: The genre and activity of the experimental article in science.* The WAC Clearinghouse. https://wac.colostate.edu/books/landmarks/bazerman-shaping/ (Originally published in 1988 by University of Wisconsin Press).

Bazerman, C. (1981). What written knowledge does: Three examples of academic discourse. *Philosophy of the Social Sciences, 11*(3), 361–388. https://doi.org/10.1177/004839318101100305.

Blair-Loy, M & Cech, E. A. (2022). *Misconceiving merit: Paradoxed of excellence and devotion in academic science and engineering.* The University Press of Chicago.

Canagarajah, S. (1996). "Nondiscursive" requirements in academic publishing, material resources of periphery scholars, and the politics of knowledge production. *Written Communication, 13*(4), 435–472. https://doi.org/10.1177/0741088396013004001.

Condon, F. & Young, V. A. (Eds.). (2016). *Performing antiracist pedagogy in rhetoric, writing, and communication.* The WAC Clearinghouse; University Press of Colorado. https://doi.org/10.37514/ATD-B.2016.0933.

Fahnestock, J. & Secor, M. (1988). The stases in scientific and literary argument. *Written Communication, 5*(4), 427–443. https://doi.org/10.1177/0741088388005004002.

Falconer, H. M. (2019a). "I think when I speak, I don't sound like that": The influence on social positioning on rhetorical skill development in science. *Written Communication, 36*(1), 9–37. https://doi.org/10.1177/0741088318804819.

Falconer, H. M. (2019b). Mentored writing at a Hispanic-serving institution: Improving student facility with scientific discourse. In I. Baca, Y.I. Hinojosa & S.W. Murphy (Eds.), *Bordered writers: Latinx identities and literacy practices at Hispanic-serving institutions* (pp. 213–230). State University of New York Press.

Finkenstaedt-Quinn, S. A., Polakowski, N., Gunderson, B., Shultz, G. V. & Gere, A. R. (2021). Utilizing peer review and revision in stem to support the development of conceptual knowledge through writing. *Written Communication, 38*(3), 351–379. https://doi.org/10.1177/07410883211006038.

Gallagher, J. R., Turnipseed, N., Yoritomo, J. Y., Elliott, C. M., Cooper, S. L., Popovics, J. S., Prior, P. & Zilles, J. L. (2020). A collaborative longitudinal design for supporting writing pedagogies of STEM faculty. *Technical Communication Quarterly, 29*(4), 411–426.

Gere, A. R., Limlamai, N., Wilson, E., MacDougall Saylor, K. & Pugh, R. (2019). Writing and conceptual learning in science: An analysis of assignments. *Written Communication, 36*(1), 99–135. https://doi.org/10.1177/0741088318804820.

Hora, M. T., Smolarek, B. B., Martin, K. N. & Scrivener, L. (2019). Exploring the situated and cultural aspects of communication in the professions: Implications for teaching, student employability, and equity in higher education. *American Educational Research Journal, 56*(6), 2221–2261. https://doi.org/10.3102/0002831219840333.

Hyland, K. (2012). *Disciplinary identities: Individuality and community in academic discourse.* Cambridge University Press.

Inoue, A. B. (2019). *Labor-based grading contracts: Building equity and inclusion in the compassionate writing classroom* (1st ed.). The WAC Clearinghouse; University Press of Colorado. https://doi.org/10.37514/PER-B.2019.0216.0.

Jeon, A-J., Kellogg, D., Khan, M. A., Tucker-Kellogg, G. (2021). Developing critical thinking in STEM education through inquiry-based writing in the laboratory classroom. *Biochemistry and Molecular Biology Education, 49*(1), 140–150. https://doi.org/10.1002/bmb.21414.

Kao, V. & Kiernan, J. (Eds.). (2022). *Writing STEAM: Composition, STEM, and the new humanities*. Taylor & Francis.

Kohn, A. & Blum, S. D. (Eds.). (2020). *Ungrading: Why rating students undermines learning (and what to do instead)*. West Virginia University Press.

Larsen, K. & Gärdebo, J. (2017). Retooling engineering for social justice: The use of explicit models for analytical thinking, critical reflection, and peer-review in Swedish engineering education. *International Journal of Engineering, Social Justice, and Peace, 5*(1–2), 13–29. https://doi.org/10.24908/ijesjp.v5i1-2.8928.

Lund, T. J. & Stains, M. (2015). The importance of context: An exploration of factors influencing the adoption of student-centered teaching among chemistry, biology, and physics faculty. *International Journal of STEM Education, 2*(13), 1–21. https://doi.org/10.1186/s40594-015-0026-8.

McGee, E. O. (2020a). Interrogating structural racism in STEM higher education. *Educational Researcher, 49*(9), 633–644. https://doi.org/10.3102/0013189X20972718.

McGee, E. O. (2020b). *Black, brown, bruised: How racialized STEM education stifles innovation*. Harvard Education Press.

McGee, E. O. & Martin, D. B. (2011). "You would not believe what i have to go through to prove my intellectual value!" Stereotype management among academically successful Black mathematics and engineering students. *American Educational Research Journal, 48*(6), 1347–1389. https://doi.org/10.3102/0002831211423972.

McGee, E. O. & Robinson, W. H. (Eds.). (2020). *Diversifying STEM: Multidisciplinary perspectives on race and gender*. Rutgers University Press.

Miner, K. N., Walker, J. M., Bergman, M. E., Jean, V. A., Carter-Sowell, A., January, S. C. & Kaunas, C. (2018). From "her" problem to "our" problem: Using an individual lens versus a social-structural lens to understand gender inequity in STEM. *Industrial and Organizational Psychology, 11*(2), 267–290. https://doi.org/10.1017/iop.2018.7.

Moon, A., Gere, A. R. & Shultz, G. V. (2018). Writing in the STEM classroom: Faculty conceptions of writing and its role in the undergraduate classroom. *Science Education, 102*(5), 1007–1028. https://doi.org/10.1002/sce.21454.

Myers, G. (1990). *Writing biology: Texts in the social construction of scientific knowledge*. The University of Wisconsin Press.

Myers, G. (1985). Texts as knowledge claims: The social construction of two biology articles. *Social Studies of Science, 15*(4): 593–630. https://doi.org/10.1177/030631285015004002.

National Research Council (2012). Discipline-based education research: Understanding and improving learning in undergraduate science and engineering. The National Academies Press. https://doi.org/10.17226/13362.

Paretti, M. C., Gustafsson, M. & Eriksson, A. (2019). Faculty and student perceptions of the impacts of communication-in-the-disciplines (CID) on students' development as engineers. *IEEE Transactions on Professional Communication, 62*(1), 1–16. https://doi.org/10.1109/TPC.2019.2893393.

Poe, M., Inoue, A. B & Elliot, N. (2018). *Writing assessment, social justice, and the advancement of opportunity.* The WAC Clearinghouse; University Press of Colorado. https://doi.org/10.37514/PER-B.2018.0155.

Poe, M., Lerner, N. & Craig, J. (2010). *Learning to communicate in science and engineering: Case studies from MIT.* MIT Press.

Reynolds, J. A., Thaiss, C., Katkin, W. & Thompson, R. J. (2012). Writing-to-learn in undergraduate science education: A community-based, conceptually driven approach. *CBE—Life Sciences Education, 11*(1), 17–25. https://doi.org/10.1187/cbe.11-08-0064.

Skopec, M., Fyfe, M., Issa, H. Ippolito, K., Anderson, M. & Harris, M. (2021). Decolonization in a higher education STEMM institution—is 'epistemic fragility' a barrier? *London Review of Education, 19*(1), 1–21. https://doi.org/10.14324/LRE.19.1.18.

Thompson, R., Finkenstaedt-Quinn, S., Shultz, G., Gere, A. & Reynolds, J. A. (2021). How faculty discipline and beliefs influence instructional uses of writing in STEM undergraduate courses at research-intensive universities. *Journal of Writing Research, 12*(3), 625–656. https://doi.org/10.17239/jowr-2021.12.03.04.

Venters, C., Groen, C., McNair, L. D. & Paretti, M. C. (2018). Using writing assignments to improve learning in statics: A mixed methods study. *International Journal of Engineering Education 34*(1), 1–13. https://doi.org/10.18260/1-2--22207.

Walton, R., Moore, K. R., and Jones, N. N. (2019). *Technical communication after the social justice turn: Building coalitions for action.* Routledge.

Wolfe, J., Olsen, B. & Wilder, L. (2014). Knowing what we know about writing in the disciplines: A new approach to teaching for transfer in FYC. *The WAC Journal, 25*, 42–77. https://doi.org/10.37514/WAC-J.2014.25.1.03.

Yu, H. & Northcut, K.M. (Eds.). (2017). *Scientific communication: Practices, theories, and pedagogies.* Taylor & Francis.

Zellmer, A. J. & Sherman, A. (2017). Culturally inclusive STEM education. *Science, 358*(6361), 312–313. https://doi.org/10.1126/science.aaq0358.

Appendix: Topic Matrix for Collection

Topic area	1. Barlow & Quave	2. Blomstedt	3. Callow & Shelton	4. Callis	5. Bitter & Oraby	6. Burry et al.	7. Riedner, Francis, & Paretti	8. Mallette	9. Seraphin	10. Newell-Caito	11. Johnson et al.	12. Fink
Ableism/Disability	x							x			x	
Critical Inquiry	x		x	x				x			x	x
Critical Literacy			x	x				x		x	x	x
Cultural Diversity		x	x					x	x			x
Decolonial Approaches	x	x	x		x			x		x		x
Disciplinary Identity		x		x		x		x	x			
Feminist Theory	x			x				x			x	
History of Science	x	x	x	x				x			x	
Indigenous Cultures	x	x		x				x		x		
Interdisciplinarity	x			x				x		x	x	x
Meaningful Writing			x		x			x	x			
Reflective Assignments	x			x				x	x	x	x	x
Translingualism		x						x		x	x	x
White Language Supremacy		x				x		x				
Community Engagement							x	x				
Course Level												
First-Year	x		x	x				x		x	x	x
Second and Third Year		x		x	x			x			x	x
Capstone							x	x		x		x
Graduate			x					x				
Assessment and Evaluation												
Ungrading & Labor-based Contracts			x				x	x	x			
Instructor Feedback						x	x	x	x			
Transparent Assignment Frameworks							x	x				x
Self-Assessment	x	x					x	x		x		

Section 1. Disrupting the Status Quo

Disrupting the status quo in STEM courses is about creating space where everyone sees themselves as members of the disciplinary community. It is about consciously recognizing that doing things the way they have always been done does not necessarily mean that those practices are the best way to move forward. We begin the collection in this disruptive space in an effort to orient readers' thinking toward the more radical possibilities challenging the status quo can lead to. It is important to note that this work is not about being confrontational or antagonistic. Rather, the end goal is to lift the curtain to show the humanity behind the systems we consciously and subconsciously reinforce through our work. It is about recognizing what no longer works when we choose to curate educational and disciplinary spaces that include those who have historically been left out or erased.

Section 1 opens with poignant student reflections from Dhatri Badri and Riya Sharma. Badri discusses her experiences as a woman in biology and the impact of a single course on helping her carve out and own a space for herself in the field. Sharma focuses on the power of a science writing course that made challenging assumptions and identifying bias the focus of the work.

The chapters in this section then take up that theme of creating space by exploring topics such as disrupting traditional ways of thinking about science in themed first-year writing courses (Barlow and Quave), interrogating English and language supremacy in STEM writing instruction (Blomstedt), illustrating how interdisciplinary collaborations can open space for developing Critical Science Literacy (Callow and Shelton), the use of real, current data related to social justice in an introductory statistics course to highlight institutionalized bias (Callis), and developing courses that highlight the ways in which cultures from around the globe have contributed to foundational elements of STEM (Bitler and Oraby). The last chapter in this section asks students in Technical and Professional Communications (TPC) courses (which often target STEM students) to take up considerations of justice and cultural difference in their work (Burry et al.).

As noted in the Introduction, these chapters are meant to generate thinking about what is possible and practical within educators' own instructional spaces. The assignments and pedagogical strategies described offer starting places—approaches to try, test, and modify as needed.

Student Vignette

Dhatri Badri
BOSTON UNIVERSITY

In several of my STEM classes, I experienced imposter syndrome—this feeling of not being good enough or fully represented as an Indian woman in my biology major. I lacked a connection between my personal identity and the STEM classes I had taken due to this absence of belonging. It was in a writing-intensive course about the history and philosophy of math and science that I finally felt a sense of belonging. This class was singular in my STEM education because I focused on an issue that resonated with me personally. Outside of this class, I would not have had the opportunity to explore the topic on which I eventually wrote a 15-page paper: "Scientific Developments in Colonial-Era India."

This class acknowledged and emphasized that several civilizations and cultures contributed to the scientific community and went underappreciated and unrecognized due to the Eurocentric nature of science. This intrigued me; I wondered if this had happened to my own culture. Through the assigned readings in this course, I realized how little other cultures' contributions to science were taught in school, and I wanted to add to the discourse by including my own culture. Consequently, my final paper focused on the scientific developments in India during British rule. The reflections written after the readings were vital to my paper—they allowed me to see scientific concepts from a historical and philosophical angle, which was often not evident in my core STEM classes. The class not only helped me to appreciate my cultural history but also to admire that my people contributed to the very field I am studying.

This paper bridged the gap between my place within the STEM fields and my experiences as a woman and a person of color. It is well established that these fields, especially their research aspect, is male-dominated. In my experience, my male classmates have typically been more assertive and confident, in contrast to some of my female classmates, who tend to be more reserved and second-guess themselves. In one of the reflections for this class, I recalled instances with my male classmates and professors in group discussions like lab meetings where I thought to myself, "What if I say something—or should I even say anything at all? What if they judge me?" While they had not meant to ignore me, I always felt like an outsider and like I needed to work much harder to make my ideas heard. This class gave me the space and confidence to communicate those very ideas.

Through this writing assignment, exploring my Indian ethnicity and culture allowed me to engineer my own inclusion in the STEM fields in spite of imposter syndrome. Due to this course and its assigned writings, I was further inspired to write an article in my college's journal about women who were overlooked in the STEM fields. It was beneficial for me to see my heritage and community represented in the field I am currently studying. I hope that my presence in the STEM fields will inspire others with my ethnic background to pursue similar careers and interests, while simultaneously paying homage to the scientific successes of our ancestors.

Student Vignette

Riya Sharma

THE GEORGE WASHINGTON UNIVERSITY

I am not a stranger to the pressures of performing well academically. Sitting at the kitchen counter at age 10, I'd feel tears form in my eyes as my mother scolded my inability to understand algebra. For what seemed like hours, I struggled to grasp the mathematical concepts necessary for success in the future STEM courses she envisioned me taking. I was further discouraged from pursuing STEM-related opportunities and careers as I heard the soft giggles of my peers echo while I failed to answer geometry questions correctly. Enjoying my education became challenging as school fostered an environment centered around competition and awards in place of students and their learning experience. Even in middle school, students began to tie their self-worth to scores on exams and boast about their ability to excel on practice SATs. While I continued to push myself and remained a relatively good student, my accomplishments felt small compared to those around me.

Growing up South Asian in a predominantly South Asian community within the US, I was surrounded by parents who lauded their children's achievements in the STEM fields and their placement into prestigious high schools, universities, and research fairs. I was and continue to be incredibly proud of my peers and their contributions. They are continuing the legacies of hardworking immigrant parents and transforming their futures. However, I also felt out of place. I felt an average student such as myself, who performed worse in STEM classes, was too stupid to continue in a STEM discipline. Too stupid to make my parents and my larger community proud. I dismissed the idea of ever engaging in STEM.

It wasn't until college that I felt included and as if I had the potential to succeed. To fulfill a requirement, I enrolled in Writing Race, Measuring Marginalization, a course on science writing. Although the course was centered on writing, its material combined the natural and social sciences and quantification. Because of my past experiences in STEM education, I was initially hesitant about this course. However, Dr. Kylie Quave, my professor, quickly helped me not only feel comfortable but enthusiastic about the material through her teaching. The course took a student-focused approach. Instead of simply feeding information to students through static slides and lectures, Dr. Quave opened the door to discussion, allowing students to share their personal experiences and perspectives without fearing judgment. Through this method, I (and my peers) felt actively involved in the learning process. It wasn't just us learning from the professor. She was learning from us as well.

Each class would focus on a new topic, from the dangerous effects of quantitative methods in craniometry to the hypertension hypothesis and the use of ancestry in biomedical research. Productive class discussions accompanied lessons on each topic. Students, including myself, would ask questions and share their thoughts here. Such talks were instrumental in creating a welcoming and positive learning environment, and many of them stuck with me. I recall a classmate describing her current struggles with the US's perception of race. Another explained how she'd experienced the effects of systemic racism firsthand. In previous courses, I hesitated to raise my hand for fear of being perceived as unintelligent or answering questions incorrectly. Not here. Hearing others openly share their points of view and being encouraged by Dr. Quave revitalized me, imbuing me with a sense of curiosity and wonder about the sciences that I thought I would never feel.

In addition to open discussion, the course showed me there was more to STEM than rigid facts and figures or competition in the classroom. I could connect with material in a new way through writing assignments and exercises. We were not simply assigned formulaic research papers and expected to regurgitate material from class. Dr. Quave worked with us one-on-one to help us develop research questions we felt interested in and passionate about while fostering collaboration through multiple peer reviews and group papers. I learned how essential discussion and writing are to student engagement, especially in STEM fields where this approach is less prevalent. Such curriculum and instruction methods significantly contributed to my decision to pursue a degree combining STEM and the social sciences, which I thought was never possible. Writing and discussing so openly with my peers not only made me feel heard but as if others wanted to hear my voice.

STEM Writing as Disruption: Views from First Year Writing

Jameta Nicole Barlow and Kylie E. Quave
THE GEORGE WASHINGTON UNIVERSITY

Sociopolitical conditions have distinctly influenced the development of scientific disciplines in the United States. These histories have promoted traditionally white, male knowledge producers as objective and reliable while sidelining others who have been deemed less neutral, objective, and authoritative in society (Kozlowski et al., 2022). Imagined hierarchies of knowledge and knowledge producers in the sciences have come at the expense of robust explanations of the world and its humans, which could otherwise impact society positively. Recent public health crises continue to highlight the disparate ways in which science and technology fall short in addressing underlying social inequities in this modern, pluralistic society. Moreover, the COVID-19 pandemic revealed widespread, keen interest in, and the need for, the critical analysis of science and its applied impact on human behavior, decision-making, and medical interactions.

This chapter is a call for STEM and STEM writing faculty to critically examine and center multiple perspectives on the roots of scientific knowledge production in our classrooms. Our objective is to explain the implications of historical and social realities within knowledge production—and their attendant epistemic injustices (Prescod-Weinstein, 2020)—within our first-year undergraduate science writing classrooms. We explain how Black Feminist, Indigenous Feminist, and other anti-colonial approaches to writing in STEM are not only the lenses of our own pedagogies, but also how these approaches can be parlayed into many kinds of STEM writing classrooms. Our theory and practice (i.e., praxis) of teaching citation are explained here as a site for making these pedagogical goals reality.

Historical and Present Challenges of STEM Knowledge Production

We approach this work by centering the tenets of Black Feminist, Indigenous Feminist, and other anti-colonial and decolonizing paradigms. Black Feminist and Womanist approaches entail collective struggles to address systemic inequities in the present and past, and the outcomes of centuries of exclusion and oppression. Following the lead of thinkers such as bell hooks and her critique of the imperialist white supremacist capitalist patriarchy (1984/2000), Black Feminism acknowledges the importance of intersectional approaches to the array of oppressions made possible by the multiple dimensions of marginalized identities (e.g., A. J. Cooper, 1892; B. Cooper, 2015, 2016; Crenshaw, 1994; Hill Collins, 1989). Black Feminist thought centers the experiences of Black women as essential to understanding the ways that multiple, interlocking oppressions operate in society. Since at least the 19th century, voices, such as Anna Julia Cooper (1892), have called for resistance to overly simplified explanations of how social inequality works. Cooper's work has been followed by a long line of Black Feminist theorists urging ways to end all oppressions by attending to the exclusions and marginalizations experienced by Black women (e.g., the Combahee River Collective Statement, 1978) and by understanding the ways systems including racism, sexism, patriarchy, capitalism, and more operate to uphold each other. More recently, theorizing these interlocking systems is referred to as "intersectionality theory" (Crenshaw, 1994).

Indigenous Feminist ways of knowing and doing prioritize Indigenous sovereignty for Indigenous lands and people. Similar to other critical feminisms, Indigenous Feminist thought emphasizes an intersectional approach to disrupting current and historical harms perpetuated on groups excluded from dominant society. Through Indigenous Feminism, decolonization is not merely a metaphor but rather demands the re-positioning of resources in just ways, with material change toward honoring self-determination (Tuck & Yang, 2012). Indigenous Feminist thinkers challenge the ways that scientists have violated Indigenous sovereignty and have unduly dismissed and sidelined Indigenous ways of knowing as non-scientific and inferior or subjective (Kolopenuk, 2020; Steeves, 2021; TallBear, 2014, 2016). Like Black Feminism, Indigenous Feminist theory similarly requires that the experiences of Native and Indigenous women be considered legitimate and be included as credible sources. Indigenous Feminist approaches require us to consider that not all women's lives are the same and do not need to be, and that our ways of knowing the world may be heterogeneous, which is a strength rather than a deficit. Indigenous Feminist thought rejects assumptions about the superiority of science and "Western" ideologies of normalcy and nature. Indigenous Feminism also centers decolonization as a process of letting go of unearned material dominance (Tuck & Yang, 2012) and calls attention to the specific forms of violence experienced by Native women (Green, 2017).

These anti-colonial ways of knowing are leveraged to overturn the outcomes of racialized and gendered oppressions (and gendered oppressions, as well as other forms of marginalization) and how they influenced dominant Western thought. In the sciences, ample research demonstrates systemic exclusion and delegitimization of Black and Indigenous women and others who did not fit an expected and purportedly normative EuroAmerican, male, middle-class identity (e.g., Bolnick et al., 2019; Mills, 2020; Rifkin, 2016; Shelton, 2020; White & Draycott, 2020).

In other words, simply existing, for Black and Brown individuals, as well as other marginalized groups, is a disruption.

Intersectional and decolonizing approaches to teaching and researching are not merely exercises in adding "diverse scholars" to one's syllabus reading list; they are not a matter of acknowledging the existence of Black women or other people historically removed from powerful institutions of knowledge production. Rather, these approaches require that teacher-scholars re-orient and re-center (Barlow & Dill, 2018) inequitable forms of knowledge production through and with STEM writing. Physicist and Black Feminist theorist, Chanda Prescod-Weinstein, points to epistemic injustices in STEM: devaluing and undermining a writer's knowledge due to their identity (2020). When epistemic injustice plays out in our STEM disciplines, it is fueled by white empiricism, which is the belief that white people are objective, neutral observers while others are biased and incapable of neutrality (Prescod-Weinstein, 2020). In a related vein, archaeologist Paulette Steeves draws upon Indigenous Feminist science and the sacred practice of burning for radical renewal to coin her original concept of pyroepistemology (2021). Steeves explains that "a practice of pyroepistemology is a ceremony that cleanses the academic landscape of discussions that misinform worldviews and fuel racism. Such literary renewal clears the way for healthy growth in academic fields of thought and centers of knowledge production" (2021, p. 20). Steeves' pyroepistemology may provide the cleansing needed to overcome white empiricisms, which may further the goal of intervening in harmful and exclusionary scientific knowledge production. This chapter offers a practice of pyroepistemology with its focus on ontologies, epistemologies, methodologies, and pedagogies that enable teacher-scholars to engage in cleansing academic practices rather than perpetuating the status quo of epistemic inequities.

The point of intervention is to decolonize methodologies by transforming the production of scientific knowledge. In fact, interrogating ontologies (i.e., What can we know? What's out there to know?) and epistemologies (i.e., How can we know? How do we know what we know?) of historical and contemporary approaches toward scientific knowledge production is how we disrupt STEM writing. As Linda Tuhiwai Smith has argued with regard to decolonizing methodologies, research is an institution marked by "its claims, its values and practices, and its relationships to power" (Smith, 1999/2021, p. 286). Smith further posits that research is "a set of ideas, practices and privileges that [are] embedded in imperial expansionism and colonization and institutionalized in academic disciplines, schools, curricula, universities and power" (Smith, 1999/2021, p. 287). As a result, re-framing and re-centering the curriculum move scientific knowledge production into the light, making the social, political, and historical contexts around research more transparent and thus closer to the scientific method's promise of empiricism and truth. As STEM knowledge has been written into existence in ways that perpetuate social inequities, so too can STEM knowledges be burnt to the ground and re-composed to overcome epistemic injustices.

Creating Inclusive Pathways Through Critical Writing Pedagogies

Writing is the primary tool by which STEM knowledge is communicated to other scholars and to the broader public. However, writing has also played an active role in reifying exclusionary ways of knowing. Teacher-scholars must be aware of and account for the historical and present challenges of STEM epistemologies discussed above and should seek out examples from their own disciplines. Knowledge in STEM fields is not neutral, though these disciplines have masqueraded as such in the Western world for at least the last five centuries. We, humans, are the ones who actually produce scientific knowledge and technological innovation: we introduce our biases, agendas, and imperfections (Marks, 2017; Smith, 1999/2021) into how we know the world and how we use writing as a technology to intervene in the world around us. An uncritical approach that treats science and technology as if they operate in a void—divorced from their cultural and social milieus—is an approach that deprives students as writers and researchers of a full picture of the human condition and what is at stake for justice and fairness in human societies.

Moreover, this does not benefit science. Scientific knowledge and technologies can, and have, improved our lives, but one need look no further than racist, sexist robots (Alaieri & Vellino, 2016; Caliskan et al., 2017) to see that social problems are only reproduced and exacerbated rather than eased or erased by scientific and technological progress. Medical algorithms and metrics (Braun, 2015; Vyas et al., 2020) are the residues of the dehumanizing histories that produced knowledge of human health (Braun & Saunders, 2017; Owens, 2017), while the technology industry can trace a throughline from legalized discrimination to de facto racism made possible through automation and apps (Benjamin, 2019). Studies of human genetic diversity have been leveraged to solidify myths about racial and ethnic inferiority, upholding white supremacy (Larsen et al., 2020; Panofsky et al., 2020). These examples from across fields taught in STEM writing courses merely scratch the surface of the wide-ranging, long-standing role of STEM disciplines and discourses that create material harm.

From the fields in which each of us was originally trained in the sciences prior to becoming writing faculty, we offer a range of case studies to students to demonstrate the harmful effects of science writing over the centuries and into the present. In anthropological archaeology and biological anthropology, for example, researchers write about the harms done to particular populations due to enduring preservation and celebration of white supremacist pasts (Carter, 2018; Mullins, 2017), some of which originates from anthropology itself (e.g., Geller, 2020; Mitchell, 2018) or through nonconsensual field research methods (Atalay, 2006; Blakey, 2020). In forensic anthropology, writers are sounding the alarm about use of the euphemism "ancestry" as a stand-in for "race," which served to uphold mythologies about biological race that cannot be supported by the sciences (DiGangi & Bethard, 2020; Tallman et al., 2021). Anthropologists are re-orienting these harmful ways of producing knowledge about the past with analysis of the language and rhetoric of anthropology (Allen & Jobson, 2016) but also by using composition itself as an instrument to disrupt (Reid, 2021).

In another disciplinary example from our courses, epistemological concerns are rarely discussed in public health and psychological science research. When they are discussed in these fields (Barlow & Dill, 2018; Bowleg, 2017; Bowleg et al., 2017), they are particularly immersed in women's and gender studies interventions (Meyer, 2007). Writing studies, in concert with the critical theories that interrogate epistemologies, offers a bridge for applied intervention by engaging philosophies of science. Community psychology (Boyd & Bright, 2007; Campbell & Murray, 2004) and community writing (Ryder, 2012) explicitly leverage agency and rhetoric to create sustainable change in communities. Research on writing as healing (Baker & Mazza, 2004; DeSalvo, 1999; Pennebaker, 1990), drawing upon the humanities and writing studies (Barlow, 2016, 2018), thus become tools for sustainable community change around healing, harm, and trauma. Public health, psychology, and composition theory are already in conversation with each other in the scholarly landscape, forging connections between knowledge production and structural inequities, and need only be explicitly presented as such in our courses.

Incorporating an equally wide range of perspectives from various disciplines, including from people of different backgrounds and positionalities, is an essential component to addressing these inequities. Anti-colonial and decolonized theoretical frameworks position teachers and learners to value multi-vocality, consent, sovereignty, differently-abled bodies, lived experience as evidence, collaboration, and a rejection of unquestioned normative categories and classifications. Including these diverse (and often excluded) scholarly and community perspectives models these values for students in the process of disrupting the standard ways of knowing in STEM disciplines.

Writing as Teaching Tool and Technology for Disrupting Inequities

In a decade in which the enduring intergenerational effects of inequity have never been clearer to more people in the US, we channel the power of writing to disrupt our pedagogies. Writing is both a tool with which to teach and a technology for disseminating scientific knowledge about the world, which can intervene in currently imperfect realizations of a pluralistic and inclusive society. Teaching STEM writing to undergraduates with anti-colonial approaches represents an ethical and more appropriate mode for this disruption to take root because students can more directly observe the impact and influence of colonialism in science as this uncovering process occurs (see Blomstedt, this volume, on the ways that White English emerged as legitimate scientific language; see Bitler and Oraby, this volume, on exclusion of non-European ways of knowing, for examples beyond our scope).

Because scientific inquiry is misunderstood as a neutral, value-free way of knowing the world (Smith, 1999/2021), instructors can use case studies showing genealogies of knowledge over time (see the models for this in more detail in Callow and Shelton, this volume) to turn that assumption on its head; tracing knowledge production reveals inconsistencies, inadequacies, and contradictions in the actual practice of science, especially as related to narrow representation of perspectives (the very problems elaborated upon by Prescod-Weinstein and Steeves, as described above). However, the lesson we ought to be extending to students is not that science is fundamentally, irredeemably flawed, but rather that the sciences are brought to life by humans working within social and individual contexts. For example, RetractionWatch.org (e.g., Marcus, 2020, 2021) offers a range of examples of the sciences failing to live up to their promise of reporting what is more true and less false about the universe. Whether studies are retracted or challenged due to the undue influence of ideology over empiricism (Larsen et al., 2020) or due to a lack of care in challenging white empiricism, the result is the same: unchecked exclusionary ideas continue to circulate under the guise of science's perceived superiority and neutrality (Nature, 2022). Intellectual and institutional barriers to marrying Black Feminist and Indigenous Feminist thought with STEM disciplines remain rigidly in place, and writing instruction can be positioned to break those barriers down. Common rhetoric poses science in opposition with these ways of knowing, but we practice pedagogies that put them into contact with each other. Our work aims to be the bridge between rhetorical writing and scientific inquiry, anchored by decolonial practices.

In our classrooms, we position students as knowledge producers themselves and empower them to use research and composition to disrupt harmful ways of knowing. Western traditions in the sciences have been forwarded in limiting and exclusionary ways, but returning to openness about what can exist (ontologies) and how we know it (epistemologies) allows us all to open up to more exhaustive ways of seeing and explaining the world. We position students to produce and to intervene in two principal ways:

1. As readers and consumers of knowledge, they learn to identify what is left unsaid and whose perspectives are overwritten in the scientific disciplines they are reading or in the selection of sources available to them (see above examples, as well as Bitler and Oraby, this volume).

2. As writers, they practice reflexive examination of their citation praxis: the ways they define sources as expert and reliable, and the ways they weigh evidence in their writing. They also learn to question how their teachers define credibility and expertise, and to seek to make those concepts more inclusive for themselves and their peers.

Our Situated Context

The authors teach in an undergraduate writing program at a historically white higher education institution in the United States. Undergraduates across the university, which is a large, private, high research activity university, are required to complete one semester in a writing and research course, and these courses emphasize disciplinary forms of writing and transfer between writing genres. About one-third of the dozens of faculty in the writing program were not trained in rhetoric and composition or related fields, but rather have come to writing instruction from other home disciplines, including STEM and the humanities. We authors are from humanistic STEM backgrounds and bring those disciplinary frames into first-year writing. Our standpoints, methodologies, praxis, and approaches are encouraged by program administration as a decolonizing practice, which we see as rooted in Steeves' pyroepistemology concept of replacing old ways of knowing and doing for more equitable futures.

Both authors are faculty in the first year writing program, considered a central component of the university's general education curriculum and a place where undergraduates are required to complete a 4.0 credit-hour introductory research and writing course. The program uses multiple disciplines and genres to prepare first year students for an academic career in writing. Both authors focus on writing in the sciences and/or health.

Course Reflection: Jameta Nicole Barlow, Ph.D., MPH

The first co-author is an unapologetic Southern Black woman and community psychologist, public health scientist, and women's health scholar teaching science and health writing. She brings her full self into the classroom—which enables her students to do the same. Her research utilizes decolonizing methodologies to disrupt cardiometabolic syndrome and structural policies adversely affecting Black girls' and women's health, intergenerational trauma, and perinatal mental health. She has spent 25 years in transdisciplinary collaborations with physicians, public health practitioners, researchers, policy administrators, activists, political appointees, and community members in diverse settings throughout the world. An alumna of the university, Barlow is deeply committed to preparing future scientists and health professionals for a future world of scientific thought and praxis, using writing as a tool of critical thought and intervention. She teaches a research-intensive first year course on writing science and health, using women's health as a point of inquiry.

This course meets any student, STEM major or not, at the door of discovery. Recent socio-political moments have attempted to sanitize science in a way that can inhibit such discovery. I aim to describe the discovery process, using STEM as our lens, in such a way that any audience could possibly replicate the experience. This method offers students space to consider multiple standpoints, interrogate their philosophy of science, and consider alternate ways of knowing—all skills critical to introducing students to university academic writing. Students practice weekly reflective responses to prompts, which may include reference reviews, current news in science, conference proceedings, non-governmental reports, and peer-reviewed manuscripts. Through this process, students also practice peer review with opinion-editorials, abstracts, elevator pitches, and academic STEM/health research mini-mock grants they will develop over the semester. Teaching students how to deconstruct research, as well as think critically about current events in STEM, encourages ongoing critical thought and practice beyond the end of the course. Moreover, I teach students to consider alternate approaches of knowledge production; thereby, introducing them to the process of interrogating both their ontologies and epistemologies and the philosophies of science in literature. This is reinforced by what I call "healthy citation practices."

Two course learning objectives central to this process are (a) critically evaluate others' research and conduct scientific research; and (b) become a thoughtful producer of research and develop a discipline of writing, editing, proofreading and "healthy citation practices." My students learn how to weave history and science to synthesize and situate a scientific topic. Through this process, students deconstruct the topic, using traditional tools Audre Lorde (1984) references as the "master's tools" (Bowleg, 2021) and develop an understanding of alternative tools, approaches, practices, and methodologies to address their scientific topic. As a result, students not only embrace critical perspectives (Bowleg, 2021), but also learn the essentiality of developing a research paradigm (Bowleg, 2021). This dynamic process involves two major steps:

1. The first week of the semester, students are tasked with writing a philosophy of science, where they engage texts (Harding, 2011; Popper, 1934/2005) and respond to prompts assessing the nature of their knowledge production (see prompts in the Appendix). Throughout the semester, students return to their philosophy of science, which inevitably expands, as their knowledge production increases through course readings, discussion, and writing exercises. Students' ability to interrogate their philosophy of science represents a necessary step in understanding science.

2. Students are tasked with placing their science/health topic within the context of history. This requires an engagement with various types of references, each offering different slices of the historical narrative. At this point in the semester, we are also deep in the engagement of critical perspectives that offer the foundation of their emerging research paradigm, which is developed through their science/health topic. This contextualization of the literature—implemented by the multiple references and critical perspectives—is augmented by healthy citation practices, where students cite the relevant primary source(s), multiple perspectives, and specifically center marginalized authors, as modeled by the hashtag movements to #CiteBlackWomen (Smith et al., 2021) and #CiteASista (Nicole & Williams, 2018). This rebalancing of the historical narratives serves to counter tunnel vision views of a science/health topic.

Course Reflection: Kylie Quave, Ph.D.

The second co-author is a white EuroAmerican woman raised in the rural south and an anthropological archaeologist teaching science writing. Her research in the South American Andes investigates how Indigenous communities prior to and during European colonization responded with resistance and persistence in the face of imperialism. Quave teaches a first-year writing and research course focused on the themes of scientific racism and racism as a public health crisis. The course brings together texts and ways of knowing from biology, anthropology, political science, economics, science, technology, and society (STS), sociology, psychology, and public health. Assignments focus on writing in different scientific genres and translating research between genres for different types of audiences.

In this science writing course, critical approaches to citation are centered in order to correct the landscape of whose research is elevated and whose research is overwritten or ignored. Citation practices—including choices about who we read, who we assign, whose ideas we deem credible, and whose work we write about—often mirror the existing inequities and exclusionary forces of the societies in which we live (Itchuaqiyaq et al., 2020). Knowledge production tends to follow the same skewed patterns of marginalization of people already pushed to the edges of societies: what this manifests as is an outsize representation of white middle-class men from EuroAmerican and European backgrounds on our bookshelves, syllabuses, and bibliographies (Craven, 2021; Edmonds, 2020; Itchuaqiyaq & Frith, 2022; Tuck et al., 2015). Ample studies across disciplines have shown this pattern to be persistent (e.g., Chakravartty et al., 2018; Hutson, 2002; Itchuaqiyaq, 2022; Mott & Cockayne, 2017).

The course on scientific racism is designed to alter attitudes and practices about citation on several overt and covert fronts. Overcoming the white empiricism that marks the status quo in many science bibliographies, I strive to channel Steeves' decolonizing method of pyroepistemology in teaching citation (2021). In Steeves' research on how deep histories of Indigenous Americans have been obscured in favor of anti-scientific denialism, she puts it as such: "Decolonizing Indigenous histories rebuilds bridges to ancestral places and times, which American archaeology burned in political fires of power and control" (2021, p. 181). Many scientific disciplines have endured such erasures, also called agnotology, which is the purposeful production of ignorance. Expertise has been ignored or cast aside to privilege the views of those already dominant in society, even when it has to do with experiences and knowledge that is not their own. Thus, in my courses, I promote a kind of anti-agnotologist way of choosing readings and citing research in our writing.

I begin my courses with critical examination of how we know what we know using the scientific method. I find that students need to be reminded of the tenets of science and the ways in which science is designed to be self-correcting. It is not a failure for scientists to err but rather is a failure when scientists do not ask about whether past knowledge production has been erroneous. For example, when Charles Darwin promoted myriad false and harmful assumptions about the nature of human races (Fuentes, 2021) while also providing researchers with the enduring theory of evolution by natural selection. Or how the "slavery hypertension hypothesis" has continued to promote the myth of an African American gene for high blood pressure, absent any evidence of such a deterministic feature (Lujan & DiCarlo, 2018).

At the heart of these and other examples is citational praxis. Students must deconstruct scientific studies and focus on writers' citational habits and how writers construct knowledge based on assumptions from their fields and elsewhere. I ask them to locate the writers' positionality as they assess the authors' epistemology (Takacs, 2003), and they work together in class to ask and answer questions that they report they have never or rarely asked about sources:

1. What is the author's background and worldview? What discipline are they from?
2. What kind of evidence is used? What is missing? What is not measured? What is excluded? How are some kinds of evidence weighted more heavily than others?
3. What kind of sources do the writers cite? What kinds of expertise have the writers prioritized and deemed credible in this study?

When students learn about citation in research writing, I steer them away from thinking of it as a legalistic matter of giving credit and instead ask them to see citation as opening up their worlds; citation creates new conversations, and I want them to see that both when they read and when they compose their own research. I urge them to view citation as a series of choices we writers are able to make and not as inevitable. Furthermore, I ask them to reconsider their assumptions about who is credible and authoritative and to question preconceptions about objectivity and bias. Instead, students are required to incorporate citations from writers of varied positionalities and backgrounds and are encouraged to think critically about what counts as "scholarship." I do not ask students to make checklists of identities or fill quotas; rather, I merely ask them to reflect on who is there in the citations, who is not, and what we might miss without greater multivocality.

Furthermore, I do not restrict student writers to a definition of scholarship that only includes peer-reviewed works but rather ask them to consider the role of peer review and what other forms of review could provide comparable outcomes beyond academic publishing. We look into examples of how peer review sometimes fails, particularly by reading cases from RetractionWatch.org and reflecting on shortcomings in our own shared process as peer reviewers in our course. When students are restricted to only citing that which has undergone peer review, they miss out on a whole universe of expertise that is excluded from academic knowledge-making.

The learning outcome in this science writing course that is informed by Black Feminist, Indigenous Feminist, and decolonizing approaches is to have a critical understanding of expertise, credibility, evidence, and authority that is demonstrated through a reflexive and inclusive citation practice. The scaffolding to support this kind of learning outcome in any STEM writing course ought to include the following:

1. A syllabus that models prioritization of voices from historically and systemically excluded experts on the course material. Those sources should be scholarly in the broadest sense of the term, in which scholarship is work that is supported by evidence, embedded in prior research, and which is produced by someone with experience or training in the research area.
2. Lessons on bibliometric inequalities in the STEM discipline one is teaching. In archaeology, for example, there are many studies demonstrating an overall underrepresentation of researchers from minoritized backgrounds (e.g., Goldstein et al., 2018; White & Draycott, 2020), and there is also a strong tradition of tracking publication and citation statistics (e.g., Heath-Stout, 2020; Hutson, 2002) to understand who is given a platform to produce knowledge in the field.
3. Scaffolding increasingly complex lessons throughout the semester that introduce citation as a practice shaped by our social and political values. Citation choices are presented as not inevitable but rather as a series of decisions we make as writers, even as our choices may be limited by previous bottlenecks in the publication pipeline that result in outsize influence from certain kinds of researchers and writers. Students encounter this problem first as readers of chosen texts, then as researchers making their own choices, and then as

writers tasked with presenting cited sources with evaluative context. In other words, students learn to offer sources not as self-evident but rather as products of knowledge production processes occurring before the student writer encounters them.

4. Citation styles are taught as functional, and alternative forms are explored to reveal what is missing in traditional scholarly citation practice. It is insufficient to teach students the formal moves of citations without helping them see the construction of citation norms. Personal communications, including oral knowledge like that used to pass on Indigenous ways of knowing in some cases, are often excluded from reference lists and thus devalued, as outlined in Lorisia MacLeod's guide to citing Indigenous oral knowledge (Kornei, 2021). Helping students to find ways to add to or defy normative structures, such as MacLeod's creation of new templates for oral knowledge, puts marginalized epistemologies on even ground with easily cited scientific journal articles.

5. Students are required to reflect on the choices they make in selecting and evaluating sources to include in STEM writing, and they take responsibility for understanding how the epistemologies of the authors they cite are shaped by their positionalities. They are moreover responsible for identifying gaps in understanding that may be introduced by privileging a limited scope of worldviews in their research. Evaluation of their research paper is partly based on how accountable they are to these values, as realistic and truthful explanations of our world through science writing are only possible when we read and cite capaciously.

Broader Contexts for Promoting More Just and Inclusive Ways of Teaching and Learning

Cultivating the next generation of STEM scholars and writers who are well positioned to contribute to innovation is the collective goal we share in our pedagogies. As scientists who teach writing, we are committed to creating a formula for this journey of decolonizing science and democratizing knowledge. Because scientific disciplines have historically sidelined the research of those already marginalized in scientists' broader societies, knowledge is and has been a structural inequity. Addressing the exclusionary ways our disciplines were formed and currently operate requires altering the fabric of how we teach; this must be enacted at multiple levels. We cannot stop at making our reading lists more inclusive; we must also hold students accountable for understanding what is at stake and how to disrupt these structures as they participate in knowledge production.

These teaching methods do not go unquestioned in our experiences. Students have usually been taught prior to their first year in college that the sciences are a place of certainty, single answers, and exactitude. They often have the sense that multiple ways of knowing must not be valid, and that questioning of established paradigms is undesirable. Some struggle to accept decolonizing ways of thinking about research, expertise, and knowledge production. However, helping them to examine case studies such as those cited here from various disciplines, and then asking them to practice citation in the ways we've outlined helps many to re-assess their relationships to the sciences.

Teaching STEM writing in a way that promotes transparency about the intellectual histories (including the successes and the failures in those histories) of STEM disciplines in order to instill sound research and communication methods harnesses the power of teaching to transform society. Doing so through critical analysis of how writers know what they know, disrupts the myth of neutral scientific ways of knowing, while composition offers a site of liberation for those marginalized and excluded by the sciences.

Acknowledgments

This chapter builds on collaboration between the authors and Dr. Cecilia Shelton, whose teaching, research, and writing have strongly influenced the authors' writing here. We are indebted to our friend and colleague. We also thank the peer reviewers for helping improve this manuscript.

References

Alaieri, F. & Vellino, A. (2016). Ethical decision making in robots: Autonomy, trust and responsibility. In A. Agah, J.-J. Cabibihan, A. M. Howard, M. A. Salichs & H. He (Eds.), *Social robotics* (pp. 159–168). Springer International Publishing. https://doi.org/10.1007/978-3-319-47437-3_16.

Allen, J. S. & Jobson, R. C. (2016). The decolonizing generation: (Race and) theory in anthropology since the eighties. *Current Anthropology, 57*(2), 129–148. https://doi.org/10.1086/685502.

Atalay, S. (2006). Indigenous archaeology as decolonizing practice. *American Indian Quarterly, 30*(3/4), 280–310. JSTOR. https://www.jstor.org/stable/4139016.

Baker, K. C. & Mazza, N. (2004). The healing power of writing: Applying the expressive/creative component of poetry therapy. *Journal of Poetry Therapy, 17*(3), 141–154. https://doi.org/10.1080/08893670412331311352.

Barlow, J. N. (2016). #WhenIFellInLoveWithMyself: Disrupting the gaze and loving our Black womanIst self as an act of political warfare. *Meridians, 15*(1), 205–217. https://doi.org/10.2979/meridians.15.1.11.

Barlow, J. N. (2018). Restoring optimal Black mental health and reversing intergenerational trauma in an era of Black Lives Matter. *Biography, 41*(4), 895–908. https://doi.org/10.1353/bio.2018.0084.

Barlow, J. N. & Dill, L. J. (2018). Speaking for ourselves: Reclaiming, redesigning, and reimagining research on Black women's health. *Meridians, 16*(2), 219–229. https://doi.org/10.2979/meridians.16.2.03.

Benjamin, R. (2019). *Race after technology: Abolitionist tools for the New Jim Code*. John Wiley & Sons.

Blakey, M. L. (2020). Archaeology under the blinding light of race. *Current Anthropology, 61*(S22), S183–S197. https://doi.org/10.1086/710357.

Bolnick, D. A., Smith, R. W. A. & Fuentes, A. (2019). How academic diversity is transforming scientific knowledge in biological anthropology. *American Anthropologist, 121*(2), 464–464. https://doi.org/10.1111/aman.13212.

Bowleg, L. (2017). Towards a critical health equity research stance: Why epistemology and methodology matter more than qualitative methods. *Health Education & Behavior, 44*(5), 677–684. https://doi.org/10.1177/1090198117728760.

Bowleg, L. (2021). "The master's tools will never dismantle the master's house": Ten critical lessons for black and other health equity researchers of color. *Health Education & Behavior, 48*(3), 237–249. https://doi.org/10.1177/10901981211007402.

Bowleg, L., Río-González, A. M. del, Holt, S. L., Pérez, C., Massie, J. S., Mandell, J. E. & Boone, C. A. (2017). Intersectional epistemologies of ignorance: How behavioral and social science research shapes what we know, think we know, and don't know about U.S. Black men's sexualities. *The Journal of Sex Research, 54*(4–5), 577–603. https://doi.org/10.1080/00224499.2017.1295300.

Boyd, N. M. & Bright, D. S. (2007). Appreciative inquiry as a mode of action research for community psychology. *Journal of Community Psychology, 35*(8), 1019–1036. https://doi.org/10.1002/jcop.20208.

Braun, L. (2015). Race, ethnicity and lung function: A brief history. *Canadian Journal of Respiratory Therapy: CJRT = Revue Canadienne de La Thérapie Respiratoire : RCTR, 51*(4), 99–101.

Braun, L. & Saunders, B. (2017). Avoiding racial essentialism in medical science curricula. *AMA Journal of Ethics, 19*(6), 518–527. https://doi.org/10.1001/journalofethics.2017.19.6.peer1-1706.

Caliskan, A., Bryson, J. J. & Narayanan, A. (2017). Semantics derived automatically from language corpora contain human-like biases. *Science, 356*(6334), 183–186. https://doi.org/10.1126/science.aal4230.

Campbell, C. & Murray, M. (2004). Community health psychology: Promoting analysis and action for social change. *Journal of Health Psychology, 9*(2), 187–195. https://doi.org/10.1177/1359105304040886.

Carter, C. R. (2018). Racist monuments are killing us. *Museum Anthropology, 41*(2), 139–141. https://doi.org/10.1111/muan.12182.

Chakravartty, P., Kuo, R., Grubbs, V. & McIlwain, C. (2018). #CommunicationSoWhite. *Journal of Communication, 68*(2), 254–266. https://doi.org/10.1093/joc/jqy003.

Combahee River Collective. (1978). A Black feminist statement. In Z. Eisenstein (Ed.), *Capitalist patriarchy and the case for socialist feminism*. Monthly Review Press.

Cooper, A. J. (1892). *A voice from the south*. Aldine Printing House.
Cooper, B. (2016). Intersectionality. In L. Disch & M. Hawkesworth (Eds.), *The Oxford handbook of feminist theory* (pp. 385–406). Oxford University Press.
Cooper, B. (2015). Love no limit: Towards a Black feminist future (in theory). *The Black Scholar, 45*(4), 7–21. https://doi.org/10.1080/00064246.2015.1080912.
Craven, C. (2021). Teaching antiracist citational politics as a project of transformation: Lessons from the Cite Black Women movement for white feminist anthropologists. *Feminist Anthropology, 2*(1), 120–129. https://doi.org/10.1002/fea2.12036.
Crenshaw, K. W. (1994). Mapping the margins: Intersectionality, identity politics, and violence against women of color. In M. A. Fineman & R. Mykitiuk (Eds.), *The public nature of private violence: The discovery of domestic abuse* (pp. 93–118). Routledge.
DeSalvo, L. A. (1999). *Writing as a way of healing: How telling our stories transforms our lives*. Beacon Press.
DiGangi, E. A. & Bethard, J. D. (2020). Uncloaking a lost cause: Decolonizing ancestry estimation in the United States. *American Journal of Physical Anthropology, 75*(2), 422–436. https://doi.org/10.1002/ajpa.24212.
Edmonds, B. (2020). The professional is political: On citational practice and the persistent problem of academic plunder. *Journal of Feminist Scholarship, 16*, 74–77. https://doi.org/10.23860/jfs.2019.16.08.
Fuentes, A. (2021). "On the races of man": Race, racism, science, and hope. In J. DeSilva (Ed.), *A most interesting problem: what Darwin's* Descent of man *got right and wrong about human evolution* (pp. 144–161). Princeton University Press. https://press.princeton.edu/books/hardcover/9780691191140/a-most-interesting-problem.
Geller, P. L. (2020). Building nation, becoming object: The bio-politics of the Samuel G. Morton crania collection. *Historical Archaeology, 54*(1), 52–70. https://doi.org/10.1007/s41636-019-00218-3.
Goldstein, L., Mills, B. J., Herr, S., Burkholder, J. E., Aiello, L. & Thornton, C. (2018). Why do fewer women than men apply for grants after their PhDs? *American Antiquity, 83*(3), 367–386. https://doi.org/10.1017/aaq.2017.73.
Green, J. A. (Ed.). (2017). *Making space for Indigenous feminism* (2nd ed.). Fernwood Publishing.
Harding, S. (2011). *The postcolonial science and technology studies reader*. Duke University Press. https://www.dukeupress.edu/the-postcolonial-science-and-technology-studies-reader.
Heath-Stout, L. E. (2020). Who writes about archaeology? An intersectional study of authorship in archaeological journals. *American Antiquity, 85*(3), 407–426. https://doi.org/10.1017/aaq.2020.28.
Hill Collins, P. (1989). The social construction of black feminist thought. *Signs: Journal of Women in Culture and Society, 14*(4), 745–773. https://www.jstor.org/stable/3174683.
hooks, b. (2000). *Feminist theory: From margin to center* (2nd ed.). Pluto Press. (Original work published in 1984).
Hutson, S. R. (2002). Gendered citation practices in American antiquity and other archaeology journals. *American Antiquity, 67*(2), 331–342. https://doi.org/10.2307/2694570.
Itchuaqiyaq, C. U. (2022). The woman who tricked the machine: Challenging the neutrality of defaults and building coalitions for marginalized scholars. In M. J. Faris & S. Holmes (Eds.), *Reprogrammable rhetoric: Critical making theories and methods in rhetoric and composition* (pp. 75–91). Utah State University Press.

Itchuaqiyaq, C. U. & Frith, J. (2022). Citational practices as a site of resistance and radical pedagogy: Positioning the multiply marginalized and underrepresented (MMU) scholar database as an infrastructural intervention. *Communication Design Quarterly*, *10*(3), 10–19. https://doi.org/10.1145/3507870.3507872.

Itchuaqiyaq, C., Litts, B., Suarez, M., Taylor, C. & Glass, C. (2020). Citation as a critical practice. *Intersections on Inclusion: Critical Conversations about the Academy*, *1*. https://digitalcommons.usu.edu/inter_inclusion/1.

Kolopenuk, J. (2020). Miskâsowin: Indigenous science, technology, and society. *Genealogy*, *4*(1), 21. https://doi.org/10.3390/genealogy4010021.

Kornei, K. (2021, November 9). Academic Citations Evolve to Include Indigenous Oral Teachings. *Eos*. http://eos.org/articles/academic-citations-evolve-to-include-indigenous-oral-teachings.

Kozlowski, D., Larivière, V., Sugimoto, C. R. & Monroe-White, T. (2022). Intersectional inequalities in science. *Proceedings of the National Academy of Sciences, 119*(2). https://doi.org/10.1073/pnas.2113067119.

Larsen, R. R., Cruz, H. D., Kaplan, J., Fuentes, A., Marks, J., Pigliucci, M., Alfano, M., Smith, D. L. & Schroeder, L. (2020). More than provocative, less than scientific: A commentary on the editorial decision to publish Cofnas (2020). *Philosophical Psychology, 33*(7), 893–898. https://doi.org/10.1080/09515089.2020.1805199.

Lorde, A. (1984). *Sister outsider: Essays and speeches*. Penguin.

Lujan, H. L. & DiCarlo, S. E. (2018). The "African gene" theory: It is time to stop teaching and promoting the slavery hypertension hypothesis. *Advances in Physiology Education, 42*(3), 412–416. https://doi.org/10.1152/advan.00070.2018.

Marcus, A. A. (2020, July 31). Springer Nature retracts paper that hundreds called "overtly racist." Retraction Watch. https://retractionwatch.com/2020/07/31/springer-nature-retracts-paper-that-hundreds-called-overtly-racist/.

Marcus, A. A. (2021, August 25). Journal retracts more articles for being "unethical, scientifically flawed, and based on racist ideas and agenda." Retraction Watch. https://retractionwatch.com/2021/08/25/journal-retracts-more-articles-for-being-unethical-scientifically-flawed-and-based-on-racist-ideas-and-agenda/.

Marks, J. (2017). *Is science racist?* John Wiley & Sons.

Meyer, M. D. E. (2007). Women speak(ing): Forty years of feminist contributions to rhetoric and an agenda for feminist rhetorical studies. *Communication Quarterly, 55*(1), 1–17. https://doi.org/10.1080/01463370600998293.

Mills, K. J. (2020). "It's systemic": Environmental racial microaggressions experienced by Black undergraduates at a predominantly white institution. *Journal of Diversity in Higher Education, 13*(1), 44–55. https://doi.org/10.1037/dhe0000121.

Mitchell, P. W. (2018). The fault in his seeds: Lost notes to the case of bias in Samuel George Morton's cranial race science. *PLOS Biology, 16*(10), e2007008. https://doi.org/10.1371/journal.pbio.2007008.

Mott, C. & Cockayne, D. (2017). Citation matters: Mobilizing the politics of citation toward a practice of 'conscientious engagement.' *Gender, Place & Culture, 24*(7), 954–973. https://doi.org/10.1080/0966369X.2017.1339022.

Mullins, P. R. (2017, April 30). The aesthetics of bliss and trauma in plantation weddings. Archaeology and Material Culture. https://paulmullins.wordpress.com/2017/04/30/the-aesthetics-of-bliss-and-trauma-in-plantation-weddings/.

Nature. (2022). How Nature contributed to science's discriminatory legacy. *Nature, 609*(7929), 875–876. https://doi.org/10.1038/d41586-022-03035-6.

Nicole, J. & Williams, B. M. (2018). #CiteASista: Today & Everyday. https://citeasista.com/.

Owens, D. C. (2017). *Medical bondage: Race, gender, and the origins of American gynecology*. University of Georgia Press.

Panofsky, A., Dasgupta, K. & Iturriaga, N. (2020). How white nationalists mobilize genetics: From genetic ancestry and human biodiversity to counterscience and metapolitics. *American Journal of Physical Anthropology, 175*(2). https://doi.org/10.1002/ajpa.24150.

Pennebaker, J. W. (1990). *Opening up: The healing power of expressing emotions*. Guilford Press.

Popper, K. (2005). *The logic of scientific discovery*. Routledge. (Original work published in 1934).

Prescod-Weinstein, C. (2020). Making Black women scientists under white empiricism: The racialization of epistemology in physics. *Signs: Journal of Women in Culture and Society, 45*(2), 421–447. https://doi.org/10.1086/704991.

Reid, L. C. (2021). "It's not about us": Exploring white-public heritage space, community, and commemoration on Jamestown Island, Virginia. *International Journal of Historical Archaeology, 26,* 25–52. https://doi.org/10.1007/s10761-021-00593-9.

Rifkin, M. (2016). Addressing underrepresentation: Physics teaching for all. *The Physics Teacher, 54*(2), 72–74. https://doi.org/10.1119/1.4940167.

Ryder, P. M. (2012). *Rhetorics for community action: Public writing and writing publics*. Lexington Books.

Shelton, C. (2020). Shifting out of neutral: Centering difference, bias, and social justice in a business writing course. *Technical Communication Quarterly, 29*(1), 18–32. https://doi.org/10.1080/10572252.2019.1640287.

Smith, C. A., Williams, E. L., Wadud, I. A., Pirtle, W. N. L. & Collective, T. C. B. W. (2021). Cite Black women: A critical praxis (a statement). *Feminist Anthropology, 2*(1), 10–17. https://doi.org/10.1002/fea2.12040.

Smith, L. T. (2021). *Decolonizing methodologies: Research and Indigenous peoples* (3rd ed.). Bloomsbury Academic & Professional. (Original work published in 1999).

Steeves, P. F. C. (2021). *The Indigenous paleolithic of the western hemisphere*. University of Nebraska Press; JSTOR. https://doi.org/10.2307/j.ctv1s5nzn7.

Takacs, D. (2003). How does your positionality bias your epistemology? *Thought & Action, 19*(1), 27–38.

TallBear, K. (2014). Standing with and speaking as faith: A Feminist-Indigenous approach to inquiry. *Journal of Research Practice, 10*(2), N17–N17.

TallBear, K. (2016, January 28). Science and whiteness. https://www.youtube.com/watch?v=pzVKVBgb4S4.

Tallman, S., Parr, N. & Winburn, A. (2021). Assumed differences; unquestioned typologies: The oversimplification of race and ancestry in forensic anthropology. *Forensic Anthropology*, Early View. https://doi.org/10.5744/fa.2020.0046.

Tuck, E. & Yang, K. W. (2012). Decolonization is not a metaphor. *Decolonization: Indigeneity, Education & Society, 1,* 1–40.

Tuck, E., Yang, K. W. & Gaztambide-Fernández R. (2015, April). Citation practices. *Critical Ethnic Studies*. http://www.criticalethnicstudiesjournal.org/citation-practices.

Vyas, D. A., Eisenstein, L. G. & Jones, D. S. (2020). Hidden in plain sight—Reconsidering the use of race correction in clinical algorithms. *New England Journal of Medicine, 383*(9), 874–882. https://doi.org/10.1056/NEJMms2004740.

White, W. & Draycott, C. (2020, July 7). Why the whiteness of archaeology is a problem. SAPIENS. https://www.sapiens.org/archaeology/archaeology-diversity/.

Appendix: Philosophy of Science Writing Prompt:

What is truth? And, how do we seek it? Is there only one truth? Are there multiple truths? Is truth necessary in science? Why or why not? How did your gender, sexuality, race, class, religion, neighborhood, nationality, personality contribute to your understanding of your world and what is meaningful? How do you begin research? What is important to you? Why do scientists rely on models and theories which are at least partially inaccurate? How is this related to implicit or unconscious bias? Explicit bias? What role does ethical research play in your approach to or thoughts about science? What is your goal in science?

The Inclusive Potential of Teaching the History of (White Mainstream) English as the International Language of Science

Elizabeth Blomstedt
UNIVERSITY OF SOUTHERN CALIFORNIA

Students whose primary language is not White Mainstream English, the version of English most valued by academia, often feel disadvantaged in classrooms where the majority of the reading, writing, and discussion is conducted in White Mainstream English. Students may view their language identities as a liability rather than an asset, which can hinder their ability to think through course concepts, read and comprehend academic publications, participate in class discussion, and complete ambitious and thoughtful writing projects (Baker-Bell, 2020). This view is reinforced when their field and their instructors implicitly or explicitly uphold the notion that White Mainstream English is the rightfully preferred English variety for the field, ignoring the historical and ongoing impacts of white language supremacy (WLS) on our everyday linguistic practices (Lee & Rice, 2007); this is why I am using "White Mainstream English," rather than terms like "Standard Written English" or "Edited American English," in this chapter: to highlight the role that white privilege plays in making certain Englishes or languages "standard" (Baker-Bell, 2020, p. 3). In STEM writing classes, this notion is codified by the concept of English as the International Language of Science (EILS) or the use of White Mainstream English in most publications, conference presentations, and other texts circulating within the field. While we ought not assume that any student who does not have a background in White Mainstream English will automatically struggle with confidence or communication in a writing course and must assume that even students from similar linguistic backgrounds will have different attitudes and experiences, acknowledging that white supremacy culture has created a barrier for many of our students is an important first step in enacting anti-racist pedagogical practices in writing instruction to create a more inclusive classroom. For STEM writing instructors, one way to do this is to teach the historical origins of EILS to illustrate that White Mainstream English's use in their field is not a rightfully-earned position but rather the result of factors related to colonialism, power, and luck (Huttner-Koros, 2015; Phillipson, 1992; Porzucki, 2014).

In this chapter, I present a series of lessons that use translingual pedagogical approaches to challenge white supremacist conceptions of linguistic ability and create a more inclusive STEM writing classroom. The lessons teach students the history of

EILS as a means of challenging embedded notions of White Mainstream English supremacy in STEM, pairing that with reflective writing on the students' own views of their linguistic practices. Table 2.1 outlines the different parts of this lesson.

Table 2.1. Overview of Assignment Sequence Detailed in This Chapter.

Activity	Topic
Individual written reflection	Personal views of writing ability, specifically looking at multiple linguistic identities or practices
Whole class discussion	Personal views of writing ability, specifically looking at multiple linguistic identities or practices
Small group discussion	Hypothesizing about how and why White Mainstream English became the language of science (EILS)
Assigned reading	History and complications of EILS, specifically: Nina Porzucki's "How did English become the language of science?" from The World Adam Huttner-Koros' "The Hidden Bias of Science's Universal Language" from The Atlantic
Individual written reflection	Identifying and challenging limiting beliefs about writing ability
Mini-lecture	Benefits of multilingualism and speaking or writing in multiple Englishes

To better understand the context and necessity of these lessons, I begin the chapter with a review of the damaging impacts of WLS on STEM students and examine how translingual pedagogies may assist in addressing those harms. I then discuss how to enact a translingual pedagogy in a STEM writing course through these lessons, discussing each part of the lesson in depth and ending with student responses, including how to respond to student resistance.

(White Mainstream) English as the International Language of Science (EILS)

In our pursuit of inclusive pedagogical practices, we must first acknowledge that the dominance of this particular variety of English, White Mainstream English, is both the product and proponent of WLS. The Conference on College Composition and Communication's (CCCC) "Statement on White Language Supremacy" defines WLS as a structural tool of white supremacy that uses "the ideology of individualism as it works with meritocracy" to position performance in White Mainstream English as a valid criterion for evaluating communication skills (*CCCC Statement on White Language Supremacy*, 2021). White Mainstream English's dominance in STEM fields is tied to the global spread of English and is largely the result of Western countries using English as a tool to dominate and control new, current, and former colonies, a tactic Robert Phillipson (1992) refers to as linguistic imperialism. While there are obvious benefits to having a global lingua franca (or common language), the dominance of White Mainstream English has created concern about its negative impacts for other languages because this linguistic imperialism places White Mainstream English above other Englishes and other languages. While the debates about the causes, costs, and benefits of EILS will continue to produce important insights, for the purposes of the set of lessons I present in this chapter, I will focus on the ways EILS's positioning of White Mainstream English as the "rightful" and "natural" language of science harms multilingual students.

Publication in STEM journals demonstrates the negative impacts EILS has on multilingual writers. Writers based in the United States enjoy a greater rate of publication and are more likely to serve on the boards of academic journals in their fields (Canagarajah, 1996, 2002; Gibbs, 1996), and more recent metanalyses from multiple STEM fields have found that academic journals feature more writing from scholars in countries where English is the dominant language (Clavero, 2011; Yen & Hung, 2018). The deep, personal impacts this has on STEM students are visible in Dhatri Badri's opening vignette (this collection), where she shares the ways in which her identity as an Indian woman prevented her from feeling a sense of belonging in STEM. Scientists whose first language is not English face a 30 percent lower chance of having their papers be accepted for publication than native English speakers (Pronskikh, 2018) and cite their own English-language research more frequently than the research written in their mother tongue (Grabe, 1988), highlighting that publishing in White Mainstream English is an important tool for gaining cultural capital. These practices, often made obvious in bylines, impact how STEM students view themselves, their linguistic abilities, and their chances of contributing knowledge in their field. A study of 45 multilingual international students found that 82 percent of respondents rated their English skills as "weak" or "adequate," and there was an inverse relationship between respondents' assessment of the strength of their English abilities and their perception of English's importance in their field (Tardy, 2004). Multilingual students may feel defeated or overwhelmed by the position of White Mainstream English as the primary language of STEM, particularly if White Mainstream English's position of supremacy is seen as innate and their language practices are seen as detracting from their ability to communicate in White Mainstream English.

Yet despite these harms, STEM (writing) instructors may not prioritize linguistic justice, which Jerry Won Lee (2016) defines as "confronting the inequitable discursive economics that afford disproportionate amounts of social capital to certain language practices over others" (p. 176). This may be because they view STEM writing as chiefly communicating objective knowledge. In the previous chapter, Jameta Nicole Barlow and Kylie Quave pinpoint how the misunderstanding of science as a "neutral, value-free way of knowing the world" often hides the influence of colonialism on STEM and the inequities embedded in the generation of scientific knowledge (this collection). This misunderstanding can also hide linguistic hierarchies. Many hold the view succinctly articulated by Vitaly Pronskikh (2018) that "[m]uch of the STEM discourse is sufficiently technical to reduce the role of natural language and linguistic injustice to a relatively minor degree" (p. 83). This avoidance is further incentivized by a false belief that writing instruction is separable from the course content (Donnell et al., 1999; Minakova & Canagarajah, 2020) and by institutions' undervaluing of teaching writing across the curriculum, which allows them to relegate writing instruction generally, and critical literacy awareness more specifically, to first-year writing programs and writing centers (Jordan & Kedrowicz, 2011). But the reality is that STEM writing's replication of linguistic injustice has major consequences. Beyond impacting students' views of their own language practices and scholars' publication rates and prestige, the linguistic injustice caused by complicity with WLS in the STEM writing class reduces our capacity to address global problems meaningfully, as explained by Ghanashyam Sharma (2018):

> These monolingual orientations are uniquely harmful for the STEM fields because scientists are among the first in line to have to cultivate a sense of global citizenship, advance knowledge, and address social challenges on global scales. Curricular and pedagogical blind spots created by monolingual worldviews can create practical challenges when STEM scholars and students are faced with the complexities of conveying specialized knowledge to outside and mixed audiences. They can also undermine academic engagement in cross-cultural and transnational communication as well as obscuring political and socioeconomic issues in academic and professional writing. (p. 44)

Just as Asao B. Inoue (2019) argued that those in the field of writing studies have power in arguments around the valid use of language and thus a responsibility to dismantle WLS, those in STEM fields have a responsibility to challenge the monolingual view of White Mainstream English as the sole valid language of communication for science. While I contend that linguistic justice does not necessarily require removing English as the international language of science, the "confrontation" it requires of us does prompt us to acknowledge and work to undo the negative impacts EILS has on writers from different linguistic and cultural backgrounds. This is owed both to the STEM community, which is made up of (largely multilingual) faculty, researchers, and students from a variety of cultural and linguistic backgrounds, and to the public, which depends on STEM's findings to shape their lives.

Contextualizing EILS through Translingual Pedagogies

Much like Barlow and Quave (this collection) advocate for an anti-colonialist pedagogical approach that examines and challenges how knowledge has been produced in STEM, I advocate for a translingual pedagogical approach that examines and challenges positioning White Mainstream English as the "language" of science in order to create a more inclusive STEM classroom and global community. Translingualism is a tool for challenging WLS's monolingual views and creating a more equitable STEM field. Like multilingual views of language, translingualism acknowledges the validity of multiple linguistic expressions; however, where multilingual views of language suggest linguistic systems are separate and compartmentalized, translingual dispositions view languages as dynamic and fluid, flowing into each other and informing each other more than we might initially think (Frost et al., 2020). As Nancy Bou Ayash (2020) explains, this approach is inherently anti-racist because it "contests a dominant monolingual English-only ideology, which propagates problematic representations and treatments of language as stable, internally uniform, and having status outside and beyond the cultural, political, economic, and ideological forces that bring about its practices" and instead "foregrounds the mutable, performed, and emergent nature of language and insists on the agency of its users and learners" (p. 14). A translingual approach to language not only advocates for those for whom English is their second language but also for those "native speakers" who primarily communicate in other varieties of English like Black English (Lee, 2016, p. 178). Challenging the idea of language as discrete and concrete opens up space for a diversity of language practices, including different varieties of a language, to be considered valuable.

Translingual pedagogical approaches are utilized in two ways in the lesson I present in this chapter: challenging how students view the powerful position of White Mainstream English and leading students to see the value of their own translingual writing processes. To practice a translingual pedagogy means we must be honest with students about the inaccuracies of White Mainstream English's assertion that it has "status outside and beyond the cultural, political, economic, and ideological forces that bring about its practices" (Ayash, 2020, p. 14). The lesson series presented in this chapter does this by teaching students the history of the perception of White Mainstream English as the "language" of STEM, illuminating the forces that brought White Mainstream English to that place of prominence alongside reflections of their own linguistic practices in order to encourage students to feel confident in their own diverse linguistic abilities. One of the most obvious ways that translingual approaches can be adopted is through the creation of multilingual writing products that utilize code-meshing, the practice of using different language varieties or languages in the same rhetorical context (like combining White Mainstream English with Black English or combining English and Spanish) which many see as a tool for promoting egalitarian language practices (Ricker

Shreiber & Watson, 2018). However, this is not the only avenue for embracing translingual writing pedagogy; truly translingual approaches to language show that multilingualism can play a pivotal role at other stages of the writing process, even for writing projects that are monolingual (Lee, 2016). Developing awareness and appreciation for translingual writing processes has the potential to help our students see the value of their own diverse language practices rather than viewing them as detracting from their ability to communicate in White Mainstream English.

Learning the History of White Mainstream English's Ascent to EILS

The lesson I outline in this chapter would be suitable for any discipline-specific STEM writing course or even a general STEM course with a significant writing component. I developed it specifically for an upper-division elective writing course titled Technical Writing for Scientists and Engineers that I taught at my previous institution, a four-year public research university where writing was primarily taught through required first-year writing courses embedded in residential colleges. This was the first upper-division interdisciplinary writing elective offered on our campus, and it grew out of conversations with STEM professors and department administrators who wanted to both prepare their undergraduate students to write in the major and provide graduate students with the opportunity to TA for the course, giving them teaching experience that might also further hone their own writing skills. The goal of the course was to acquaint students from different STEM fields with some of the most common genres of STEM writing, including research articles, review articles, research posters, and conference proposals. STEM professors I collaborated with on the course expressed a desire for students to develop confidence in these genres and in their writing, reading, and critical thinking skills.

I've primarily taught this lesson in courses where the majority of students were multilingual, though, as I'll share later in the chapter, students whose home language closely resembled White Mainstream English also benefit from this series of lessons. Because the course was writing-focused, we did have explicit conversations about WLS from the very first day of class. On the first day of all of my writing classes, I show Jamila Lyiscott's spoken word poem "3 Ways to Speak English" (https://www.ted.com/talks/jamila_lyiscott_3_ways_to_speak_english) as a way to introduce thinking about text rhetorically and linguistic justice. In it, she challenges the ways white people react to her performance of White Mainstream English as a Black woman and highlights the ways her being a "tri-lingual orator" is both shaped by violent historical forces of colonialism and works to make her a better communicator (Lyiscott, 2014). While explicit classroom conversations about WLS in academia will support the lesson presented here, it can stand on its own as an introduction to EILS.

I find it most impactful to teach this early in the semester or quarter, though it may also be helpful at the start of units focused on style or editing.

Beginning to Examine Personal View of Writing Ability

I begin with an in-class reflective assignment where students freewrite (record one's thoughts in writing without stopping to consider structure, composition, or grammar) answers to the following questions about their writing development:

1. How would you assess your overall writing ability? What do you think are your strengths? Your weaknesses?
2. What types of feedback have you gotten from instructors in the past? What negative feedback sticks out in your mind? What positive feedback can you remember?
3. Do you speak different languages with your family, friends, or in other contexts? Or do you speak other types of English than the type of English we read in STEM journal articles? Describe the different languages and/or Englishes you speak in different contexts.
4. Do you think the different languages or Englishes you speak help or hurt your ability to write in academic settings? Why or why not? Do you ever use your other languages or Englishes when thinking about, planning, or drafting writing assignments? If so, how?

Though these can be assigned for homework or presented to students all at once, I prefer to put one set of questions up on the board or projector screen, allow a few minutes for students to write, and then put up the next set, telling students that they can take more time on questions they find more generative. This approach also gives me the opportunity to explain what I mean by "other Englishes" and list examples of how they might use different languages and Englishes in their composing processes. I then have a fifteen-minute whole-class discussion, asking for student volunteers to share their responses to questions 3 and 4. This discussion can alleviate students' sense of alienation in their language practices or beliefs about them and introduce them to new ideas or insights about linguistic practices. We also begin to discuss concepts like White Mainstream English and WLS toward the end of this discussion if we have not covered them in previous classes. I also collect students' freewriting so that I get to know students' individual experiences and viewpoints of their linguistic practices.

Brainstorming then Reading about the Causes of EILS

I then put students into groups of 3 or 4 to hypothesize about why White Mainstream English became the international language of science, asking them to appoint one student notetaker to record their ideas. For interdisciplinary courses, I put students in groups according to major (or similar majors) so that they can discuss both EILS broadly and why they think White Mainstream English is the language most often used for their specific discipline. This leads to the next question I give them to discuss: "Is English, specifically White Mainstream English, a 'good' language for science? Why or why not?" After groups have discussed for about ten minutes, I have each group share the ideas that came up in their discussion. Some are able to pinpoint specific historical phenomena that contributed to White Mainstream English becoming a default language, like computers and coding languages being created in the United States. Other students may have knowledge about other languages (Latin, French, or German) being used in their field in past decades and centuries. This discussion, especially sharing ideas about whether or not English is a "good" language for STEM, is an important first step toward examining the belief that White Mainstream English's higher status is innate and separate from those of other Englishes or languages.

I typically end class here and assign students two brief readings from popular journalism sources that explain the history and consequences of EILS. The first is an audio broadcast (also available as a text article) produced by Nina Porzucki (2014) for *The World*, titled "How did English become the language of science?" The piece covers the rise of EILS in the 20th century, attributing it largely to German falling out of favor after World War I. Understanding how EILS is culturally and historically influenced enables students "to develop an understanding of the discipline as culturally situated" and challenge the "Eurocentric perspective" that STEM is taught from, as Alicia Bitler and Ebtissam Oraby elaborate on in their chapter later in this section (this collection). The second piece I assign, "The Hidden Bias of Science's Universal Language," by Adam Huttner-Koros (2015) in *The Atlantic*, focuses on the ways EILS harms scientists whose native language is not White Mainstream English, the production and circulation of scientific knowledge, and other languages. I like to start the next class period with an open discussion in which students share their reactions, experiences, and questions they have about the content presented in the readings. If discussion stalls, I'll ask students to take a few minutes to review the pieces and write two discussion questions, then ask them to read their questions aloud. For especially quiet classes, taking five minutes for students to freewrite a reflection can help them gather their thoughts to begin a class discussion, or students can be paired off to share their freewritten thoughts with a partner.

Challenging Limiting Beliefs about Writing and Multilingualism

After this discussion, I transition back to where this series of activities started: students reflecting on their views of themselves as writers. I ask students to revisit their writing from the previous class, specifically their thoughts about their weaknesses as a writer. I explain the concept of limiting beliefs, which are ideas that people hold to be true about themselves that hold them back in some way; they often begin with statements like "I can't" or "I don't" and are typically adopted from our experiences, education, faulty logic, or out of fear. I then prompt students to write out some limiting beliefs they have about themselves as a writer, sharing some examples like: "I don't have anything original to write about," "Nobody will care what I have to say," "I'm not a good enough writer to do this topic justice," or "I am bad at grammar." I share research with students about the negative impacts of limiting beliefs, of which there are many; Tamlin Conner and Lisa Feldman Barrett (2005), for example, found that unconscious beliefs can hinder our ability to embrace challenges, and thus simply identifying our limiting beliefs can be an important first step to becoming more capable and confident writers. To continue this process, I ask students to do the following with their limiting beliefs:

1. Write out where you think these beliefs originate from. Are you extrapolating one piece of feedback you received once? Are you using this as an excuse to not try something new or something that scares you?
2. Challenge your limiting beliefs. Write out evidence that contradicts or challenges your limiting beliefs. What positive feedback have you received about these aspects of your writing or thinking? Alternatively, how might simply identifying these limiting beliefs serve you in your journey to dismantle them?

I conclude by sharing overviews of research on the writing skills of multilingual students and students who speak different versions of English in order to help students see their language practices as assets rather than liabilities before prompting them to reframe their own limiting beliefs in light of this information. There are numerous pieces of research supporting the assertion that multilingualism improves creativity, critical thinking, and cultural awareness, but I like to show students the American Academy of Arts & Science's 2017 report from the Commission on Language Learning's executive summary, which highlights the specific positive impacts speaking multiple languages has on one's cognitive abilities, cultural sensitivity, and even on preventing or slowing negative health impacts associated with aging (Commission on Language Learning, 2017). I invite multilingual students who have examples of their multilingualism giving them greater rhetorical knowledge and flexibility to share those experiences with the class, providing concrete examples of how these strengths manifest themselves. Once we've discussed the specific benefits of multilingualism, I prompt students to think about how this might cause them to rethink their limiting beliefs. I then ask students who are comfortable to submit their written reflection on limiting beliefs to me so that I can be aware of the areas they are working on gaining confidence; given that all students have different experiences and perspectives, these are helpful for me as I continue to deepen my understanding of my specific students' attitudes toward writing and their linguistic abilities, especially those who are less willing to share in discussion. I can also refer back to these before commenting on students' writing to tailor my feedback to their specific concerns.

This lesson combines both instruction on the history of EILS with personal reflection in order to help students not only understand how EILS has been shaped by WLS but how those practices, in turn, impact their conceptions of themselves as people and as writers and thinkers in the STEM community. Combining this with small-group and whole-class discussions gives students the opportunity to learn from their peers' application of the material to their own writing lives.

While the type of reflective writing described here may be less popular in STEM writing courses than in expository writing courses, it is essential for transferring writing skills from one writing situation to another and for facilitating the types of attitudinal shifts around linguistic practices necessary to enact an inclusive pedagogy (Hendricks, 2018; Herrington & Stassen, 2016; Yancey et al., 2014). In her study on how college students develop as writers, Lee Ann Carroll (2002) advocates for writing instructors to help students gain awareness of their own development through "self-reflection that learns a new knowledge or skill by unlearning and revising old knowledge or skill" (p. 131). If we want students to question their internalization of White Mainstream English as supreme language, we must guide them to question their own internalized views of it before introducing new translingual views of writing. This reflection helps students develop a wider, more flexible approach to the writing process, and it is central to aiding in their development of the types of critical thinking skills that will enable them to question the impacts WLS has had on academic writing and STEM writing in particular. Ideas about these and other potential benefits of metacognitive and reflective assignments in STEM writing courses can be found throughout this collection (e.g., Badri; Barlow and Quave; Callow and Shelton; Bitler and Oraby).

Before turning to how students react to this lesson, I'd like to consider how translingualism is at play here. While I do tell students they are welcome to write in other Englishes or utilize code-meshing to write in other languages in their free-written self-reflections (a practice I welcome whenever students do reflective writing), for many students, the readings they did, the discussions we had, and the writing they produced were in White Mainstream English. The reason I label this pedagogical practice as translingual is because it runs counter to the monolingual perspectives upheld by WLS by directly challenging the notion that White Mainstream English is innately superior to other Englishes and languages through teaching the history of the social, political, and economic forces at play in crowning White Mainstream English the international language of science. The final piece of the lesson, teaching students about the benefits of multilingualism and speaking multiple Englishes, begins the work that I will continue throughout the semester of making room for students' linguistic backgrounds and practices to be seen as having a positive impact on their language abilities. This work includes Lyiscott's "3 Ways to Speak English" from the first day of class and also includes instructional texts on science writing that acknowledge the ways power and privilege play a role in shaping what are often assumed to be "objective" scientific texts. In my STEM writing course, I rely on excerpts from *The Scientists' Guide to Writing* by Stephen Heard (2016) because it acknowledges the cultural and historical factors that influence norms of scientific writing, like this excerpt from his chapter on sentences that contextualizes the use of passive voice in scientific writing:

> Early scientific writing was predominantly active-voice (Gross et al. 2002). This fit well with science done by respected gentlemen and with authority derived from virtual witnessing (Box 11.1): vivid description of the actors and the action conferred rhetorical strength. As science became professionalized in the nineteenth century, however, scientists looked for objectivity in prose—with objectivity meaning "knowledge that bears no trace of the knower" (Daston and Gallison 2007). The passive voice let writers suppress any mention of the person who actually conducted the experiment, analyzed the data, or drew the conclusion. This is odd, though, because we all know it's only pretense: trees don't fell themselves! Authority in modern science comes from our adoption of appropriate conduct and techniques—not from pretending we don't exist. (p.165)

Instructors can also assign texts from the field of writing studies on the rhetorical choices made in scientific texts, including Wayne C. Booth's (2004) discussion of the decisions made by Watson and Crick to humbly present their double-helix findings (pp. 57–59). These texts and ideas help students continue to hone their rhetorical skills by considering the ways in which context, purpose, and audience shape even seemingly objective scientific texts, and they elucidate the reality that our conventions around language are culturally produced.

Student Reactions and Growth Opportunities

In this section, I present the most common types of student responses to this sequence of lessons, both to demonstrate their value for students and to prepare instructors for discussions about WLS and potential resistance points. My students' reactions to this assignment vary. I've had a student come to my office hours to tell me they decided to teach their two-year-old child their family's home language after learning there are benefits to multilingualism, and I've also had a student in class discussion accuse me of not doing my job by presenting this "distracting" lesson when I could be teaching them White Mainstream English. I've also been met with silence in the whole-class discussion portions of this lesson, even if their freewriting shows engagement with the ideas and readings. Even those who are skeptical of the lesson early in the semester sometimes cite it in their end-of-semester reflective portfolios and course evaluations as an important spark for shifting the way they thought about writing. Because no two individuals have the exact same attitude toward language difference, it is important to meet students where they are and to be mindful of their experiences and attitudes in individual conferences and written feedback with them. This section, thus, is not meant to generalize how different student "types" may react to this series of lessons but rather to help prepare instructors for a few broad categories of responses.

Students who respond positively to the lesson often demonstrate how these ideas impacted their own attitudes and approaches to their writing, thinking, and composing processes in their final freewriting on their limiting beliefs on writing, though insights also develop in small and large-group discussions. Multilingual students, particularly multilingual international students, sometimes express relief that their struggles with learning the expectations of communicating in White Mainstream English are not a personal failing and are instead influenced by a specific set of historic and cultural factors. One student wrote in his written reflection identifying and challenging limiting beliefs about his writing ability that he was relieved to hear from other native speakers of White Mainstream English that they, too, struggled with early drafts, and the historical context for EILS helped him see that he was not doomed to be a bad scientist because he struggled with White Mainstream English. In recognizing this historical context, white students who communicate primarily in White Mainstream English are often shocked by the prospect that they could be reading or writing in French or German if not for the circumstances that led to EILS. These students sharing their surprise or commenting on how ill-prepared they would be to read biology articles in French is a stark example of WLS, and it often softens their disposition toward the translingual pedagogical approach I adopt in my classes. For many students, this lesson sequence contextualizes the prominence of White Mainstream English in STEM communication in a way that helps them develop greater empathy toward themselves and their classmates; I see this empathy toward others most clearly in peer review and class workshops later in the semester.

Some students take this awareness a step further and become interested in linguistic justice or in building a more audience-aware communication style in STEM. Building on the earlier lesson using Lyiscott's "3 Ways to Speak English" to illustrate the ways power plays a role in our reaction to different Englishes, and Black English specifically, students who communicate in different varieties of

English often connect the role that power played in establishing EILS to the ways their own varieties of English are treated in the United States. These students are often the most willing to begin to challenge conventions of STEM writing, like adopting the #BetterPoster[1] model over the traditional scientific research poster, and embrace a translingual disposition in their drafting process that helps them integrate their different linguistic practices in a way that feels productive rather than detracting from their ability to communicate effectively. Linguistic justice and shifting STEM audiences' expectations for STEM communication become a secondary interest for some students, whether they are multilingual, speak multiple varieties of English, or adhere fairly closely to White Mainstream English in the majority of their communication. These interests are encouraged by ideas in course readings and discussions, including the Heard (2016) and Booth (2004) ideas referenced earlier, and they can also be explored through class discussions and workshops of sample student texts that continually call attention to audience responses to texts.

Some students are, of course, more resistant, objecting to the pedagogical approach or the content presented in the lesson sequence. For students in any of my writing classes who are resistant to my broader pedagogical practices, specifically utilizing reflection and encouraging a translingual disposition, I share research from the field of writing studies that supports these practices, including many of the texts cited in this chapter. However, some students are so deeply rooted in the idea that writing is a skill that can be easily transmitted to them using the banking model of education that they expect me to be focused solely on sharing knowledge about "good writing" with them rather than leading them to reflect on their own writing processes and the cultural and historical forces influencing them. Though this attitude about the banking model of education is present in most writing courses, in my experience, it is more common in STEM writing courses where students appear to have already internalized the idea voiced by Pronskikh (2018) that the "technical" nature of STEM communication makes it more objective and thus makes linguistic justice less relevant. Multilingual international students who have been working on their White Mainstream English skills for years in order to prepare them for success in their education and future careers, for example, may have little interest in performing inquiry into their own language practices, particularly if they see this as detracting from the time I have to teach them White Mainstream English. When these attitudes appear in class discussion, I like to offer more context for the impossibility of a static "Standard English" existing. One way I do this is by reciting or showing students the prologue to the Canterbury Tales,

---

1   #BetterPoster is a template for scientific research posters created by Mike Morrison that minimizes text on the research poster in favor of making it easier for people to read the main finding of a research project when perusing poster sessions. It focuses on meeting the needs of the genre while also encouraging discussion about research during a poster session rather than silent reading.

which is incomprehensible to modern-day English speakers, to show students how drastically the English language has changed in just 600 years. I may also introduce the concept of World Englishes, asking students to consider what is needed from a lingua franca and why we punish writers who effectively communicate meaning but do not perform White Mainstream English flawlessly. I also express my commitment to helping students work on the literacy skills they want to develop while highlighting the benefits of viewing texts as being shaped rhetorically and influenced by the context in which they were written and approaching grammatical issues by performing inquiry in their own patterns of error. While this does not win over all the skeptics in my class, it tends to soften many students' dispositions toward my approach. I am careful to encourage audience awareness, though, teaching students to be mindful that some professors or journals will be much stricter about adherence to proper grammatical or syntactical conventions; in our class, though, we will focus most heavily on the necessary thinking and composing skills that tend to be more difficult to develop than proper proofreading.

Toward a More Inclusive STEM Writing Class and Community

In this chapter, I've presented a lesson that combines written self-reflection, small-group and whole-class discussion, and instruction on the history of EILS and the benefits of multilingualism and communicating in multiple Englishes aimed at helping students better understand the historical roots of and impact of WLS on STEM writing. I contend that this translingual pedagogical approach challenges assumptions about the "rightful" place of White Mainstream English as the language of STEM, assumptions that can harm the attitudes and practices of students who are multilingual or who communicate in other Englishes. This lesson has taken on additional importance in recent years with the rise of text-generating generative artificial intelligence systems like ChatGPT; these tools are produced by scholars in STEM fields and currently produce text that largely adheres to these standards dictated by WLS. I often follow this assignment with readings about biases in ChatGPT, giving my STEM writing students the opportunity to examine how language bias in STEM can have broader impacts on our world.

Of course, this short lesson sequence is not the only necessary step in creating an inclusive STEM classroom; this approach must continue throughout the semester in order to truly help students feel that their linguistic practices and backgrounds are welcome in the STEM writing classroom. For example, I mentioned earlier that this lesson helped students gain empathy that manifested itself in their response to their peers' papers in group conferencing and peer review. I did not rely on this singular lesson sequence to shape peer review practices, though; I modeled a respectful form of peer response and specifically asked students to focus on higher-order concerns and avoid making judgments of a student's grammar, punctuation, or other minor language issues. In other words, this assignment sequence is an important foundational lesson in creating an atmosphere that is accepting of language difference and, more specifically, the translingual nature of writing, but this attitude must be affirmed in other teaching practices as well.

My primary goal in this lesson is to create a classroom environment where all students, regardless of their home languages or cultural backgrounds, can learn, and challenging WLS is central to achieving this goal in any writing course. But a necessary and related outcome of these kinds of pedagogical shifts is creating an attitude in academia and in STEM specifically that is more welcoming of so-called "language differences." If English is to be a lingua franca for the scientific community, why can't we loosen our understanding of what "correct" English is? What is lost in the refusal to be accepting of World Englishes? Seeing students embrace these ideas about challenging WLS gives me hope that these kinds of lessons will equip the next generation of researchers and scientists to make productive changes to the expectations of language practices in STEM journals, classrooms, and conferences, but we cannot shirk our own responsibility to push for greater acceptance of language difference. As scholars in writing studies and as scholars in STEM fields, we have the power to influence our colleagues, colleges, and publications to recognize the white supremacist roots of the adherence to a strict form of White Mainstream English. We ought to use it to create a more inclusive field.

References

Ayash, N. B. (2020). Developing translingual language representations: Implications for writing pedagogy. In A. Frost, J. Kiernan & S. B. Malley (Eds.), *Translingual dispositions: Globalized approaches to the teaching of writing* (pp. 13–31). The WAC Clearinghouse; University Press of Colorado. https://doi.org/10.37514/INT-B.2020.0438.

Baker-Bell, A. (2020). *Linguistic justice: Black language, literacy, identity, and pedagogy.* NCTE/Routledge. https://doi.org/10.4324/9781315147383.

Booth, W. C. (2004). *The rhetoric of rhetoric: The quest for effective communication.* Wiley-Blackwell.

Canagarajah, S. (1996). "Nondiscursive" requirements in academic publishing, material resources of periphery scholars, and the politics of knowledge production. *Written Communication, 13*(4), 435–472. https://doi.org/10.1177/0741088396013004001.

Canagarajah, S. (2002). Multilingual writers and the academic community: Towards a critical relationship. *Journal of English for Academic Purposes, 1*(1), 29–44. https://doi.org/10.1016/S1475-1585(02)00007-3.

Carroll, L. A. (2002). *Rehearsing new roles: How college students develop as writers*. Southern Illinois University Press.

Clavero, M. (2011). Language bias in ecological journals. *Frontiers in Ecology and the Environment, 9*(2), 93–94. https://doi.org/10.1890/11.WB.001.

Commission on Language Learning. (2017). *America's languages: Investing in language education for the 21st century*. American Academy of Arts & Sciences. https://tinyurl.com/ycknvhk7.

CCCC statement on White Language Supremacy. (2021). Conference on College Composition and Communication. https://cccc.ncte.org/cccc/white-language-supremacy.

Conner, T. & Barrett, L. F. (2005). Implicit self-attitudes predict spontaneous affect in daily life. *Emotion, 5*(4), 476–488. https://doi.org/10.1037/1528-3542.5.4.476.

Donnell, J. A., Petraglia-Bahri, J. & Gable, A. C. (1999). Writing vs. content, skills vs. rhetoric: More and less false dichotomies. *Language and Learning Across the Disciplines, 3*(2), 113–117. https://doi.org/10.37514/lld-j.1999.3.2.09.

Frost, A., Kiernan, J. & Malley, S. B. (2020). *Translingual dispositions: Globalized approaches to the teaching of writing*. The WAC Clearinghouse; University Press of Colorado. https://doi.org/10.37514/INT-B.2020.0438.

Gibbs, W. (1996). Lost science in the Third World. *Scientific American, 273*(2), 92–99. https://doi.org/10.1038/scientificamerican0895-92.

Grabe, W. (1988). What every EFL teacher should know about reading in English. *Anglo-American Journal, 7*, 177–200.

Heard, S. B. (2016). *The scientist's guide to writing*. Princeton University Press.

Hendricks, C. C. (2018). WAC/WID and transfer: Towards a transdisciplinary view of academic writing. *Across the Disciplines, 15*(3), 1–15. https://doi.org/10.37514/atd-j.2018.15.3.11.

Herrington, A. J. & Stassen, M. L. A. (2016). Intersections of writing, reflection, and integration. *Across the Disciplines, 13*(4), 1–11. https://doi.org/10.37514/atd-j.2016.13.4.21.

Huttner-Koros, A. (2015, August 21). The hidden bias of science's universal language. *The Atlantic*. https://www.theatlantic.com/science/archive/2015/08/english-universal-language-science-research/400919/.

Inoue, A. B. (2019, March 14). How do we language so people stop killing each other, or what do we do about white language supremacy? https://docs.google.com/document/d/11ACklcUmqGvTzCMPlETChBwS-Ic3t2BOLi13u8IUEp4/.

Jordan, J. & Kedrowicz, A. (2011). Attitudes about graduate L2 Writing in engineering: Possibilities for more integrated instruction. *Across the Disciplines, 8*(4), 1–17. https://doi.org/10.37514/atd-j.2011.8.4.24.

Lee, J. J. & Rice, C. (2007). Welcome to America? International student perceptions of discrimination. *Education, 53*(3), 381–409. https://doi.org/10.1007/s10734-005-4508-3.

Lee, J. W. (2016). Beyond translingual writing. *College English, 79*(2), 174–195.

Lyiscott, J. (2014, February). 3 ways to speak English. *TED*. https://www.ted.com/talks/jamila_lyiscott_3_ways_to_speak_english.

Minakova, V. & Canagarajah, S. (2020). Monolingual ideologies versus spatial repertoires: Language beliefs and writing practices of an international STEM scholar. *International*

Journal of Bilingual Education and Bilingualism, 26(6), 708–721. https://doi.org/10.10 80/13670050.2020.1768210.

Phillipson, R. (1992). *Linguistic imperialism*. Oxford University Press.

Porzucki, N. (2014, October 6). How did English become the language of science? *The World*. https://theworld.org/stories/2014-10-06/how-did-english-become-language -science.

Pronskikh, V. (2018). Linguistic privilege and justice: What can we learn from STEM? *Philosophical Papers, 47*(1), 71–92. https://doi.org/10.1080/05568641.2018.1429739.

Ricker Schreiber, B. & Watson, M. (2018). Translingualism ≠ code-meshing: A response to Gevers' "Translingualism revisited" (2018). *Journal of Second Language Writing, 42,* 94–97. https://doi.org/10.1016/j.jslw.2018.10.007.

Sharma, G. (2018). Internalizing writing in the STEM disciplines. *Across the Disciplines, 15*(1), 1–21. https://doi.org/10.37514/ATD-J.2018.15.4.19.

Tardy, C. (2004). The role of English in scientific communication: Lingua franca or Tyrannosaurus Rex? *Journal of English for Academic Purposes, 3*(3), 247–269. https:// doi.org/10.1016/j.jeap.2003.10.001.

Yancey, K. B., Robertson, L. & Taczak, K. (2014). *Writing across contexts: Transfer, composition, and sites of writing*. Utah State University Press.

Yen, C.-P. & Hung, T.-W. (2018). New data on the linguistic diversity of authorship in philosophy journals. *Erkenntnis, 84*(4), 953–974. https://doi.org/10.1007/s10670 -018-9989-4.

"Science has always been about asking questions": Critical Science Literacy in STEM Writing

Megan Callow
University of Washington–Seattle

Holly Shelton
George Fox University

During my (Holly's) first teaching job in Chile, *mi hermana anfitriona* introduced me to a Spanish idiom, "*tener buen lejos,*" which means something or someone looks good from a distance but becomes less attractive once you get a closer look. For many diverse students, STEM fields are initially attractive, but the actual experience of entering a STEM field often results in underrepresented students being disparately turned off and turned away. Biology instructors at our own university are aware of opportunity gaps in their introductory survey courses, which have impacted access to STEM fields for students who are historically underrepresented in STEM majors, and these instructors have even actively researched and implemented opportunities for change. One of their published studies specifically recognizes the following:

> Underrepresented minority (URM) students in the United States . . . start college with the same level of interest in STEM majors as their overrepresented peers, but 6-[year] STEM completion rates drop from 52% for Asian Americans and 43% for Caucasians to 22% for African Americans, 29% for Latinx, and 25% for Native Americans (Theobald et al., 2020, p. 6476).

While STEM faculty have focused on mitigating disparate impacts on student attrition based on *achievement* to provide greater access and opportunities in STEM, we focus on *experience* for the same purposes. Both Dhatri Badri's vignette and a journal entry from Alicia Bitler and Ebtissam Oraby's chapter in this volume have highlighted the role that a single course can have for students in situating themselves and their chosen STEM field. Badri related how bridging the gap between her place in STEM and her experience as a woman of color helped her to engineer her own inclusion, while the journal writer in Bitler and Oraby said that recognizing cultural situatedness led to understanding her discipline better. To these insights, we add a chorus of student voices on how and to what extent these

DOI: https://doi.org/10.37514/ATD-B.2024.2364.2.05

acts of bridging, engineering, situating, and recognizing knowledge and identity within a single course can impact student experience.

This chapter describes our journey as writing instructors situated within a disciplinary writing program to propose and teach a course entitled Critical Science Literacy in the Natural Sciences. This chapter attempts to synthesize some of the theoretical strands that feed into the critical science literacy (CSL) framework and describe the course we designed to enhance students' capacities for critical inquiry into their own disciplines. Such capacities can facilitate students' identities as scientific practitioners, which can then enhance their agency and feelings of legitimacy as communicators and can also enable them to critique and disrupt from within disciplinary structures that have caused historical exclusion or harm. Consequently, CSL can better support students across diverse backgrounds who are working toward a degree, and eventually a career, in the natural sciences. In this chapter, we describe the theoretical underpinnings of CSL and how our institution and program contexts revealed a need for CSL-based curriculum. We then offer an overview of the course itself and finally reflect on its successes and possibilities based on student survey data.

The Framework: Critical Science Literacy

We have adapted the concept of CSL from Susanna Priest (2013) and Maria E. Gigante (2014), who developed the concept in response to the social turn in the natural sciences and to increasing lay publics' participation in science. Priest highlights the social and cultural nature of science, writing that scientific knowledge includes "the kind of everyday, tacit knowledge of 'how things work' that members of a culture take for granted but outsiders can find mystifying" (p. 138). The other key piece of Priest's definition is the civic importance of scientific knowledge: citizens need to be able to "sort out which truths should be relied on in any given moment" (p. 138). These "moments" of reliance on science have proliferated in part because of the grave stakes of issues like climate change and global pandemics and because of the multitude of public-facing platforms on which scientific debates now play out.

Gigante (2014) takes up Priest's version of CSL but expands its role in undergraduate science education. Gigante calls on specialists in writing studies, rhetoric of science, and science communication to help introduce CSL because undergraduate survey science courses do not often focus on how "scientific culture operates," nor are students typically "prompted to take responsibility for communicating their research to nonexpert publics" (p. 78). Personally and professionally, we can attest to a growing necessity for scientists to be able to communicate with different kinds of publics. Gigante asserts that a key element of CSL is rhetorical knowledge and suggests that "a rhetorically grounded science writing course can assist science majors with understanding the intricacies of scientific communication" within and beyond science (p. 79). We argue that CSL can help expose what Jennifer Mallette describes in this volume as "tacit requirements and expectations" in science, which "can serve to widen gaps between students with and without access to stronger preparation in writing, better mentoring, or effective peer educational networks" (this collection). Additionally, CSL can help students recognize and, we hope, disrupt what Jameta Barlow and Kylie Quave (this volume) term "white empiricism" and other "inequitable forms of knowledge production" in STEM (this collection).

Whether or not they do so deliberately in their claims that science is social, civic, and rhetorical, Priest and Gigante evoke "threshold concepts" in biology. Jan H.F. Meyer and Ray Land (2012) define a threshold concept as "a transformed way of understanding, or interpreting, or viewing something without which the learner cannot progress" (p. 3). Threshold concepts in a discipline have certain characteristics: they transform understanding, are irreversible (once you know, you cannot unknow), are integrative (that is, they reveal interrelated ideas), are bounded, and are potentially troublesome. "Troublesome" knowledge can refer to the conceptually difficult or tacit, which is why grasping a threshold concept can often be accompanied by an "Ah ha!" moment.

Charlotte Taylor (2012) notes that threshold concepts in biology can be difficult to generalize because there are so many far-flung subfields that comprise the biological sciences, all with different foci and methods: environmental, cellular/molecular, marine, biochemistry, physiology, evolutionary, genetics, etc. Speaking broadly, however, she found that threshold concepts in biology tend to be interrelated, and deeper understanding comes from link-building (examples include the complexity of living systems, probability and uncertainty, consequences of meiosis, and creating hypotheses). The webbed, cumulative nature of scientific knowledge creates challenges for teaching undergraduate biology courses, and these challenges may lead to pedagogies that are more focused on "fact-transmission" than on the contexts, cultures, and processes of science (Taylor, 2012, p. 90; see also Weinstein, 2009).

Threshold concepts from writing studies relevant to CSL are that writing is a social and rhetorical activity and that writing enacts and creates identities and ideologies (Adler-Kassner & Wardle, 2015). These understandings have been called for and cultivated by every contributor to this volume as we recognize the situatedness of science knowledge across time, space, and identity. Of particular interest to us as writing instructors is the role of language in learning science because, first, many scientific discourses can be technical and specialized and difficult to learn, and second, because scientific language varies so much as its audiences and stakeholders shift. Meyer and Land's (2012) characterization of the role of language in the threshold concepts framework is helpful:

> Specific discourses have developed within disciplines to represent (and simultaneously privilege) particular understandings and ways of seeing and thinking. Such discourses distinguish individual communities of practice and are necessarily less familiar to new entrants (p. 14).

Not only is language characterized as a layer of difficulty (an aspect of "troublesomeness") in achieving a threshold concept, but the passage also alludes to power relationships inherent in learning disciplinary discourse. We propose that CSL acknowledges (the learning of) scientific discourse(s) as its own threshold concept, particularly in the sense that discourse is something "without which [students] cannot progress" (p. 14) within the discipline. Facility with discourse should not be limited to mastery of terminology but should include recognition of the highly rhetorical and hierarchical nature of scientific discourses. To that end, CSL includes the ability to move across multiple discursive channels because communicating in science often involves acts of translation and accommodation (Fahnestock, 1986).

Learning disciplinary discourse also involves moving through certain kinds of communities and cultures. To learn science—to learn to *be a scientist*—is in large part a process of language learning, but one that includes "not merely knowledge of form, but knowledge of the rhetorical requirements of that form and of the writing behaviors common to professional scientists" (Poe et al., 2010, p., 23). Cultivating what Mya Poe, Neal Lerner, and Jennifer Craig call a "scientific discursive identity" can be challenging, given that it requires understanding "the *social significance* of appropriating scientific discourse in interpersonal and intrapersonal contexts" (Brown et al., 2005, p. 781; emphasis added). Scientific texts *have* social significance to the degree that they conform to certain tenets, such as objectivity, certainty, and authority (or, read another way, exclusion). The more technical or specialized a discourse, the more difficult it may be to enter into because, often, these discursive practices are so different from the kinds of literacy practiced in personal lives and communities (Adler-Kassner & Wardle, 2015). This leaves us with questions about the ways that scientific discourse occludes subjectivity, uncertainty, and exclusion based on difference.

As we developed our own CSL course, we wished to focus not only on the rhetorical and social aspects of science and its discourses but also on the ways that these knowledges have traditionally harmed or excluded certain groups; the ways that scientific texts have suppressed, sterilized, or rationalized that harm; and the ways that scientific practice and communication have discouraged a sense of belonging among those with ostensibly conflicting cultural frameworks and literacy practices. That is, we wanted to lean into the *criticality* of CSL. To build a more capacious CSL, we borrow from Allan Luke's (2012) work on critical literacy (distinct from

CSL and from the phrase "critical thinking" commonly used in K–12 teaching and learning contexts), a framework that entails:

> a) a focus on ideology critique and cultural analysis as a key element of education against cultural exclusion and marginalization; b) a commitment to the inclusion of working class, linguistic, and cultural minorities, indigenous learners, and others marginalized on the basis of gender, sexuality, and other forms of difference; and c) an engagement with the significance of text, ideology, and discourse in the construction of social and material relations, everyday cultural and political life (p. 6).

Many in the various but interlocking fields of Science and Technology Studies (STS) have written of the ways that scientific practice has lent itself to the kinds of critiques Luke enumerates here (Epstein, 2007; Nelson, 2016; Roberts, 2011; and TallBear, 2013 are but a few examples). We try to feature, whenever possible, these critiques in our courses.

In sum, we embrace Priest's emphasis on the social and civic nature of science and Gigante's emphasis on science as rhetorical. In our own development of a CSL course, we also focus on the ways that scientific discourse can suppress human bias and the very real repercussions that bias can have on material conditions and embodied experience. Because acquiring a new discourse is like taking on a new identity, we aim to teach skills in rhetorical analysis, but in critical terms, and ask constantly how scientists' experiences and positions shape how they conceive of and communicate science. Our ultimate goal is to work against the "weed out" culture so commonly seen in undergraduate science programs; just as advocates are pushing for more human-centered approaches for the stakeholders of science (patients, consumers, citizens), this course seeks to create communities of belonging for young scientists themselves.

The Context: The Institution and the Program

The University of Washington is a public Research I university located in Seattle. As of autumn 2022 (UW is on the quarter system), the flagship Seattle campus enrolls over 48,000 students, including 33,000 undergraduates. About 26 percent of its students hail from Washington State, and 11 percent are international students; 36 percent of students identify as Caucasian, 24 percent as Asian, 9 percent as Hispanic/Latino, 5 percent African American, and about 1 percent each as Hawaiian/Pacific Islanders and American Indian. The most popular majors include computer science, business administration, psychology, biochemistry, and electrical engineering, and 48% of students occupy STEM majors (University of Washington, 2022).

Competition for admission into many STEM majors at UW can be intense, which makes for a high-stakes experience. Demand for STEM majors has increased so much in the last decade that students must apply for acceptance to the most popular majors, though acceptance rates are highly variable; for example, nearly 100 percent of the 550–600 students who apply to the biology major are accepted, whereas over 7,500 students named computer science as their top choice for a major, but there are only 550 spaces per year (Stiffler, 2022). While the numbers for the biology major seem reassuring on their face, one staffer from the department admitted that the introductory courses serve a "weed out" function, which prompts questions about the forms of gatekeeping students experience as they progress toward (and through) their majors. Additionally, persistently low funding from the state and increased reliance on funding from industry (UW is among the top ten universities receiving industrial support in the US) have led to a STEM-oriented culture on campus that is felt across disciplines and departments.

The Program for Writing Across Campus (PWAC)[1] offers disciplinary writing seminars in two different formats, linked and unlinked. The linked courses (the majority of the courses we offer) are each connected with large lecture courses in a wide variety of disciplines that range from philosophy to biology (we offer writing courses linked with several of the institution's most popular majors listed above). Only students enrolled in the lecture course are eligible to take the linked writing course. In recent years, PWAC has been offering a growing number of unlinked discipline-specific courses: writing in the humanities, writing in the social sciences, technical writing, and CSL. While all our courses satisfy the general education Composition and Additional Writing requirements, they are not required for any major or program. Students seeking to satisfy their writing requirements in discipline-specific contexts tend to gravitate toward our program rather than taking general composition courses.

All PWAC writing assignments and scaffolding activities are designed to cultivate students' knowledge of the topics, genres, and methodologies of the lecture course (or discipline) with which it is linked. PWAC instructors, as writing rather than content experts, help students orient themselves in a particular field and orienteer themselves toward developing disciplinary writing identities. This specificity of writing contexts enables PWAC instructors to develop inquiries surrounding students' writing and learning that are infrequently addressed in other institutional contexts.

The unlinked CSL course we present in this chapter was first conceived based on the observations, experiences, and critical conversations among three biology-linked writing instructors (one of whom is Holly) and, later, with the PWAC director (Megan). BIOL 180 is the first of a three-sequence introductory set of biology courses,

---

1   The program was formerly called The Interdisciplinary Writing Program, but the name was changed in 2022. Data for this chapter were collected prior to the name change, but for practical purposes, we use the new name here.

which regularly enrolls 600–800 students, mostly sophomores (Brownell et al., 2014; Haak et al., 2011). Students must pass these courses to continue major coursework in biology. The department has taken steps to fill opportunity gaps based on class structures/activities and measuring student achievement scores quantitatively; however, qualitative student experience has not been considered during these studies.

As an anecdotal insight into potential student experience, two PWAC instructors (Holly and Hsinmei Lin) observed the first day of class in BIOL 180. One slide from the lecture dealt with defining science in contrast to pseudo-science and religion. In the scenario, someone had crashed on a bicycle, and the wounds were now healing. Students were asked to explain the healing process according to science ("The blood clotted to form a scab, and now my cells are regenerating skin"), pseudo-science ("I was wearing my lucky shirt, so I made it through"), and religion ("I prayed, and a higher power healed me").

For the PWAC instructors, the scenario established a straw-man argument where pseudo-science and religion could be simplified, essentialized, caricatured, and knocked down easily in contrast to science. Hsinmei noted that pseudo-science is often used as a colonizing proxy term for Indigenous Traditional Knowledge (TK) and/or Traditional Chinese Medicine (TCM), which was part of her own cultural background. Holly reflected on her own extended identity crisis when she was told by her undergraduate science instructors that there was no room for her faith tradition if she were pursuing science and how she later discovered that many famous scientists did, in fact, draw on their faith traditions to animate their work in science. Underrepresented students could feel that important and complex aspects of their identities are not welcomed in disciplinary coursework even when active learning components are implemented structurally.

Student identity and experience are important for understanding inclusion but were not addressed by the Biology Department studies. It is worth noting that some of the structural changes implemented by the Biology Department still provided hope for greater access in STEM courses, so we will briefly overview some of the key findings based on their framing around "achievement" (even though "opportunity" may be a better framework for placing responsibility on the program, rather than students, for disparate results and access). Theobald et al. (2020) note that grades in STEM courses are heavily dependent on exam scores, so exam score "achievement gaps" (in the language of the Biology Department) create a barrier for URM students when grades fall below the threshold that allows for continued study or if students decide themselves to withdraw. The research team was able to show that active learning reduced "achievement gaps" in exam scores by 33 percent and passing rates by 45 percent. Overall, they propose that deliberate practice and a culture of inclusion are the two key elements to reversing achievement gaps in STEM courses.

If active learning increases opportunities for URM student success, what does this active learning look like? David C. Haak et al. (2011) show that high-structure classes with pre-reading quizzes, informal class group work, and multiple-choice clicker or random call questions were able to halve the "achievement gap" observed in traditional lecture-based class structures. Questions that prompted students to engage higher levels of Bloom's taxonomy and opportunities for group work where students could co-construct knowledge were crucial. Sara E. Brownell et al. (2014) explain that experimental design is a fundamental skill and an important aspect of science literacy and critical thinking. Asking students to analyze an experiment OR design an experiment using worksheet prompts as a group for 30 minutes during class led to more accurate conceptions of sample size and repeating an experiment. In addition, Sarah L. Eddy et al. (2015) identify several dimensions that can be used to evaluate a class's overall active learning elements for students: does the class create opportunities for student practice, logic development, accountability, and reducing student apprehension?

All of these studies identify aspects of STEM content courses that can reduce differential passing rates for underrepresented students based on exam scores. However, constraints remain in the learning system overall related to class size, time, content, instructor training, etc., and as noted before, the studies do not collect or address actual student experience.

To return to the anecdotal example of defining science, a more complex and inclusive understanding of science would mean shifting away from a framework of "evidence-based teaching" and toward a pedagogical and curricular emphasis on belonging and equity (see Mallette, this volume). To do so means examining the underlying values and activities of a scientific community, such as asking questions and making systematic observations, rather than setting science in conflict with other knowledge systems or ways of knowing. It would also mean asking students to identify what parts of their own backgrounds contribute to the ways they approach or participate in science, which is how Holly and Hsinmei, as writing instructors, expanded this conversation in their own linked writing classes.

The authors of this chapter recognize that writing courses have unique potential to emphasize elements identified as important for underrepresented student success in BIOL 180 research, such as practice, logic development, and reducing student apprehension and feelings of exclusion. By foregrounding the nature of science and the positionality of its practitioners, CSL can enhance students' feelings of inclusion in STEM courses (and fields).

The Course: Critical Science Literacy in the Natural Sciences (ENGL 296)

After observing the ways that undergraduates were moving toward and through natural science majors, we saw the need for a course that supplemented (and, ideally, reframed) students' apprenticeships in the sciences. We named the course "Critical Literacy in the Natural Sciences," hoping to capture the theoretical investments we have described and also hoping to market the course to (current or aspiring) science majors looking to satisfy a composition requirement. In planning discussions, the authors (Megan and Holly), along with a doctoral student in biology, Aric Rininger, developed a curriculum where students would learn to become confident, authoritative participants in science and scientific discourse while at the same time becoming familiar with the ways that Western values are embedded and centered (often invisibly) in the sciences and its related institutions. Through course content and culturally responsive, anti-oppressive pedagogies, the course aims to help students interrogate these values as they enter advanced study.

Like all PWAC courses, our CSL course implements pedagogical approaches that our program holds dear: students work with a variety of texts, including primary and secondary scientific texts as well as their own and their peers' writing. Throughout, they move through cycles of reading, discussion, reflective and formal writing, peer review, conferencing, revision, and intensive instructor interaction and feedback.

The course is organized around three broad learning goals, listed below, with narrower, orienting questions following each goal. In this course, and as we spell out explicitly in the syllabus, students will work toward the following:

Understanding the nature of science as contingent, contested, and situated.

- What purpose does science serve? Does it have social or moral responsibility?
- How are questions formulated and answered in the sciences? What kinds of questions can science answer? Why do people choose particular questions in science, and how do they develop hypotheses? What sociopolitical and ethical values underlie scientific assumptions, questions, and hypotheses?

Engaging a diversity of ways of knowing and doing in science across cultures and nations, including identifying strengths and limitations of different approaches.

- What are essentialist vs. holistic ways of knowing in science?
- How do scientists situate the self in relation to various communities (academic, professional, disciplinary, cultural, national, indigenous, etc.) and ecologies (environmental, institutional, research contexts, topics/objects of inquiry)?
- By what means can students locate themselves within scientific practice and discourse? What kinds of cultural and intellectual capital do they bring to the course, and might they bring to scientific inquiry?
- How can students deploy a critical lens as they navigate scientific fields first as apprentices and then as professionals?

Tracing the genealogies of ideas in circulation as information moves through pipelines and networks.

- How do scientific concepts and "discoveries" get reified as they are communicated across various platforms? How do reified concepts privilege or harm certain groups? How does "reality" differ across those groups?
- In what ways can novice scientists use transformative communication practices within a realm where the language of Western mainstream science is dominant?
- How can inquiry into scientific content provide occasions for writing to learn as a form of reflection and engagement? Equally important, how can communicating scientific content provide occasions for learning to write in order to share that knowledge with particular audiences?
- How can intertextual connections across time and space provide greater insights into particular "facts"?
- In what ways can scientific communication practices (both traditional and transformative) serve as a vehicle for responding to all of the above questions?

According to the American Association for the Advancement of Science's Benchmarks for Science Literacy (1993), "Students should learn that all sorts of people, indeed, people like themselves, have done and continue to do science" (n.p.). These students bring a huge diversity of cultural and intellectual experience to their studies but may not always feel that non-Western ways of knowing and practicing are embraced (or even acknowledged). UW's STEM education is seen internationally as second to none, but our undergraduate and postsecondary programs encounter the same pipeline issues that the sciences face globally: women, people of color, and indigenous populations are dropping off of scientific career pathways because scientific fields are not sufficiently culturally responsive (Grunspan et al., 2016). This course aims to help students become aware of these hierarchical traditions and to enable them to critique and transform traditional approaches to doing science.

After developing the course in 2019 and submitting it for the formal course approval process, the English department chair informed us that, while he loved the course, we would need to present the syllabus at a meeting of department chairs in the Natural Sciences division of our College. This was a courtesy expected of any department developing a course out of their disciplinary wheelhouse: as English faculty, we needed to get the science department chairs' blessing to offer a course situated in the sciences.

As program director, Megan prepared to present at that meeting and did so somewhat apprehensively. After all, the idea for the course first came about after we had been observing survey science courses with some concern. While our course contains no explicit critique of the science curriculum at our institution, it *does* critique some of the traditional methods and assumptions embedded in science generally. We were surprised and delighted, however, to hear rave responses from the science chairs. They explained that their own major curricula were so packed with content that they had no time to explore issues like the nature of science or uncertainty, subjectivity, or racism in science. The only concern they brought up was that there was no way we would be able to serve hundreds of natural science majors in a given year. Indeed, with classes capped at 23 (as most composition courses are at UW) and at three sections a year, we are lucky if we can serve 70. Still, we were pleased to move forward with the course, and we have been offering about one section per quarter, three to four quarters per year, since 2019. Below, we describe our three major assignment sequences, along with reflections on the assignments' strengths and suggestions for adapting them.

Project 1: Tracing the Life of a Scientific Fact

This assignment draws its inspiration from Jeanne Fahnestock's classic (1986) article, "Accommodating Science: The Rhetorical Life of Scientific Facts," in which Fahnestock analyzes rhetorical shifts as scientific articles are translated from specialized research reports to public-facing summary articles. In this project, students select a topic or event (the Rover landing on Mars and the U.S. arrival of the Asian giant hornet, aka "murder hornet," are two recently selected topics), typically finding the item on a social media platform like Twitter and collecting publications translating the study on different platforms until they trace it back to the source, a peer-reviewed research article.

Students analyzed the ways that authors' rhetorical strategies (e.g., tone, voice, use of images and layout, etc.) changed across publications and then presented their analysis in Adobe Creative Cloud Express (formerly called Spark), a free online program for producing multimedia content. The program is accessible and user-friendly and offers a platform for integrating images, text, and other media, which in turn encourages deep analysis of visual, textual, and digital rhetorics. We have found that students not only become more able to demonstrate how the writing occasion and audience expectations shape the ways that "facts" get represented in texts, but they also come to question the very nature of a scientific fact. If a tweet attempts to capture attention and persuade, then so does an article in *Nature*, in its own way. The assignment also has the added benefit of cultivating media literacy and research skills through the "tracing" process (see Megan's version of the prompt in the Appendix).

Students report through assignment debriefs, mid-quarter check-ins, and course evaluations that learning about the ways that information changes as it is translated across platforms is transformative for them. We find it to be an especially successful assignment at the start of the quarter because it sets up a foundation for students' understanding of accommodation for different rhetorical situations and the stability (or rather dynamic nature) of scientific knowledge, which informs their future thinking and writing. Additionally, it is a creative, low-stakes "writing to learn" assignment that eases students into the next, higher-stakes project.

Project 2: Generating a Scientific Question

The second assignment sequence asks students to consider the scientific hypothesis and research question.[2] Where do research questions and hypotheses come from? What assumptions about reality are embedded within them? How does the formulation of a research question select certain realities and deflect others? We hoped, through this project, to pull back the curtain on the ideological dimensions that are presumed to be neutralized in scientific research.

In the first part of this project, students read and discuss an excerpt from Robin Wall Kimmerer's book *Braiding Sweetgrass*, where the author, a botany professor and member of the Citizen Potawatomi Nation, describes the question that brought her to a botany major as an undergraduate. She wrote, "I wanted to learn about why goldenrod and asters look so beautiful together"—to which her adviser responded, "That is not science," and "If you want to study beauty, you should go to art school" (2013, p. 39–41).

After guided discussion about what counts as science and what kinds of questions are legitimate, students then describe their own curiosities about the natural world (which are often the inquiries that drove them to science in the first place). Finally, they convert these questions into a formal research design proposal, sometimes writing toward a real-life targeted audience, such as the UW Mary Gates Research Scholarship. In the process of drafting, students examine real-life proposals and consider how their own inquiries must be (re)framed for the conventions of formal, institutional contexts.

While students report that this assignment has a lot of practical value—for many it is their first opportunity to compose in a genre that they will use professionally—we ourselves sometimes feel challenged to support it as fully as we can. Neither of us is a scientist. While we do guide students to think through tenets of experimental design like internal and external validity, replicability, etc., we are not able to guide them as well as we would wish (dilemmas about instructor expertise and authority are not uncommon in disciplinary writing programs). However, it gives students more time, space, and authority to apply principles of experimental design than during large survey courses and offers an opportunity to "learn to write" in a disciplinary genre.

---

2   UW quarters are approximately ten weeks long, and each major project is spaced roughly three weeks apart.

Our limitations in Western science give way to even greater limitations in other scientific traditions, such as indigenous science, holistic or "folk" approaches, or TCM. Some students bring with them immensely valuable cultural experience that they can juxtapose against Western frameworks, gaining a deeper understanding of each; other students do not necessarily possess such backgrounds. In future versions of the course, we hope to provide richer introductions to non-Western scientific frameworks. This might include guest speakers or field trips and certainly should include the hiring of instructors who can bring multi- or transcultural perspectives on science. Perhaps most important, we need to better facilitate cross-pollination among students—for example, by sharing former or current student writing—so they can benefit from each other's immense intellectual and cultural knowledges.

During revisions of this chapter, we learned that PWAC instructor Christopher Chan, who has been teaching ENGL 296 in recent quarters, has updated this assignment sequence. He assigns an "ethnographic vignette" in which students are asked to "answer a question about how people conduct, live with, think about, talk about, experience, or contribute to science: a short, descriptive essay that illustrates a scene that you observed, and then uses this observation to explicate an answer for your question." We love this adaptation and believe it responds directly to the course's orienting question about how scientists situate the self within various communities.

Project 3: Science Literacy Narrative

All scientists have intellectual, cultural, and linguistic histories. For the (ostensible) sake of neutrality and objectivity, apprentices are trained to divorce themselves from these histories, especially when they are doing and communicating research. This assignment asks students to read examples of different types of scientists' narratives, which reveal how personal history and professional practice can interface. Examples of such narratives include a 2019 op-ed by Katherine Hayhoe, a climate scientist and self-described evangelical Christian; Nobel Prize winner Youyou Tu's (2011) narrative about how her cultural grounding in TCM enabled her to develop the anti-malaria drug that has saved millions of lives; and/or the documentary film, *Oliver Sacks: His Own Life* (Burns, 2020), about the famous neurologist and his struggles with homophobia and addiction.

Students also read the chapter "How Does Rhetoric Work in Multimodal Projects?" in Ball, Sheppard, and Arola's 2018 textbook *Writer/Designer* to help them ground their choice of a genre that best serves their own scientific narratives. In their narratives, which have included podcasts, films, comic strips, photo essays, and medical school personal statements, students explore how their identities, investments, and intellectual interests have shaped their trajectories as scientists. Holly has used digital stories as a particular genre for the class in which students share their videos at a class showcase event. This assignment is a form of reflection and an orientation to/within the desired scientific field, but we also hope that it will serve as a form of self-advocacy—a confident declaration of intention to participate.

In Their Own Words: Student Survey Feedback

We wanted to be able to assess how we were achieving our learning goals, so we have administered a survey to each course section since we first offered it in the winter of 2020. We obtained IRB approval and collected responses anonymously at the end of the quarter, explaining that responses might be used in scholarly publication, that participation was completely optional, and that it had no bearing on course grades. Given its voluntary nature, we acknowledge selection bias may be at work in the responses. However, while generalizable patterns are helpful, even disparate responses are informative and offer possibilities for differentiated instruction. From the survey, which consists of seven open-ended questions, we hoped to gain a deeper understanding of the following broad questions, which organize the themes we discuss below.

- How did students' understanding of science as contingent, contested, and uncertain change while taking a CSL course?
- How did the course affect students' sense of belonging in the discipline?
- Did the course encourage students to make connections to other courses/learning occasions in the field?

Shifting Understandings

Of 32 responses to the question, "How has your belief in scientific certainty shifted?" 18 said that they had come to understand scientific knowledge as dynamic. As one representative response stated, "I think this class has pushed me to be willing to question the extent and limits of science." Of the 12 who said that their understanding of scientific certainty had *not* shifted, the belief went both ways: some students maintained their belief in scientific certainty; others (about three) maintained their prior belief in *uncertainty* (the course only affirmed it). These two responses stood out in their juxtaposition:

> My belief hasn't shifted because I'm queer, disabled and neurodivergent- bias in science affects me and my people to an extremely dramatic degree.

> No, I still believe that science is absolute. Although there may be bias, as long as the evidence is there, it is ok.

The first response shows awareness of the ways that identity and positionality affect beliefs about the nature of science. It is plausible that those who occupy traditionally marginalized identity groups are better positioned to embrace scientific contingency, just as those with non-normative cultural backgrounds seem better able to integrate multiple scientific frameworks (as we discussed above). However, it is hard to apply the logic in reverse since those who affirmed their belief in scientific certainty did not (could not?) attribute their beliefs to their own positioning. These observations only affirm to us the importance of asking students to share their own experiences and identities with the class community.

In addition to students' shifting perspectives on scientific certainty we observed a shift in their overall metacognition about the nature of science and scientific communication. In response to the multi-part question, "What were your goals for taking this class? Did these goals change over time? To what extent did the course align with your goals?" we noticed a pattern in which students said they initially wanted to "gain skills for scientific writing" (33 of 39 responses) and to "fulfill [the] general education writing requirement" (7 responses). A number of respondents (about 10) expressed a pivot not just in their scientific understanding but also in the kind of knowledge they viewed as important, some noting the unexpectedness of the shift.

> Over time, I became interested in the person behind the publication; what they're [sic] motives are or beliefs, etc. Having finished the course, I can say that I did align with the types of things that I wanted to learn. I want to be a more capable scientist, obviously, and I think that this course helped me more carefully define what "capable" means.

> To be honest, my main goal for taking this class was to receive my English composition credit. However, as the class continued, I realized that I really enjoyed the readings and discussions. This class has taught me to think in slightly different ways, and to—above all—ask questions.

Other students did not report such a shift, though we were satisfied to see that the course still met the practical goals students arrived with. We are reminded here of the writing to learn/learning to write duality of disciplinary writing instruction. The phrase "writing to learn" is shorthand for the idea that the very act of writing stimulates a deeper understanding of a given concept. "Learning to write" (or "learning to communicate") involves practicing composing discipline-specific genres (e.g., a research proposal), largely for the purposes of application in real-world contexts.

Good disciplinary writing courses involve both learning goals, but we were unsurprised that students may value "writing to learn" and "learning to write" differently, depending on their larger goals. The following responses show such different goals—one was self-understanding and intellectual challenge, the other professional preparation.

> I wanted to continue to understand who I was as a writer and develop my writing skills as a scientist. This class definitely challenged me to try things that I have never tried before, and I thought it really pushed me out of my comfort zone.

> I wanted to get better at reading and writing critical science literature to help prepare me for nursing school as well as a career in nursing.

Sense of Belonging in the Sciences

As our course goals indicate, an essential element of CSL is a sense of belonging in the sciences. Enhancing students' feeling of belonging in their future field helps address pipeline issues for URM students, but our hope, too, is that our students' sense of their own humanity as scientists will support their understanding that science is a social, human-driven enterprise.

Of 37 responses to the question, "Do you feel like you have a greater sense of belonging in the sciences after taking this course? Why or why not?" 22 said they felt a greater sense of belonging, five said no, and four gave a qualified response. Some of the explanations students gave for a greater sense of belonging include an expanded exposure to/interest in other classmates' areas of research, increased confidence in "involving myself in more opportunities related to science," an ability to read and write about science in new ways, feeling "more connected with science than before," and overcoming a "sort of 'imposter syndrome'." Some students responded that even though their sense of belonging was not affected, their knowledge increased. For example:

> I don't necessarily feel a greater sense of belonging, but I do feel a greater sense of understanding regarding scientific publications and communication in general.

> I'm not sure, I am more aware though of how exclusionary the current and especially the past fields of science have been towards anyone outside of white males.

One key finding that relates to students' belonging is their ability to choose their own assignment topics—in survey responses and course evaluation responses alike, this is one of the most highly lauded aspects of the course (as it is in our other science writing courses). Encouraging students to select their own writing topics may have had the unanticipated effect of cultivating a sense of belonging in the sciences for students, perhaps more than any other aspect of the curriculum, as this response illustrates: "Every single project I did directly related to something I was truly interested in. It made it so easy to write about something you care about and not just some essay prompt." We urge educators to allow student choice whenever possible in order to encourage student ownership of their own work.

Making Connections

The final question in our survey was, "What connections did you notice between your learning in this course and other courses? Are there frictions or tensions with learning in other courses?" As the subquestion shows, we considered it a very real possibility that CSL could challenge what students were learning in other science courses. The responses revealed a much more interesting picture. In fact, not only did some students see alignment between CSL and other science courses (6 of 33 responses), but many others (14 of 33) spoke about how this course and their other studies in science were mutually reinforcing. We were surprised and delighted to see that, for these students, learning transfer was so immediate and so visible. A couple representative responses include:

> This course helped me be more aware of the rhetorical situation and the audience my writing is addressing. This awareness helped me write to my audience better in other science classes. In addition, reading and researching more about scientific fields that interest me in this class helps me gain more interest and motivation in work for other science classes.

> Science has always been about asking questions, and this class has helped me ask myself more questions when I read a source.

We find responses like these heartening and hope to enhance these connections by encouraging students to pursue topics in the CSL course that they are simultaneously learning about in other science courses. We also surmise that if CSL can help science students better contextualize and create connections across traditionally siloed training milestones (e.g., courses in the major) in science, then this may have meaningful implications for students' sense of belonging in their desired fields.

Other students (about 13 of 33) did not see alignments or even saw active tensions between the CSL and other science courses. Interestingly, students did not cite differences in content but rather characterized the tensions as pedagogical or, in a couple cases, epistemological in nature. A couple of students pointed explicitly to grading systems as key factors, as this one did:

> Some frictions in other courses is the grading system. I really enjoyed that this course had a grading contract so I was responsible for my own grade. We were graded on improvement and completion, not right or wrong.

As another student pointed out, the nature of knowledge tends to be of less concern in other science courses, which may have inhibited their ability to draw connections:

> There was more wanting to know about where information came from in this course than others I have taken. Some other professors don't seem to care much about those things.

While the student above noted that the nature of scientific knowing is a matter of value, the student below observes that it is a matter of time and the imperative of content coverage in science courses:

> Science courses don't have the time to teach you the things that this class does. That's understandable, as there is enough to go over in a basic STEM class. Where else am I supposed to learn this except from a more humanities-centered class?

What the variety of responses to this question tells us is that some students are more open to connections between CSL and science courses, in spite of pedagogical and epistemological differences, and some view them as (even justifiably) siloed (i.e., the sciences and the humanities should each stay in their lane). From these findings, we take away an increased motivation to help students bring inquiries from other areas of their studies into our course. A question for future research is to explore the ways science education is structured, pedagogically and institutionally, since those structures likely have direct implications for how students perceive course content. How do high-stakes exams and grade curves, for example, shape students' knowledge frameworks?

Final Thoughts

As the last excerpt above notes, the humanities may have a meaningful role in science education. We continue our efforts to that end through our Critical Science Literacy in the Natural Sciences course, to which students have responded overwhelmingly positively. Demand for the course increases every quarter. But aside from the findings we describe above, it remains unclear how or whether students are bringing CSL forward into their further studies in science. Certainly, we have no evidence that science instructors are aware of the course, let alone aligning their courses with ours.

There are some glimmers of change, however. Just in the few years elapsed between the introduction of this course and the completion of this chapter, there has been a visible shift in the Biology Department's curricular emphases on the situated nature of science and its harmful histories. The department has even begun to assign more writing in its intro series courses (which is no small feat in a course with 700+ students enrolled) and has invited Megan, in her new role as campus-wide Director of Writing, to come speak at the Biology Teaching and Learning Group about teaching writing in science. We view these as promising shifts, though their incrementality reminds us of the many institutional barriers in place to major curricular reform.

For most STEM majors on campus, writing requirements are not "baked in" to the major pathway. Rather, students must "forage" for courses to satisfy those requirements. Some of them do find their way to our course, but even three or four full sections a year serve only a small fraction of the total number of natural sciences majors on our campus. As we have said, the chairs of the natural sciences departments appeared wildly enthusiastic about the CSL course, but none have signaled a commitment to supporting (with funding or labor) a more systematic offering of this curriculum. We aspire to even greater collaboration with and buy-in from science departments that may have the resources to support the broader reach of CSL curriculum. The siloed nature of university disciplines (and department budgets) certainly discourages that kind of transdisciplinary collaboration, but we hope to leverage the growing interest in social justice to foster a stronger exigence for institutional change. Meanwhile, we offer our colleagues the findings in this chapter and in this volume as a testament to the possibilities for inviting rather than suppressing various identities, experiences, and knowledge traditions as a means to belonging in the sciences. After getting a closer look at the sciences, can all students, in fact, see themselves reflected?

References

American Association for the Advancement of Science (1993). *Project 2061: Benchmarks for science literacy*. Oxford University Press.
Adler-Kassner, L. & Wardle, E. (Eds.). (2015). *Naming what we know: Threshold concepts of writing studies*. Utah State University Press.
Ball, C. E., Sheppard, J. & Arola, K. L. (2018). *Writer/designer: A guide to making multimodal projects*. Bedford/St. Martin's.
Brown, B. A., Reveles, J. M. & Kelly, G. J. (2005). Scientific literacy and discursive identity: A theoretical framework for understanding science learning. *Science Education, 89*(5), 779–802. https://doi.org/10.1002/sce.20069.
Brownell, S., Wendcroth, M., Theobald, R., Okoroafor, N., Koval, M., Freeman, S., Walcher-Chevillet, C. L. & Crowe, A. (2014). How students think about experimental

design. *Bioscience, 64*(2), 125–137. https://doi.org/10.1093/biosci/bit016.

Burns, R. (Director). (2020). *Oliver Sacks: His own life* [Film]. Zeitgeist Films.

Eddy, S., Converse, M. & Wenderoth, M. (2015). PORTAAL: A classroom observation tool assessing evidence-based teaching practices for active learning in large science, technology, engineering, and mathematics classes. *CBE Life Sciences Education, 14*(2), 14:ar23. https://doi.org/10.1187/cbe.14-06-0095.

Epstein, S. (2007). *Inclusion: The politics of difference in medical research*. University of Chicago Press.

Fahnestock, J. (1986). Accommodating science: The rhetorical life of scientific facts. *Written Communication, 3*(3), 275–296. https://doi.org/10.1177/0741088386003003001.

Gigante, M. E. (2014). Critical science literacy for science majors: Introducing future scientists to the communicative arts. *Bulletin of Science, Technology & Society, 34*(3–4), 77–86. https://doi.org/10.1177/0270467614556090.

Grunspan, D. Z., Eddy, S. L., Brownell, S. E., Wiggins, B. L., Crowe, A. J. & Goodreau, S. M. (2016). Males under-estimate academic performance of their female peers in undergraduate biology classrooms. *PLoS ONE, 11*(2), 1–16. https://doi.org/10.1371/journal.pone.0148405.

Haak, D., Hille Ris Lambers, J., Pitre, E. & Freeman, S. (2011). Increased structure and active learning reduce the achievement gap in introductory biology. *Science (American Association for the Advancement of Science), 332*(6034), 1213–1216. https://www.doi.org/10.1126/science.1204820.

Hayhoe, K. (2019, Oct. 31). I'm a climate scientist who believes in God. Hear me out. *The New York Times*. https://www.nytimes.com/2019/10/31/opinion/sunday/climate-change-evangelical-christian.html.

Kimmerer, R. W. (2013). *Braiding sweetgrass: Indigenous wisdom, scientific knowledge, and the teaching of plants*. Milkweed Editions.

Luke, A. (2012). Critical literacy: Foundational notes. *Theory into Practice, 51*(1), 4–11. https://doi.org/10.1080/00405841.2012.636324.

Meyer, J. H. F & Land, R. (2012). Threshold concepts and troublesome knowledge: An introduction. In J. H. F. Meyer & R. Land (Eds.), *Overcoming barriers to student understanding* (pp. 3–18). Routledge.

Nelson, A. (2016). *The social life of DNA: Race, reparation, and reconciliation after the genome*. Beacon Press.

Poe, M., Lerner, N. & Craig, J. (2010). *Learning to communicate in science and engineering: Case studies from MIT*. The MIT Press.

Priest, S. (2013). Critical science literacy: What citizens and journalists need to know to make sense of science. *Bulletin of Science, Technology & Society, 33*(5–6), 138–145. https://doi.org/10.1177/0270467614529707.

Roberts, D. (2011). *Fatal invention: How science, politics, and big business re-create race in the 21st century*. The New Press.

Stiffler, L. (2022). 'It is not acceptable': UW computer science program can't keep up with record demand from undergrads. *Geek Wire*. https://tinyurl.com/bdhp5ber.

TallBear, K. (2013). *Native American DNA: Tribal belonging and the false promise of genetic science*. University of Minnesota Press.

Taylor, C. (2012). Threshold concepts in biology: Do they fit the definition? In J. H. F. Meyer & R. Land (Eds.), *Overcoming barriers to student understanding* (pp. 87–99). Routledge.

Theobald, E., Hill, M., Tran, E., Agrawal, S., Arroyo, E., Behling, S., Chambwe, N., Laboy Cintrón, D., Cooper, J. D., Dunster, G., Grummer, J., Hennessey, K., Hsaio, J., Iranon, N. Jones, L. II, Jordt, H., Keller, M., . . . Freeman, S. (2020). Active learning narrows achievement gaps for underrepresented students in undergraduate science, technology, engineering, and math. *Proceedings of the National Academy of Sciences—PNAS, 117*(12), 6476–6483. https://doi.org/10.1073/pnas.191690311.

Tu, Y. (2011). The discovery of artemisinin (qinghaosu) and gifts from Chinese medicine. *Nature Medicine, 17*(10), 1217–1220. https://doi.org/10.1038/nm.2471.

University of Washington. (2022). *Fast Facts—Seattle Campus.* https://www.washington.edu/opb/uw-data/fast-facts/.

Weinstein, M. (2009). Critical science literacy: Identifying inscription in lives of resistance. *Journal for Activist Science & Technology Education, 1*(2), 1–11.

Appendix: Project 1: Tracing the Life of a Scientific Fact

Background

The study of a scientific phenomenon triggers a wide variety of forms of communication, like ripples in a pond. Most published scientific research starts as a peer-reviewed article, but it can get portrayed in a press release, a journalistic article, a social media post, and even make its way into mainstream media like movies or late-night talk shows. The nature of research findings can change, and sometimes get distorted or simplified, as it is translated across communication platforms.

Adjustments in communication are often appropriate according to changing audiences and purposes, such as research reported to a general scientific audience, a more specialized field within science, or to K–12 students or general public audiences. However these adjustments still have an impact on what is understood (or understandable), in what ways, and by whom. (One example of such a cluster of texts is here: https://www.nps.gov/whsa/learn/nature/white-animals.htm). In what ways do these acts of translation affect the actual information?

Assignment Overview

For this project you will select one particular phenomenon and study the "genealogy" of that phenomenon as it moves through various stages of translation: from grant proposal to research article to press release to mainstream media coverage to social media to textbook (to documentary, to literature, to . . .). Not all research articles get translated into other venues, so you will need to find one that has been described in multiple venues. The best way to go about this is to work your way backward: start with some kind of communication in the mainstream media or in social media and trace your way back to the original study.

For this assignment please find a minimum of four different pieces of science communication that all relate to the same research findings. Your sources should have no more than two of the same genre in any analysis (e.g., no more than two tweets, or research articles, etc.). This document gives some pointers on tracing a piece of scientific information across sources (see https://tinyurl.com/3x26a8un).

Assignment Requirements

This project will be composed as a presentation in Adobe Spark, which we got some practice with when we created our own introductions during Week 1 (see https://spark.adobe.com/sp/). I hope you feel some confidence using the platform now, but please do refresh yourself with this tutorial if you need to (see https://tinyurl.com/9y2nsxue). In your Spark presentation you should include the following elements:

- Inclusion of at least four related sources, as described above.
- A written and visual representation of each source. The visual representation is most likely to be a screenshot, though you are welcome to include web links as well. For the textual description, you should consider describing the source in such a way that that the author and context is fully clear, e.g., "Dina Smith is a geneticist at Harvard and she published an article on her research about mice brains entitled 'Mice Brains and DNA . . .' etc.". You don't need a formal citation in the body of the presentation, though you will need them in the references (see below).
- A central claim or "takeaway" about how or whether the information has shifted across sources (e.g., the information was distorted or simplified; the findings were clearly portrayed and remained faithful to the original claims; the findings became more emotional, etc.).
- Written analysis that amounts to a minimum of about 1200 words. It is perfectly appropriate if chunks of text are broken up by images, external links, or other media. Your analysis should be rhetorical in nature—feel free to attend to forensic, epideictic, and deliberative rhetoric and to the ways that particular vocabulary, titles, images, and tone shift across the translations.
- A list of references at the end of the Spark presentation, properly formatted according to APA format (use the menu on the left for info on formatting different kinds of sources: journal articles, websites, etc.)

Timeline

- Library instruction with biological sciences librarian: Friday, 4/8, 1:30pm
- Rough draft due in Canvas and to peer conference groups: Wed., 4/14, 5pm
- Conference: 4/15 and 4/16 (see Canvas for your assigned time)
- Final draft: Wed., 4/21, 10pm

Integrating Social Justice Data and Scaffolded Writing with Universal Design Principles Into Introductory Statistics

Laura Kyser Callis
CURRY COLLEGE

The very history of modern statistics is entrenched in racism (Clayton, 2020). The famous statisticians whose names grace the methods students learn in introductory statistics, such as Karl Pearson and Ronald Fisher, were overt white supremacists and eugenicists (Clayton, 2020). To be redundantly clear: the methods that we teach our students were developed for the expressed purpose of having a seemingly objective measure to demonstrate the inferiority of specific populations of people (Clayton, 2020). Moreover, mathematics and statistics have often been used as weapons of oppression historically, in living memory, and today—redlining, algorithmic policing, and even, closer to home for our students, the use of mathematics placement metrics to limit who is allowed access to high-quality mathematics instruction and STEM careers (D'Ignazio & Klein, 2020; Ngo & Velasquez, 2020). Undergraduates can and should have these conversations—and, indeed, in my (elective) History of Mathematical Inquiry class, students engage in these topics with passion. However, most college students do not take math electives. Many students end their mathematics study with a required quantitative reasoning course, such as college algebra or, increasingly, statistics. It is understandable that statistics faculty may be reluctant to include the racist history of statistics in their courses. Statistics is often compulsory for students and a service course for faculty. Students may already be disengaged due to the compulsory nature, and faculty feel pressure to "cover" an ambitious list of topics in a syllabus to support the needs of other departments. Even if instructors were aware of this history, sharing it with students would, first, be time away from helping students develop the statistical mindset and skills necessary for their careers and civic participation and, second, could risk students' disengagement from the field altogether.

One solution is to integrate social justice topics and data sets into the syllabus, to use data sets about racism, sexism, and injustice to teach the statistical topics addressed in introductory statistics. This approach frames statistics as a powerful tool for highlighting the reality of injustice; it is not just a tool of the oppressor. This method can broaden the range of students who are interested in statistics and mathematics by demonstrating that statistics can be used for purposes other than finance or science; statistics can be used to help others and understand the scope of real-world problems.

Giving students the tools to understand and quantify real-world problems with statistics is only part of the equation, however. To truly be able to use statistics for change, students need practice in communicating as well. Just like "Standard" English (see Blomstedt, this collection), mathematics and statistics are languages of power. People listen when statistics are used. Statistics instructors apprentice students into thinking statistically and communicating quantitative information.

This chapter describes two assignments that integrate real data about social justice topics and prompt students to write about quantitative information about social justice topics. The assignments, one used as an in-class lesson or discussion board prompt and the other used as a midterm project, were both used in an introductory statistics course. In this chapter, I first articulate the conceptions of equity that inform this work. Next, I detail the contexts of the college and the course in which the assignments were used. Then, I describe each assignment and the students' responses to the assignment, explaining how concepts of Universal Design for Learning (UDL) can be seen in the assignments. Last, I identify challenges that arise in designing and enacting these assignments and similar ones.

Theoretical Framework

Rochelle Gutiérrez (2012) described different conceptions of equity used by mathematics educators and researchers. She identified two dimensions of equity: the *dominant axis,* which addresses access and achievement, and the critical axis, which is concerned with identity and power. Through the *dominant axis* lens, educators consider whether students have opportunities to learn mathematics at a high level and whether these opportunities result in equitable outcomes. Educators and researchers ask, do policies result in learners from historically excluded populations enrolling in higher-level mathematics courses at the same rate as their peers? Does the curriculum support them in understanding key ideas? Do students from marginalized backgrounds learn as much and persist as long in STEM courses as their peers? The assumption of this lens is that educators are working to include students in the dominant system. However, the dominant axis does not prompt educators to consider if the system itself is a just one.

In teaching literature, Rudine Sims Bishop (1990) wrote about how books are mirrors and windows, allowing one to see both oneself and to learn about others. Gutiérrez (2012) similarly used this metaphor when explaining the role of identity in equity in mathematics education—part of the *critical lens*. If the only mathematicians highlighted in mathematics textbooks are white and male, female students of color may not see themselves in mathematics. If mathematics instruction only rewards fast fact recall and symbolic manipulation, students with processing difficulties or dyscalculia may not see themselves in mathematics. If the discourse around mathematics ability is about ranking or if classroom activities valorize competition, women, who are more likely to be socialized to cooperate, will be less likely to see themselves in mathematics. If placement policies consistently keep students of minoritized racial or linguistic groups in courses that repeat previously learned content, students may not see themselves in mathematics (Larnell, 2014; Ngo & Velasquez, 2020; Rios, 2023). If practice problems are primarily about optimization and consumer finance, learners who are more concerned with improving the world around them may not see themselves in mathematics. As Muna Abdi (2021) noted, "It is not inclusion if you are inviting people into a space you are unwilling to change."

The critical axis also addresses the idea of power. The power dimension is multilevel. At a micro level, there are power differentials in classrooms. At a macro level, there are power differentials in what counts as knowledge or proof, or who is asked to provide data to support their claims (D'Ignazio & Klein, 2020). The assignments described in this chapter are most aligned with the power dimension in that they support students in using mathematics to examine social injustice. In this way, they are aligned with the work of Eric Gutstein (2006) in *Reading and Writing the World with Mathematics*. Building on Paulo Freire's work, Gutstein defined reading the world with mathematics: "to use mathematics to understand relations of power, resource inequities, and disparate opportunities between different social groups and to understand explicit discrimination based on race, class, gender, language and other differences" (p. 25–26; see also Fink, this collection). To write the world with mathematics is to use mathematics to change the world. Gutstein's work offers up cases where students and communities have used mathematics to work for change, as have others (e.g., D'Ignazio & Klein, 2020; Turner & Font Strawhun, 2013).

The assignments I describe here were designed to address issues of identity and power. They were written in a context infused with efforts to address achievement and access through policy changes, curriculum examination, and professional development efforts around UDL. Admittedly, the assignments are focused more on understanding statistics than on understanding the social dynamics that lead to injustice. They are focused more on learning to write in the field of statistics than writing to change the world with mathematics. They are not social justice mathematics lessons; they are statistics lessons that use social justice contexts. However, this approach may be one that is easier for statistics instructors who are new to social justice education to implement within their introductory statistics courses without sacrificing class time for addressing statistical concepts.

Context

Curry College is a proudly neurodiverse college that welcomes students with a wide range of academic preparedness. A small private college south of Boston, Curry has a nationally recognized Program for Advancement of Learning (PAL), an optional fee-based program that supports students diagnosed with specific learning differences, executive function challenges, and attention deficit disorder. About 20 percent of incoming students enroll in the program. A recent survey of Statistics 1 sections estimated that between 29.0 percent and 43.5 percent of students identify as having a learning or attention disability or having an individualized education plan in secondary school (McNally, 2024). Because of the prevalence of students with learning differences and PAL faculty, the culture at Curry College may be very different from other colleges. Students talk more freely about their learning disabilities (LD), and many faculty understand LD through a *difference* versus a *deficit* lens. There is also familiarity with UDL principles. In this chapter, I will call attention to the elements of UDL or UDL-Math that are used in the assignments (CAST, 2011; Lambert, 2021).

Curry College also has a moderately racially diverse student body. Although we are a primarily white institution, our racial diversity mirrors that of our location, Massachusetts. Two-thirds (66.3 percent) of the students identify as White, 12.30 percent are African American, and 7.58 percent are Hispanic or Latino (*Curry College | Data USA*, 2021). In Massachusetts, 79.4 percent of residents are White, 9.5 percent are African American, and 13.1 percent are Hispanic or Latino (*U.S. Census Bureau QuickFacts*, 2022). Like other colleges across the country, Curry is not immune to the increasingly overtly racist and sexist rhetoric. In 2022, Curry experienced a bias incident the College took very seriously, canceling classes to host a mandatory community meeting of students, staff, and faculty. During that meeting, students asked faculty to include more opportunities to learn about and talk about racism, sexism, and social justice in their courses. Some of the materials described in this chapter were developed in response to students' requests.

With an acceptance rate of 88 percent in academic year (AY) 2023–2024, students in our courses vary widely in their academic preparation. Surveys of our students in Statistics 1 show that some students have taken AP Statistics or another statistics course in high school, while other students stopped at Algebra 2 in their penultimate year in secondary school. Students typically take Statistics 1 in their first or second year at Curry to meet their general education quantitative requirement or a prerequisite requirement for their major. There is often a significant amount of time between this course and students' previous mathematics courses. In our in-take assessments, student responses vary widely in regard to their perceived skill and interest in mathematics. Given the diversity in our courses, the faculty have agreed that class size for mathematics courses be capped at 26; statistics courses are usually at capacity, though summer courses may be smaller.

The course uses a simulation-based inference (SBI) curriculum rather than the consensus curriculum, Introduction to Statistical Investigations (ISI) by Nathan Tintle and colleagues (2016). In the consensus curriculum, students first study descriptive statistics and then learn the formal rules for when they can apply different distributions—primarily the binomial and normal distributions—to conduct different tests of statistical inference. Typically, there is a focus on formalism, notation, and hand calculation. In SBI, students study statistical inference at the beginning of the semester. They create chance models with real and digital dice, spinners, cards, and other probability devices to represent the null hypothesis and sampling. The curriculum we use has freely available applets for this purpose. Nationwide, researchers have found that the SBI approach has helped students to learn and retain more statistical concepts, particularly students with weaker procedural fluency (Chance et al., 2018; Tintle et al., 2012, 2018).

In our pilot of this curriculum at Curry College, we found that students with learning disabilities performed better both in terms of their final grades and in terms of a low-stakes concept inventory under the SBI curriculum, compared to the previously used consensus curriculum, even when active learning techniques were used with the consensus curriculum by an experienced instructor (Callis, 2022b; Callis & McNally, 2021; McNally, 2024). In our early investigations of instructional practices that relate to students' development of conceptual understanding, opportunities to talk with each other and with the professor about their thinking, in real time in the classroom or over Zoom, seemed to be a key component. Thus, the class uses small group work, whole class discussion, and immediate feedback systems through Desmos Activity Builder during class time.

The Assignments

Discussion Board: School Discipline

In most learning management systems, there are tools for creating discussion board assignments, where students and the instructor can write and react to each other's responses. In a statistics course, this tool allows students to consider their peers' methods and conclusions, some of which are legitimate and some of which are not, and to benefit from the feedback that the professor gives to their peers. A discussion board can be a space that is in between the informality of a student's own notebook, which only needs to be understood by one individual, and a formal paper, which is expected to provide enough detail to be understandable by the academic community and to follow the conventions and formalism of a particular discipline.

As part of addressing students' desire to have more opportunities to engage in social justice issues in their other courses, I created three discussion board prompts that ask them to apply the methods they have learned to scenarios addressing racism and sexism. I piloted these prompts in a summer 2023 course. Some of the students completed the course asynchronously, and some joined a synchronous session on Zoom, so there were likely varying degrees of feelings of belonging among different students. Students were graded on the prompts and received ongoing feedback from their peers and from me. At the end of the semester, students chose one of their responses to write up more formally for an ePortfolio, a requirement of the General Education curriculum. One of the prompts is presented below.

Discussion Board: The Chance Model & Pre-School Discipline

There are a lot of data that suggests that African American children are more likely to receive exclusionary discipline consequences, like suspension and expulsion, than White American children in public schools, even for the same behavior. One group of researchers at the Yale Child Center began to investigate whether teachers might be more likely to unconsciously expect African American children to misbehave [see https://tinyurl.com/yzut382r]. They asked teachers to watch a video of four children together in a preschool setting: an African American boy, an African American girl, a White boy, and a White girl. The teachers were told that the researchers were interested in teachers' perception of misbehavior, but they did not tell the teachers they were investigating race. Teachers were told to press a bar if they saw misbehavior in the video, but no misbehavior occurred. The teachers were then asked which child they were the most "concerned" about in terms of needing to watch for potential misbehavior. There were 132 teachers in the study. 42 percent of the teachers chose the African American boy as the child they were most concerned about in terms of potential misbehavior.

In a 2–3 paragraph response, address all of the questions below.

1. Why might this issue be worth investigating?
2. What is the research question?
3. What is the parameter of interest?
4. What percent of the teachers would you expect to choose the African American boy if race and gender were not factors?
5. What are the null and alternative hypotheses? Write them in words and symbols.
6. What is the value of the sample statistic?
7. How would you set up a simulation to determine if the sample statistics could have occurred due to chance alone?
8. Run the simulation using spinners in the applet [see https://tinyurl.com/rt585wxk]. Report your p-value.
9. Make a conclusion about this study. Does race and/or gender seem to impact teachers' anticipation of preschool students' misbehavior?

Student Response. Eight of the students who wrote responses to this prompt correctly answered all parts of the question and engaged in a concerned way about the topic itself—children and racism. They explained the null model conceptually—if racism were not a factor, we would expect the African American boy to be picked 1 out of 4, or 25 percent, of the time. This could be modeled by spinning a spinner with four equal parts, each part labeled with one of the children, and spinning the spinner 132 times to represent the 132 teachers. These eight students were also able to articulate conclusions that reflected their understanding of the simulation and the resulting p-value. These students were able to write the hypotheses correctly with formal notation, $H_0: \pi = 0.25$ and $H_A: \pi > 0.25$, where π represents the long-run proportion of times the African American boy would be chosen if it was random chance.

Four of the students set up their null hypotheses "incorrectly": $H_0: \pi = 0.5$ and $H_A: \pi > 0.5$, but these were sensical "mistakes"—in their less formal writing, they explained that they were testing a hypothesis that African American and White American children were equally likely to be identified by the teachers; gender was not taken into account. Indeed, one of the students herself pointed out this difference in method. Their conclusions would have been valid if they had included the African American girls in their sample statistic. The linked article notes that 11 percent of the teachers chose the African American girls, so the sample proportion of teachers choosing an African American student would be 42% + 11% = 53%. However, this "mistake" led to very rich discussions. For instance, it does not make sense to test the one-sided alternative hypothesis $H_A: \pi > 0.5$ if our sample proportion is smaller than the null hypothesis parameter value, as 0.42 is compared to 0.5; this reason is why these four students were getting such unusual p-values, over 0.50. A one-sided alternative hypothesis needs to align with the direction of the data. The "mistake" also highlights the importance of multivariable thinking. If we ignore gender and use the sample statistic of 0.53, the proportion of teachers who chose an African American student, we find that the p-value is quite large, around 0.28. The sample proportion of 0.53 is so close to our null hypothesis parameter value of 0.50; a difference of 0.03 could be due to chance. However, when we consider gender *and* race, when we use the sample proportion of 0.42 teachers choosing an African American boy as concerning and the null hypothesis parameter value of 0.25, the p-value is very close to 0. The sample proportion of 0.42 is so far away from what we would expect under the chance model, 0.25, that it is not reasonable to think this difference would be due to chance. These are major topics in statistical inference—multivariable thinking, the role of chance, factors that impact the strength of evidence—and these issues were able to surface in the first two weeks of the semester through a writing assignment that gave students access to their peers' thinking and the instructors' thoughts on their thinking. These opportunities might have occurred during a traditional class, but the writing assignment allowed there to be a public record of these topics.

The affective impact of writing for others about social topics is also evident here. Students commented on the unfairness of African American preschoolers being expected to misbehave more by teachers. They cheered each other on in their responses as well. Often, mathematics is portrayed as cold and unemotional, but educational researchers know that emotion is an important part of learning. One student commented on the prompt's impact on her perception of mathematics in a response to a peer's post: "It's interesting how statistics can be used to solve societal and racial issues, I feel like that's often brushed past and it is mainly only viewed as something used solely for math-based problems." Addressing social issues in community writing assignments like discussion boards may be a way to bring emotion back to mathematics.

Mid Term Project: Deaths in Police Interactions

One of the challenges instructors face in introductory statistics courses is giving students the opportunity to explore all parts of the statistical inquiry cycle. In our curriculum, students identify a research question from an existing study, but they do not create their own research question. Developing a research question, collecting or finding a data set, and cleaning the data can all take well over a semester, leaving little to no time to explore and analyze the data itself. In addition, learning how to conduct the methods using technology itself can be a challenge. To give students an opportunity to engage in more parts of the statistical inquiry cycle, for the past three years, I have assigned a controlled research project; students must use the data set I choose in Google Sheets so that I can support them; being familiar with 30 to 50 different data sets that students find would take too much time away from the work that I know has an immediate impact on student learning.

The data set we use is found at www.fatalencounters.org (Burghart, 2021). Housed in a Google Sheet, it attempts to document every death that has occurred in a police interaction. Paid and volunteer researchers use media reports, but they also comb through files from Freedom of Information Act requests and other data sources. There are over 30,000 cases, but the author is clear that this is not exhaustive; there is no way to know how many people actually die in police presence. There are several variables, some well researched and others still developing, from the intended and actual use of force, the location, the race, recorded gender, and age of the deceased. Students are tasked with identifying a research question *that can be answered with this data set* and working through the statistical inquiry process to answer that question in a final, short paper.

Another challenge that instructors face with final papers such as these is receiving poor quality products that are difficult to grade. One solution to this is breaking a project into steps and giving feedback to students throughout the semester on each step. For this project, each step is a step in the statistical inquiry process. First, in the second week of class, I orient students to the data. Students have some time to talk in groups about the data set and develop three research questions that they think could be answered by this data set. They post their potential questions on a discussion board post. I give them feedback on their questions. For example, *Does the race of the police officer matter?* is a very important question but one that cannot be answered by this data set, as there is no information about the officers. Students also benefit from support in developing research questions that lead to a plan of action. For example, one student suggested, *Does the location matter?* While an excellent question, it does not have an obvious plan of action. Instead, we might ask something like, *Are people disproportionately likely to be killed in a police interaction in a city or suburb? In a state with gun control laws or without? In higher poverty neighborhoods or lower poverty neighborhoods?* Students might ask questions that leave very little to write about, questions that are *mathematical* questions and not *statistical* questions—questions that do not anticipate variability. For example, students have suggested *How many victims have been shot by police?* This question can be answered with a single number; it is a mathematical question, not a statistical question.

In the fourth week of class, students review my comments and the comments of their peers to choose a research question. On the discussion board, after a brief conversation with peers in class, they brainstorm some graphs and methods that could help them answer the research question. This step is another opportunity for me to give them early guidance. For example, students are often interested in the number of deaths that occur in different states. However, states vary significantly in their populations; a bar graph of the *number* of deaths in each state would look very similar to a graph of the state populations and add very little insight. On the discussion board, I coach them through thinking about rates per 1000 people, very similar to a percentage, that could help us understand the relationship between states and deaths that occur in police presence in a deeper way. As another example, students are often surprised to find that there are more instances of White people dying in a police interaction than African Americans—14,731 White people, or 46.78 percent of the data set, compared to 8,545 African Americans, or 27.13 percent of the data set. However, there are more White people in the United States than there are African Americans. The United States is 75.5 percent White and 13.6 percent African Americans; 27.13 percent is a lot higher than 13.6 percent, and this is what is meant by the idea that African Americans are disproportionately likely to be killed in a police interaction.

The discussion boards are also a way for the instructor to gain data to inform instruction. Based on the questions I see students asking, I create videos using their lines of inquiry to show them different spreadsheet techniques, such as pivot tables. Watching these videos and trying some techniques is their pre-homework for the two class sessions that we spend working on the project, either independently or in groups, as I circulate to help. For efficiency, during this class period, I also group students together who have similar lines of investigation so that they can support each other, and I can troubleshoot the same technique fewer times.

The discussion boards also serve as a way for me to address students' use of language before their final product. There are language challenges that are both mathematical and contextual. Contextually, for example, students sometimes refer to the deceased in the spreadsheet as "criminals" or "suspects," but they are not all criminals or suspects. Indeed, one of the deceased has the age listed as 0.25—a three-month-old baby who was killed in a vehicle pursuit, the linked story explains. A three-month-old baby cannot be a criminal or a suspect. As a mathematical example, the issue of the denominators used in percentages often appears when students are studying two categorical variables. For example, students are often interested in the relationship between the highest use of force and gender. In a pivot table, students can find the total number of women who died in a vehicle pursuit. Dividing by the total number of women would lead to a different percentage than dividing by the total number of people who died in a vehicle pursuit. The sentence construction must match the calculations so that we understand which "whole" or divisor was used. Some guidance can be provided on the discussion board, and some points are worthy of addressing with the whole class during class time. This feedback is more likely to be picked up by students than comments on an end-of-term paper.

The discussion boards also allow students to interact in a hybrid social/academic way that may be limited during class time due to seating habits. On the discussion board, the instructor can direct students to others who have similar ideas for research projects, enabling them to find peers to work together with—either to work together on one common project or to submit separate projects but support each other with technology during class time. In interviews with students with learning disabilities or negative prior mathematics experiences, students reported that these structured times to interact socially and academically helped them learn more both during and outside of class time (Callis, 2022b).

Student Response. Statistics 1 is an introductory class; it introduces key ideas. In mid- and upper-level classes, these ideas are reinforced and mastered. The Fatal Encounters project is also a midterm project submitted in the seventh week of class; it is more a learning opportunity than a formative assessment. Given these stipulations, Tables 4.1 and 4.2 show some data from student papers.

The first finding is that, while the project allows students to choose variables other than race, students are willing to investigate race. Table 4.1 lists the topics students chose during the 2022–2023 academic year. Students were also very interested in multivariable thinking. For instance, some wondered if the relationship between gender and the highest level of force used was different for different racial groups. The asterisk indicates that projects in this topic may overlap with other topics.

Second, students were able to engage with the learning outcomes for the course. Table 4.2 describes the learning outcomes and the degree to which they were achieved by students through the project.

Among the fall 2022 papers, 21 students communicated clearly about the meaning of the quantitative information and connected the numbers with the context. In contrast, 22 students communicated and made connections but required clarification on some points on the final paper. These numbers are not necessarily comparable. Students varied in the complexity of their questions and the complexity of the spreadsheet mechanics to answer their questions. More complexity results in a higher challenge to precisely articulate conclusions. In short, this is not a standardized assessment.

Table 4.1. Topics Chosen by Students in Fatal Encounters in Police Interactions Project

Topic	Number of student projects spring 2022	Number of student projects fall 2022
Race*	17 (53%)	11 (37%)
Age*	6 (19%)	8 (27%)
Gender*	9 (28%)	5 (17%)
Geography	8 (25%)	2 (7%)
Time	0	1 (3%)
Mental Health	1 (3%)	1 (3%)
Other	1 (3%)	1 (3%)
Total	32	29

* Projects may overlap in topic.

Table 4.2. Learning Outcomes Addressed by the Fatal Encounters Project

Learning Outcome	Level
Engage in regular discussion of quantitative information or results, with special emphasis on the context of the problem and general, real-world knowledge.	Achieved by all who submitted paper
Utilize statistics software to perform data analyses to interpret and compare multiple representations of quantitative information and draw inferences from them.	With support
Organize, summarize, interpret, and compare single-variable data using descriptive methods of statistics.	With support
Recognize and apply the different representations of quantitative information (e.g., symbolic, visual, numerical, verbal) when describing relationships between two variables.	With support
Communicate quantitative information effectively, incorporating symbolic, numeric, and/or graphical representations and appropriate syntax within verbal and written communication.	Varies; majority met with support & feedback

The purpose of the project is not just to master learning outcomes, however. The project also provides the instructor with ways to see students' capabilities beyond mathematics. Through mentoring students in this project each semester, I witness their writing skills, their perseverance, their ability to act on feedback, and other skills needed in their workforce and civic life. I call upon what I notice about individual students in many situations. I invite students with quality projects to present at the college academic forum. Two of my students, both women, have presented, and one of them went on to present at the Joint Mathematics Meeting, the largest mathematics conference in the world, on a panel on undergraduate research (Conley et al., 2023). Inspired by Talithia Williams' introduction to her book *Power in Numbers* (2018), where she describes a moment when a teacher suggested she study mathematics, I personally invite students to take more mathematics, and my observations of their work on the project or in class directly impact the invitations. At the end of the semester, I look at all students who have received a grade of B- or higher. I write them emails, personalized based on their major and the skills I've noticed that they bring to the class, to suggest they consider a math or data analytics minor. Students often write me back with grateful emails, surprised to think of themselves as a "math person." When mathematics is done in a social, supportive environment with a real purpose that matters, many more people can think of themselves as "math people."

Universal Design for Learning

Based on learning sciences, the UDL Guidelines provide a framework for designing instruction to support students with learning differences (CAST, 2011). Like universal design in architecture, the elements of design in UDL often end up benefiting all learners, including neurotypical learners. UDL calls for engagement in the three neural networks used in learning: the affective network, the recognition network, and the strategic network. These networks are activated across the process of learning: accessing, building, and internalizing knowledge. A full discussion of the Guidelines is outside the scope of this chapter; more detail can be found at https://udlguidelines.cast.org/. Here, I give a few examples of how the assignments described above demonstrate elements of UDL. I do not claim that these assignments are exemplars of universal design. Instead, I offer an example of how statistics instructors might begin to think through elements of their assignments with UDL in mind.

Under "Engagement," to activate the affective network and provide access to the content for students, Guideline 7 calls for recruiting interest by optimizing individual choice and autonomy (Checkpoint 7.1) and relevance, value, and authenticity (Checkpoint 7.2). The authors of the Guidelines recognize it is not appropriate to give choice over every component of an assignment; as an example, I limit students to a particular data set rather than allowing them to choose a data set so that we can better support each other. Students do, however, have a choice in their research question. They are allowed to choose whether they work independently or with a group; even if they work with others, they are allowed to choose whether they will submit separate individual papers or one final group product. They have some choice in their analysis methods, though this is highly guided by the instructor through feedback so that they are making sensical choices. The context of these two assignments is also socially relevant. Students had asked for more opportunities to discuss racism in their other courses; they lived with racism through the bias incident. They themselves remember being students in preK–12 schools and have thoughts and opinions on teachers' perception of misbehavior, or lack thereof. Students are also learning skills, such as manipulation of data through Google Sheets, that they can use in their classes and their daily lives. While many students will go on to learn other data analysis software, because Google Sheets is free, widely accessible, and shareable, it or Microsoft Excel, which behaves similarly, is likely to be highly used in their careers or personal lives.

Under "Representation," to activate the recognition network and build understanding, the Guidelines recommend attending to vocabulary, symbols, syntax, and structure (Checkpoints 2.1 and 2.2). Statistics is a particularly language- and symbol-heavy discipline. It is not just a matter of new vocabulary and symbols; familiar words and symbols are also repurposed. For example, the word "mean" now signifies average; the word "variable" is no longer just a placeholder for an unknown numerical value; the symbol π now represents a population parameter instead of a familiar irrational constant. There is also a high level of precision required, both when explaining common quantities like percentages, as described earlier, and when explaining new concepts, like a p-value. For example, the p-value is not the likelihood that the null hypothesis is true; it is the likelihood of getting results similar to the observed statistic *if* the null hypothesis were true. To novices, these seem like equivalent descriptions, but they have very different meanings and consequences. The two assignments highlighted here provided repeated opportunities, through the discussion boards, for students to use both more casual language and to try out the more formal language of the discipline. The discussion board allowed me, as the instructor, to give students feedback on their developing use of the language of the discipline. Instruction for these assignments was also provided with multiple media (Checkpoint 2.5). Students had access to closed-captioned videos demonstrating how to use the applets and execute techniques in Google Sheets. Students also had static handouts with annotated pictures. In addition, I demonstrated to students in class and during office hours in real time and coached them as they tried the techniques themselves.

Under "Action & Expression," to activate the strategic network to internalize learning, the Guidelines suggest supporting planning and strategy development (Checkpoint 6.2). For example, they suggest providing "prompts to stop and think before acting," "checklists and project planning templates," and "guides for breaking long-term goals into reachable short-term objectives."(CAST, 2011, p. 26) In the Fatal Encounters project, the discussion board prompts served as a guide to break the long-term project into short-term objectives: identifying a research question, planning an analysis method, conducting the analysis, and writing the findings. There was also a template for completing the assignment, with a model response, so that students could envision the final product. The prompt for analyzing the preschool teachers' data was also scaffolded with individual prompts to work through the process. Eventually, the hope is that students will be able to work through the process on their own, but in an introductory statistics course, repeated, explicit modeling of the inquiry process can support students to internalize it.

There are other ways in which these assignments incorporate the UDL Guidelines and ways in which the assignments may fail to incorporate the Guidelines. Awareness of the Guidelines and the way they inform assignments helps me, as an instructor, think more carefully about what I emphasize and make explicit to students and how I can continue to improve the assignments over time.

Challenges in the Design and Enactment of the Assignments

All teaching has challenges. Some challenges are meaningful and interesting. In enacting these assignments, I have mostly found meaningful and interesting challenges. In designing additional assignments that integrate social justice data sets, I continue to face challenges and wonder whether there are challenges that are still unknown.

With the Fatal Encounters project, the first challenge is helping students navigate discomfort with an open-ended, ambiguous project. In most school contexts, students are not typically asked to develop their own research question, and many are unaware of what a worthwhile research question would be. This can cause frustration and confusion among some students. However, I believe that sitting with this frustration, confusion, and anxiety is a life skill they will call upon in their careers, and so it is worth the time, effort, and discomfort to help them build this skill. At the end of the project, I teach them how to talk about this experience in interview questions.

Whenever a project is incorporated into a course, the first question other faculty often ask is, how do you find the time? This question especially comes up when technology is used because both the content and the technology are new to students. In my experience, introductory statistics students typically know very little about spreadsheets; even understanding the impact of clicking on a cell, column, or row is new. I front load many issues by using the discussion board to give feedback, giving students a few minutes in class to read the feedback and brainstorm with their peers, and providing videos and handouts as resources. Instead of a midterm exam and review period, we work for two 75-minute periods in class on the project. If students have not completed the project by then and still need support, I meet with them during office hours, either in person or over Zoom. During these weeks, I schedule extra time for office hours, which is a challenge with a full course load, advising duties, and service commitments. However, this challenge is meaningful and interesting: meeting with students individually is rewarding and positively impacts my relationship with them. It helps them develop a habit of meeting with their professors, which will support them in their other courses.

In designing lessons that incorporate social justice contexts, the biggest challenge is finding studies and data sets. Racism, sexism, and injustice are complex, multivariable issues. Therefore, studies investigating these issues use complex methods for multivariable analysis, which introductory statistics students have not yet been exposed to. In addition, our curriculum uses simulation-based inference (SBI), which often requires that we have close-to-raw data, which is typically not provided in papers or available by contacting the authors. Creating new lessons requires a significant amount of time reading research studies to understand the phenomenon oneself and searching for data that would not only be accessible to students at a given point in the semester but also highlights a particular mathematical idea, such as factors that influence the width of a confidence interval or size of the p-value.

There may be challenges and difficulties that I am not aware of. I recognize that for many students in my classes, issues of racism, sexism, and injustice could be painful issues to consider. I am a white woman. While I have experienced sexism, I have not been afraid for my safety in a police interaction because of my race. I have not felt like a teacher expected I would misbehave. Before we begin a lesson, I check in with students using a Desmos slide to see how they feel about the topic. This lets me assess if we are in a somewhat safe space. For instance, one beginning slide asks students what they think about deaths that occur in police interactions. So far, students have indicated that it is upsetting, wrong, tragic, and important to address. A secondary reason I have for allowing students to choose their own research question is so that students can choose to examine something other than race. I don't ask students to present their findings or to share their personal experiences. I do not include questions that ask about racism on high-stakes summative assessments. I thank students for being willing to engage with difficult topics. I hope that these choices mitigate the discomfort the right amount. Bringing emotion back into mathematics, rehumanizing mathematics, is not without risk.

Conclusion

Often, the discourse around equity in STEM is about increasing the number of students who go further in STEM pathways and careers. The outcomes of interest are participation rates, pass rates, and persistence rates. Even in introductory courses that serve non-STEM majors, questions of success for introductory quantitative courses are often framed around success within a student's major or career. However, preparing students for careers is only one goal of higher education. We are also responsible for preparing students—of all majors—to become active, caring citizens. To this end, introductory courses that serve the general student body can be a place where students can learn to consider, discuss, investigate, and write about challenging and important issues affecting our world. Courses can attend to both the dominant axis, to access and achievement, as well as to the critical axis, to identity and power, and hopefully, the work on both dimensions will be mutually reinforcing.

The Curry College mathematics faculty has a commitment to equity. As a department, we worked together to pilot a curriculum and examine its impact on students with learning disabilities and African American students, not just on the aggregate learning gains (Callis, 2022b; McNally, 2024). We examined and changed our assignments, our policies, and our course offerings using the UDL frameworks (Callis, 2022a). As a result, we have seen positive outcomes in the access and achievement dimension. The diversity of students taking upper-level mathematics courses has outpaced the growing diversity of the college (McNally & Callis, 2021). The gap between students with learning disabilities and neurotypical students and between BIPOC students and white students has narrowed in multiple measures in Statistics 1 (McNally, 2024).

On the critical axis, I hope that the assignments I design, like the ones described here, address issues of identity and power. I hope that students are able to see themselves as mathematical because mathematics is about collaboration, about trying things out, about caring relationships, and about addressing issues that matter. I hope, too, that they learn that mathematics is a tool for critiquing injustice; it is not just a tool for science and finance. Though it is necessary to critique the disciplines of mathematics and statistics, which enjoy their own unearned privilege as "objective," many instructors, myself included, dare not broach this topic with introductory statistics students, who may already be suspicious of a subject that has been unkind to them in their K–12 schooling. Perhaps, however, by using data sets that highlight injustice to teach the statistical topics we are charged with teaching anyway, we can recruit a wider range of students to the quantitative fields. Then, perhaps they can join us in the critique—or they can stand on our shoulders, as Newton stood on the shoulders of his forebearers, and see new possibilities for mathematics and statistics based on justice.

References

Abdi, M. [@Muna_Abdi_Phd]. (2021, June 19). *It is not inclusion if you invite people into a space you are unwilling to change.* [Tweet]. X. https://tinyurl.com/3hrf2787.

Bishop, R. S. (1990). Mirrors, windows, and sliding glass doors. *Perspectives: Choosing and Using Books for the Classroom, 6*(3). https://tinyurl.com/y2wwkprt.

Burghart, D. B. (2021). *Fatal encounters.* https://fatalencounters.org/.

Callis, L. (2022a, November 4). *Departmental application of Universal Design for Learning—Mathematics.* Association of American Colleges and Universities STEM Conference, Washington, D.C.

Callis, L. (2022b, November 5). *In their own words: Instructional practices that support students with learning disabilities or negative prior mathematics experiences in simulation based inference introductory statistics classes.* Association of American Colleges and Universities STEM Conference, Washington, D.C.

Callis, L. & McNally, J. (2021, July 1). *Beyond active learning: Preliminary hypotheses on the factors that impact gains in introductory statistics: Comparing simulation vs consensus curriculum, face to face, hyflex, and online synchronous modality, and instructor factors.* United States Conference on Teaching Statistics, virtual.

CAST. (2011). *Universal design for learning guidelines version 2.0. Wakefield, MA.* https://tinyurl.com/2jzb2jcd.

Chance, B., Mendoza, S. & Tintle, N. L. (2018). Student gains in conceptual understanding in introductory statistics with and without a curriculum focused on simulation-based inference. In M. Sorto, A. White & L. Guyot (Eds.), *Looking back, looking forward: Proceedings of the Tenth International Conference on Teaching Statistics (ICOTS10, July, 2018),* Kyoto, Japan. International Statistical Institute. https://tinyurl.com/yyybwcwk.

Clayton, A. (2020, October 27). *How eugenics shaped statistics.* Nautilus. https://nautil.us/how-eugenics-shaped-statistics-238014/.

Conley, T., Jordan, H. & McDonald, C. (2023, January 6). *Undergraduate research in the scholarship of teaching and learning statistics.* Joint Mathematics Meeting, Boston, MA.

Curry College | Data USA. (n.d.). Retrieved November 3, 2023, from https://tinyurl.com/289anre8.

D'Ignazio, C. & Klein, L. F. (2020). *Data feminism.* The MIT Press. https://doi.org/10.7551/mitpress/11805.001.0001.

Gutiérrez, R. (2012). Context matters: How should we conceptualize equity in mathematics education? In B. Herbel-Eisenmann, J. Choppin, D. Wagner & D. Pimm (Eds.), *Equity in discourse for mathematics education* (pp. 17–33). Springer Netherlands. https://doi.org/10.1007/978-94-007-2813-4_2.

Gutstein, E. (2006). *Reading and writing the world with mathematics: Toward a pedagogy for social justice.* Routledge.

Lambert, R. (2021). The magic is in the margins: UDL Math. *Mathematics Teacher: Learning and Teaching PK–12, 114*(9), 660–669. https://doi.org/10.5951/MTLT.2020.0282.

Larnell, G. (2014). The stuff of stereotypes: Toward unpacking identity threats amid African American students' learning experiences. *Journal of Education, 194*(1), 49–57. https://doi.org/10.1177/002205741419400107.

McNally, J. (2024). A design-based research study of department-wide curriculum change to support equitable performance in a first undergraduate course in statistics. [Unpublished Manuscript] Science and Math, Curry College.

McNally, J. & Callis, L. K. (2021, January 7). *Removing barriers to participating in upper level mathematics*. Massachusetts PKAL Regional Network Winter Meeting, Virtual.

Ngo, F. J. & Velasquez, D. (2020). Inside the math trap: Chronic math tracking from high school to community college. *Urban Education*, 0042085920908912. https://doi.org/10.1177/0042085920908912.

Rios, J. N. (2023, January 6). *Bond or barrier: Exploring the role of language in the undergraduate mathematics classroom*. 2023 Joint Mathematics Meetings (JMM 2023). https://meetings.ams.org/math/jmm2023/meetingapp.cgi/Paper/17628.

Tintle, N., Chance, B., Cobb, G. W., Rossman, A., Roy, S., Swanson, T. M. & Vanderstoep, J. L. (2016). *Introduction to statistical investigations*. Wiley.

Tintle, N., Clark, J., Fischer, K., Chance, B., Cobb, G., Roy, S., Swanson, T. & VanderStoep, J. (2018). Assessing the association between precourse metrics of student preparation and student performance in introductory statistics: Results from early eata on simulation-based inference vs. nonsimulation-based inference. *Journal of Statistics Education*, *26*(2), 103–109. https://doi.org/10.1080/10691898.2018.1473061.

Tintle, N., Topliff, K., Vanderstoep, J. L., Holmes, V. & Swanson, T. (2012). Retention of statistical concepts in a preliminary randomization-based introductory statistics curriculum. *Statistics Education Research Journal*, *11*(1), 21–40.

Turner, E. & Font Strawhun, B. (2013). "With math, it's like you have more defense": Students investigate overcrowding at their school. In E. Gutstein & B. Peterson (Eds.), *Rethinking mathematics: Teaching social justice by the numbers* (2nd ed.). Rethinking Schools.

U.S. Census Bureau QuickFacts: Massachusetts. (n.d.). Retrieved November 3, 2023, from https://www.census.gov/quickfacts/fact/table/MA/PST045222.

Williams, T. (2018). *Power in numbers: The rebel women of mathematics*. Race Point Publishing.

A Curriculum Exploring Arab and Muslim Science: Opening Space for Other Epistemologies of Science

Alicia Bitler
MAGRUDER HIGH SCHOOL

Ebtissam Oraby
THE GEORGE WASHINGTON UNIVERSITY

In this chapter, we will share about a course that brings together history, philosophy, cultural studies, and science.[1] In the winter of 2020–2021, our four-year research university announced an initiative for interdisciplinary teaching teams to propose courses that would bring together science and humanities. According to the call for proposals, the intention of these courses was to enrich studies for both science and humanities students:

> For students in STEM, they understand STEM as fields that must interact with and be informed by fields outside of STEM, and they learn to draw upon the knowledge and methodologies of these fields. For students in the Humanities, Arts, Social Sciences, they learn that their own fields can have an impact on knowledge in key areas of STEM, and they learn about methodologies and knowledge of STEM that they can draw on in their studies.

The authors, Ebtissam and Alicia, responded with a proposed course on Arab and Muslim science. Alicia is a STEM (science, technology, engineering, and math) education instructor whose research interests include a utilization of multiple epistemologies of science in classrooms (Bang & Medin, 2010). Ebtissam is an Arabic language instructor whose research interests include the decoloniality of knowledge and Arab and Muslim ways of knowing. We have been writing and thinking together for many years and often discussed our overlapping interests in Arab and Muslim epistemologies of science. Arab and Muslim cultures are inseparable and entwined as Arab refers to the culture shaped by language, and Muslim refers to the culture shaped by religion. These two cultures exist side by side in the Middle East, being informed by and changed by each other.

---

1   At the time of this project, Bitler was an instructor at The George Washington University.

When planning the course, we thought about three audiences: the future science educators whom Alicia teaches, the Middle Eastern Studies students whom Ebtissam teaches, and other science and humanities students who were the target audience of the initiative. Science students learn their discipline within a Eurocentric perspective, implicitly learning that science is apolitical and acultural. Through the referencing of only Western scholars, students also begin to view science as a historically European enterprise, ignorant of the innumerable contributions of individuals from other cultures. These beliefs about what science is shape how science is conducted and, therefore, what can be known (Harding, 1998). By ignoring different epistemological underpinnings of science, the field is likely limited in its ability to solve the many problems facing the world (Elby et al., 2016). Therefore, we believe that it is vital for science students to develop an understanding of the discipline as culturally situated. It is potentially even more vital for future science teachers to develop this understanding, as the way that they teach can have an exponential impact on future scientists. Middle Eastern Studies students, like individuals worldwide, are often unaware of the complexity of Arab and Muslim cultures that shaped science and the world. They likely have heard references to important Arab and Muslim scholars, but they learn about them separate from understanding how culture shaped those thinkers and how those thinkers shaped the world as we know it.

The course we proposed focused on the history of Arab and Muslim science and the distinct ways science is thought about in various cultures. We engaged in the inaugural offering of the course in the fall of 2021 with students from diverse majors, such as political science, classics, and public health. One of the students had taken several Arabic language classes, and one of them was a native Arabic speaker and a practicing Muslim. Although we had hoped the course would be of interest to science students, none were able to join for the first semester that it was offered.

Our course centered on exploring the fundamental beliefs that form the basis of Arab and Muslim scientific thought juxtaposed with Western and other non-Western perspectives on science. We investigated how cultural influences shape the comprehension, implementation, objectives, and recognition of science across various societies. While we acknowledged scholarship discussing Arab and Muslim scientific achievements that influenced the European Renaissance (Al-Khalili, 2010, 2011; Nasr, 1976, 1984, 1988, 2001, 2003, 2010), our primary focus wasn't on covering this content. Instead, we aimed to analyze the varied manifestations of science in different cultures and civilizations, challenging the Eurocentric monopoly on the history, philosophy, and teaching of science. Our course underscored the significance of recognizing that all knowledge, including science, is culturally situated. Although our course's approach may not directly align with traditional science education, we contend that the critical perspectives and inquiries we explored are essential for engaging with any form of knowledge production. This chapter emphasizes the crucial role that reflective writing played in fostering the critical analysis of these epistemological frameworks within our course.

Course Intent and Goals

To destabilize and challenge the prevailing view of science as Western, male, and white, we created a course to provide space for students to explore a non-Western epistemology of science and think of science as diverse and inclusive. Throughout the course, students explore Arab and Muslim science history and culture as part of a globally shared human heritage to open a space for other ways of thinking about and doing science. Arab and Muslim scholars have contributed to science in meaningful and often unacknowledged ways, founding disciplines like chemistry, algebra, modern surgery, and optics—shaping science as we know it.

The course did not intend to stop at learning *about* Arab and Muslim science but focused more on learning *from* it. Learning *about* may promote a diverse representation in curriculum, but it does not challenge the dominant Western epistemology. Learning *from*, on the other hand, means to be challenged, to listen attentively to others, and to respond to and be changed through the encounter (Todd, 2003). Our course involved examining scientific achievements in Arab and Muslim civilizations. However, our focus wasn't on mere historical facts. Rather, we aimed to use Muslim perspectives on science to analyze their underlying epistemology and contrast them with modern views. For instance, when studying al-Jazari's (1136–1206) inventions (Hill, 1991, 2020), we paid more attention to understanding his approach to science rather than just the specifics of his creations or their relation to European literature on mechanical engineering and other fields. Specifically, we explored his book, *The Book of Knowledge of Ingenious Mechanical Devices* (1206), where he detailed his inventions with descriptions, drawings, and instructions for design, manufacture, and assembly. We questioned what drove al-Jazari to share such detailed information and discussed the Islamic concept of knowledge sharing (Anand & Walsh, 2016), comparing it with the modern patent system and considerations of research funding and profitability. Furthermore, we examined al-Jazari's integration of art, humor, and theater into his inventions, emphasizing the interconnectedness of arts and sciences in his work. This inquiry allowed us to reflect on broader themes of interdisciplinary collaboration and the holistic nature of knowledge creation.

By focusing on the cultural underpinnings of Western science (Harding, 1998), we examined the situatedness of knowledge. In alignment with the work of Megan Callow and Holly Shelton (this volume), our course aimed at critically analyzing Western values often embedded in science fields. Students learned how claims of objectivity and value neutrality in Western science stem from European culture and its Christian origins. By examining the historical erasure of Arab and Muslim science (Lyons, 2014), we highlighted the political and colonial aspects of knowledge. Mapping other epistemologies of science—Buddhist, New Confucius, indigenous, postcolonial, and feminist epistemologies—provided opportunities for students to question the foundations of science. Throughout the course, our primary aim was to establish a platform for critical analytical comparison and encourage students to reflect on the cultural foundations of their commonly accepted notions about science. We sought to inspire students by exposing them to non-Western perspectives on science, encouraging them to reconsider their understanding of science and its purposes. This invited the students to see how culture shapes the purposes of producing scientific knowledge, the ways of knowing, and the questions that guide scientific research. Our introduction of non-Western scientific frameworks, like Arab and Muslim science and other non-Western sciences, offered a much-needed multicultural perspective on science and provided important background essential for students to develop critical literacy of science (Callow & Shelton, this volume).

The shape of the course was parabolic, starting and ending with a specific Arab and Muslim context while thinking more globally in between. This pattern is apparent in Appendix A, which outlines all the course texts and essential questions. The course started by exploring some of the vast history of Arab and Muslim science, focusing on the Golden Age of Islam. We then moved to exploring the concept of epistemologies of science, not only within Arab and Muslim culture but also within other cultures. We finally returned to the history and Arab and Muslim science with a new in-depth ability to analyze its epistemological underpinnings. This arch allowed the students to self-reflect and critique their taken-for-granted thinking about science, therefore inviting a destabilization of the students' Western perspectives on science to which they have been continually exposed through educational institutions and society.

Writing Assignments and Responses

In order to think critically about the course subject, the students read complex texts, wrote reflection journal entries prior to each class, discussed emerging understandings with partners and with the whole class, completed three group research projects, and wrote two papers. By engaging with seminal readings, students expanded their knowledge of Arab and Muslim history and culture, an often misrepresented culture in the United States and a mostly unrepresented history in science classrooms. Through in-class discussions, students and teachers challenged each other's understandings of the readings and each other's cultural beliefs that shape their thinking. Through group projects, students explored topics addressed in class in more depth, synthesized multiple sources, formulated arguments about specific topics, and communicated what they learned to their classmates.

In addition to reading and discussing with others, we believe in the concept of writing to learn (Gere, 1985), that writing is an essential sense-making process that is vital to deeper engagement with ideas and self-exploration. Research supports that this is important in science classes as well as more humanities-oriented classes (Prain & Hand, 2016). The writing assignments in our course provide opportunities for students to reflect on the cultural situatedness of science. Where the projects were designed to build deeper knowledge of Arabic and Muslim science, the writing assignments were for students to make sense of what they read and learned through their own unique perspectives and voices. To open space for students to reexamine their existing modes of thinking and engage with new ideas, the course involved three major writing assignments, which will be explored in the following sections. The first and ongoing assignment is a weekly journal reflection working through their emerging thinking in response to the course texts. The second assignment, completed at the beginning of the course, is a reflective narrative essay through which students explore their personal experiences in and perceptions of science. Near the end of the course, students complete the third writing assignment in which they reflect on their emerging understanding of perspectives of science and how those understandings may impact their studies, work, and life.

Journal Reflections

For all class sessions, students write journal reflections (see Appendix B) representing their study and emerging understanding of the readings and videos for the day. The reflections are intended to help the students put the specific texts into a bigger picture and to allow the students to think about how science is viewed in different places and at different times. They require the students to be self-reflective regarding their own beliefs and experiences and how these beliefs are challenged by the readings. Additionally, the reflections enable them to participate well in class discussions and to document the development of their thoughts about key ideas throughout the course. The journal reflections are open in terms of structure, with no specific prompts given for each reflection. They are low-stakes opportunities with no end goal of preparing something for an outside audience (Gere, 1985). To reiterate the low-stakes nature, each reflection is worth a small number of points, which students earn based on assignment completion. The only requirement is that they include an analysis that goes beyond a summary of the assigned texts. If we feel a student did not meet this requirement, we provide written response questions to guide deeper analysis but do not deduct points unless it is an ongoing issue.

In an end-of-semester course evaluation during the fall of 2021, students mentioned that they found the weekly journal reflections helpful as a space to construct their emerging ideas about the texts and prepare for discussing those ideas with their peers. We believe that the open-ended structure allowed their own understanding to emerge. Throughout the course, students said that they were challenged to make sense of what they read and were experiencing both inside and outside of class in an ongoing way. This extended interpretive labor placed students in the mindset of continual meaning-making. Evidence of this mindset was particularly clear in class discussions, as students would frequently refer to conversations they had with friends or family related to ideas in the assigned readings. From the instructor perspective, the journals were also a space for one-on-one asynchronous dialogues with the students. Additionally, reading these journals before class provided us a preview of how students were making meaning of the texts, what points they highlighted, and what questions they had. This allowed for the discussion prompts to be tailored to and built on the insights of the students.

In addition to students commenting that the journal reflections were helpful and us finding that they provided useful insight, the journal reflections showed clear evidence of students making connections across time and across disciplines. In the weekly journal reflections, students connected the readings with their previous educational experiences. For example, one student wrote the following in a journal reflection:

> I read a book in my seventh grade humanities class that I thought of throughout the "Narrative as Inquiry" (Hendry, 2010) reading. The book was called *The Things They Carried* by Tim O'Brien (2009), and a major theme we discussed in class was the idea that, as the book progressed, we realized that the narrative that the narrator was creating might not have exactly matched what had objectively taken place. Despite the narrator's memory of and feelings surrounding incidents occasionally being different than the factual order of events, what mattered most in the book was the way the narrator experienced and remembered things. This was mind blowing to me in seventh grade—the idea that truth is not one objective thing, or that no one person or being decides what truth is, but instead, that truth is dependent on perspective. It was the first time I remember considering that a plurality of truths can exist at the same time, and I remember that for weeks after that class discussion, I approached many conversations imagining what truths others might be holding that could complement or contradict my truths. Only during the "Narratives" reading did I remember that experience, and realize that that is a way I was encouraged to employ a multiplicity of truths (as opposed to one Truth) in some pockets of my education. I did not see those same ideas presented in my science classes though; I almost entirely was educated to believe

> in one scientific Truth and think in the way that science should be presented in a dry, "vacuum" style.

This quote demonstrates a student making sense of past experiences in a humanities course, in past science classes, and in our course. It shows her making sense of her perception of truth and reconciling past enlightenments with the present. Her use of lowercase and capital "T" in truth also indicated to us that she was making connections to a past course reading, *Sushi, Science, and Spirituality* (Kasulis, 1995). She spoke at large in class about her appreciation for the capital and lowercase use of the word when we read that text. Although not always described to this level of detail, this type of moment was common in student reflection journal entries and in class discussions.

In addition to students showing evidence of working on connecting their past understandings with their emerging understandings in the journals, they also actively worked to reconcile their emerging understandings of science with previously gained expertise in their primary disciplines. The quote below is from a senior whose studies focused on history and classics:

> In history, especially ancient history, we often discuss the impact of creation stories and conceptions of the afterlife on political and cultural norms, and it came as no surprise that it also impacted the study of science. It emphasized the huge role that religion plays in every aspect of life, whether we like to admit that or not. Seeing how science and history are both heavily influenced by the dominant religion—and seeing how that changes how we treat the world we live in—was so intriguing to me because it finally helped me understand just how interdisciplinary the two subjects are.

This student was able to better understand her own discipline through an in-depth exploration of the cultural situatedness of science. This is just one example of using writing to actively make cross-curricular connections, a skill that is vitally needed in college students (Bear & Skorton, 2019). There were many such moments in the students' journals.

Writing provides a modality for the students to actively connect different aspects of their identities as shaped by their past, their educational experiences, exposure to new ideas in texts, and critical dialogue. We do not grade or even comment on the accuracy of ideas. Rather, the journals create a dialog where we ask thought-provoking questions, provide additional resources, and give personal responses based on our own experiences and beliefs. The journals are not a place for polished ideas to be judged by an outside audience. They are an incubator for ideas (Pearse, 1985), a low-stakes informal platform that provides a safe space for students to try new ideas and ways of thinking that might feel risky and potentially transformative.

Essay on Personal Experiences and Perceptions of Science

Near the beginning of the course, students write a self-reflective essay (see Appendix C) on their personal experiences and perceptions of science. The goal of this paper is to make space for them to explore their own histories in and beliefs about science. They are prompted to base their writing on reflection questions such as what their first memory of science was, what science activities they have engaged in throughout their life, what science means to them, and how their experiences shaped their beliefs about science. It is vital to reflect on one's own understanding in order to be able to form comparisons to the perspectives of others. Through remembering significant experiences they had with science and reflecting on their own perspectives of science, the students examine where their preconceived notions come from. This prepares the students to be able to be more open to learning from other perspectives, including those of the authors of assigned readings and others in the class.

At the end of the semester, students were asked to provide feedback on the course and all of the assignments. They did not talk a great deal about any of the large writing assignments. However, they did talk extensively about how valuable and open the class conversations were and how open they were to learning from their classmates. We believe that the self-reflection facilitated through the first paper assignment contributed greatly to students' openness to others. By examining the backgrounds that shaped their personal ideas about science and the roots of these perspectives, students were able to culturally situate their own understandings. Without self-examination, individuals are likely to believe their own perspectives are truth rather than recognizing that experiences and culture shape beliefs (Singh, 2021). In class conversations, the students showed that they were continually self-reflective and open to self-critique. We observed that the students were willing to not only hear other students' ideas but also to deeply think about the different perspectives and contemplate integrating new perspectives into their own worldviews.

In the first paper, which focused on students' personal experiences in and perceptions of science, students found a space to explore how they view science and where those ideas came from. One student wrote about how science was always presented as an objective truth:

> School always taught me about science as a logical realm with
> one correct answer that accords with the laws of nature, whatever
> they may be. Overall, then, I had always learned that there was

one law governing all the universe, an answer for every question, a path to be taken to an enlightened end all thanks to science.

This student spoke in class about the fact that she always enjoyed science but lost interest when science was not able to cure an ill family member, even though she was taught in school that science had all the answers. She also indicated that if science was talked about in a more tentative manner, she may have maintained a greater interest in the discipline. Her life experiences did not match what she was being taught, which had a detrimental effect on her interest. We believe that having a space to reflect on how her experiences shaped her beliefs about science invited her to begin to reconcile these tensions.

Another student described her perception of science and "scientist" in a way that, according to Miller et al. (2018), is very common:

> Growing up, my preconceived notion of a scientist was an old white cisgender male pouring brightly colored vials into beakers. This image was usually accompanied by scientific equations and complicated work that would make any young elementary school student rethink a career in the sciences. I was only taught the names and legacies of these men, but never of those who looked like me or my classmates.

As a female person of color who was in her first year of college and was interested in pursuing a science-related major, the acknowledgment of her perception of a scientist as very different from her own identity is vital to reconciling those differences (Barton & Tan, 2010; Johnson, 2007). She did not talk more in class about the difference between herself and the stereotypical scientist. She was the least outspoken student. It was clear to us that she used the assignment to think through something that she may not have felt comfortable openly discussing with others.

Not all the students had a typical Western perspective of science, however. For example, one of the students reflected on her lifelong experience of science as an act of creation that is deeply connected with the arts:

> How I viewed science throughout my life; [is as] a tool for bettering my life and the lives of those around me through creative power. This is why I have been so excited about our conversations in class about the deep connection between art and science; to me, the pursuit of science has always fueled my pursuit of art, and vice versa. . . . I think of science as the pursuit of knowledge for the sake of creating a better world. And it is in that image of a "better world" that we see discrepancies; that image does not look the same for everyone. To some, "better" might mean more

equitable, but to others, it might mean more profitable, more aesthetically beautiful, or healthier.

In the paper, she shared her childhood experiences of mixing flavors in the kitchen, creating toys, and creating a music-activated light-up skirt for a Cinderella performance. In the assignment, she identified her situatedness. Prior to this paper she may have approached texts with an unconscious belief that everyone viewed science the same way that she did. Through engaging with different readings in the course, she was able to openly and consciously encounter different conceptualizations of science that were not utility-based.

The initial paper invited students to reflect on what their beliefs were and where those beliefs came from. Without this grounding, students likely would have encountered all new ideas with an undercurrent of unconscious beliefs rather than an active reflection and reconciliation with their conscious beliefs. The assignment began the work of giving voice to the dialogue between their identities and new ideas.

Similar to our first assignment, in their STEM writing course, Barlow and Quave (this volume) also started their course by requiring their students to write a philosophy of science reflecting on the nature of their knowledge production. As Barlow and Quave highlighted, students starting a course with reflecting on their personal philosophies of science was an important step for understanding other philosophies of science. For us, too, this assignment was very crucial for students to understand the underpinnings of their beliefs and become open to other perspectives on science.

Essay on Emerging Understandings and Perspectives of Science

Upon completion of the course, as the final assignment, students wrote a self-reflective essay (see Appendix D) on their emerging perspectives of science and how they foresaw those emerging understandings and perspectives shaping their future scholarship, work, and being in the world. The goal of the paper was to make space for students to explore how their thinking shifted due to experiences, readings, and discussions both in and outside of the class. We believe that it is vital to reflect on one's own beliefs in order to be able to continually critique societal representations of science, with which we are inundated. Examining the changes in their perspectives of science was a critical step for the students to then reflect on how they can have a voice calling for change in the larger community.

In their final reflection on their journey of learning throughout the course, students mapped the transformation of their thinking about science. For example, one student wrote:

> There was never a divide between humanities, philosophy, sciences, and religion during the Islamic golden age, so why does the Western world strive so hard to push this divide? This course educated me to not see education as a linear concept, but more so an exploration of how different disciplines interact. There was no shame in having equal interests in fine arts and hard sciences.

Reflecting on how the course broadened her horizons, this student started to question how much knowledge she was really missing by receiving a Eurocentric science education. Changing her perspective of science, that it contains indisputable answers, she pledged to perceive science with a critical eye as much as she does other disciplines. Later in this paper, she described her future career in public health, in which she intended to bring her understanding of the importance of actively listening and having respectful conversations in order to embrace and welcome people's different ways of thinking and belief systems. As a young woman of color studying science, she expressed her growing confidence in her voice and expanded perspective on science.

Another student wrote about how her ideas on the purposes for science changed throughout the course:

> Even though I already had a more complicated view of science than many in the Western world, the biggest shift in my thinking occurred with learning about the purpose of science in the Arab World, and also encompassed how other societies viewed the idea of the truth. No matter how open-minded I was about science not being omniscient, I always saw it as something driven by profit. This class, however, and especially the discussions we had over the course of the semester, showed me how we can still make scientific progress even without profit motivation. . . . The shift from for-profit to satiating curiosity was immense for me, and it forced me to rethink so many other Western scientific achievements, even the ones of which I was already critical.

Despite claiming that she never saw science as omniscient prior to the class, she did put the study of science on a pedestal. She went on in the paper to discuss that, as a gifted student, she always felt like she was wasting her abilities by not studying science. Throughout the course, she came to understand that science is not better than humanities but is just different than humanities and can and does often occur in conjunction with humanities. She explained that since science can exist outside of economic motivation, being a STEM major does not equate to prestige. This understanding made the student feel more confident in her decision to study humanities. It also made her feel more confident exploring her interests in science despite not studying science in any official capacity. This aligns with the findings of Callow and Shelton (this volume), where students found a sense of belonging in science after a critical science literacy course.

Throughout this course, students encountered science as developed by people around the world and understood that science may be thought of differently in non-Western cultures. This allowed students from diverse backgrounds, beliefs, and interests to better see themselves involved in science and reconcile who they are and want to be.

Lessons Learned, Challenges, and Implications for Science Teachers

This course is an example of working across disciplinary lines to engage with scholarship from history, philosophy, cultural studies, and science. The writing assignments described were created for the class in order for students to examine the overlap of culture and science and to write to learn (Gere, 1985). Although this course is unique, similar writing activities could be utilized in any science content class to help students negotiate their identities and expand their thinking. At a minimum, students could be assigned reflections on their understanding of what science is at the beginning and end of science courses, with intentional exposure to varied ways of thinking about and doing science throughout a course. Students could also be encouraged to keep a personal journal, continually reflecting on the representations of science that they are exposed to and how they are making meaning of the content they are learning.

From the experience of this course, the format and requirements of writing assignments impact the students' openness toward the assignments. As previously described, students found the writing assignments in this course helpful for their intended goals. One point of feedback they provided was about the word requirements of the journal reflections. Because it is important to only grade for the intended purpose of an assignment, all writing assignments had grading criteria that highlighted the quality and depth of thought rather than aspects such as grammar and editing (see Appendices B–D). However, there was a required word limit for each assignment. For the journal reflections, some students found the word requirement to be a little long, given the twice-weekly expectation. They felt that it impeded their ability to write for learning by shifting the focus to writing for a requirement. This further highlights that the more stringent and criteria-driven grading is, the less free students will feel to do the self-reflective work that is the purpose of the assignment. Along these same lines, we believe that if our comments on the journals were more judgmental and less reflective, students would not have felt as comfortable expressing their emerging ideas.

Overall, despite years of science classes in elementary and secondary school and several college science courses, students in this class repeatedly commented that they had never been exposed to ways of thinking about science that went beyond the empirical, rational modality, which is the norm in Western approaches to science. Students reflected that being exposed to culturally situated ideas about scientific knowledge—knowledge for its own sake, the overlap between science and aesthetic arts, and various conceptions of what can and cannot be known—transformed their thinking. More than transforming their thinking, for several students, the new perceptions increased their interest in the discipline and made them feel more confident that their voices matter.

One of the major challenges that we face for our specific course is getting students interested and registered. The course was a response to an institutional initiative to bridge STEM, humanities, arts, and social sciences in a co-teaching opportunity involving a STEM and a humanities professor. Although our course received institutional approval to run, its continuation after the first year was hindered by the lack of departmental support for interdisciplinary scholarship. The course does not fulfill any requirements, and many of the ideas highlighted in the course description and syllabus are disarmingly foreign to students and professors. We hope for this course to continue as a disruptive space to interrogate the status quo in STEM education (Introduction, this volume) and impact students in STEM fields in a way that makes them question dominant paradigms in their disciplines and examine their cultural underpinnings. We hope that it allows students to go beyond learning and mastering the contents of their subjects to inquiring about the types of questions and methodologies that drive their field of study, where they stem from, whose interests they serve, and what beliefs about the world and humanity lie beneath them. We hope to create space within academia for courses that invite students to reflect on the production of knowledge as a culturally situated political act. Solutions for us include making the course fulfill general education requirements or count toward majors or minors. This would involve educating department heads on the importance of the ideas. It also involves advocating for more institutional integration of humanities and arts with STEM. More importantly, the lack of interest in this topic highlights the necessity for science teachers (elementary, secondary, and post-secondary) to become knowledgeable about and imbed ideas about the culture of sciences in their courses, which also brings into question teacher preparation in STEM fields.

As clarified earlier, the focus of this chapter is on highlighting the role of reflective writing in facilitating critical comparative examinations of epistemologies of science rather than providing information on how to include Arab and Muslim science in different science classes. Nevertheless, it's crucial to underscore the importance for science educators to integrate Arab, Muslim, and other non-Western achievements into their curricula. For example, they can include Ibn Al-Haytham's work when teaching the scientific method, alongside other significant contributions from this historical period. Resources and videos showcasing such achievements are available on the web at https://www.1001inventions.com. Supporting this integration are works from Bala & Duara (2016), Barnard (2011), Ganchy (2009), Saliba (2007), Tibi (2006), and Gutas (1998). By incorporating diverse perspectives and historical achievements into their teaching, science educators can provide students with a more comprehensive understanding of the development of scientific thought. Our course highlights the importance of examining the history and geopolitics of science in a way that goes beyond listing names of culturally diverse scientists. This approach allows students to see science as a human enterprise of exploration of the natural world and inquiry toward it rather than as a collection of absolute facts, laws, and equations to study or apply. Through learning about how science is approached differently in various cultures, students can develop critical thinking skills and learn to question and evaluate the underlying assumptions and biases embedded within scientific theories and practices. Ultimately, this approach fosters a more inclusive and equitable approach to science education, encouraging students to engage with science as a dynamic and culturally situated endeavor.

Some researchers are arguing for a shift toward an appreciation for multiple and often culturally situated ways of understanding science (Bang & Medin, 2010; Hammer & Elby, 2002; Russ, 2014). Meghan Bang advocates supporting a learner's navigation of what she calls multiple epistemologies. In a 2010 article, Bang and Medin discussed the impact of a three-week summer camp on Native American students. The curriculum of the summer camp was designed around culturally-based practices and aimed to support the navigation of multiple epistemologies of science. They showed that respecting multiple epistemologies can have a positive impact on student learning. Elby, Macrander, and Hammer's 2016 study showed that unique focuses with different communities lead to distinctive observations, distinctive questions, and distinctive interpretations of findings. Additionally, there is research showing that the Western culture of science makes it difficult for many minority students to assimilate into scientific communities (Brown et al., 2016), particularly women of color (Carlone & Johnson, 2007).

Thus, although science content courses typically have a set of stringent objectives that must be met during the course in order to prepare students to be well-versed in their discipline, a sole concentration on those objectives, to the exclusion of consideration of cultural underpinnings, may discourage students from remaining in the field. It may particularly discourage students who are already underrepresented in science and who may bring unique ways of thinking that are vital to approaching the problems facing the world. Therefore, it cannot be up to elective courses to expose students to other ways of thinking about science. In every science course, there should be space for examining the social underpinning of science. One way of implementing these much-needed reflective practices into science courses is through reflective student writings such as those described in this chapter.

References

Al-Khalili, J. (2010). *Pathfinders: the golden age of Arabic science*. Penguin UK.
Al-Khalili, J. (2011). *The house of wisdom: How Arabic science saved ancient knowledge and gave us the Renaissance*. Penguin.
Anand, A. & Walsh, I. (2016). Should knowledge be shared generously? Tracing insights from past to present and describing a model. *Journal of Knowledge Management, 20*(4), 713–730.
Bala, A. & Duara, P. (Eds.). (2016). *The bright dark ages: Comparative and connective perspectives* (Vol. 5). Brill.
Bang, M. & Medin, D. (2010). Cultural processes in science education: Supporting the navigation of multiple epistemologies. *Science Education, 94*(6) 1008–1026. https://doi.org/10.1002/sce.20392.
Barnard, B. (2011). *The genius of Islam: How Muslims made the modern world*. Knopf Books for Young Readers.
Barton, A. C. & Tan, E. (2010). 'We Be Burnin'! Agency, identity, and science learning. *Journal of the Learning Sciences, 19*(2), 187–229. https://doi.org/10.1080/10508400903530044.
Bear, A. & Skorton, D. (2019). The world needs students with interdisciplinary education. *Issues in Science and Technology, 35*(2), 60–62. https://issues.org/the-world-needs-students-with-interdisciplinary-education/.
Brown, B. A., Henderson, B., Gray, S., Donovan, B., Sullivan, S., Patterson, A. & Waggstaff, W. (2016). From description to exploration: An empirical exploration of the African-American pipeline problem in STEM. *Journal of Research in Science Teaching, 53*(1) 146–177. https://doi.org/10.1002/tea.21249.
Carlone, H. B. & Johnson, A. (2007). Understanding the science experiences of successful women of color: Science identity as an analytic lens. *Journal of Research in Science Teaching, 44*(8), 1187–1218. https://doi.org/10.1002/tea.20237.
Elby, A., Macrander, C. & Hammer, D. (2016). Epistemic cognition in science. In J. Greene, W. Sandoval & I. Braten (Eds.), *Handbook of epistemic cognition,* (pp. 113–127). Routledge.
Ganchy, S. (2009). *Islam and science, medicine, and technology*. The Rosen Publishing Group, Inc.
Gere, A. R. (2012). *Roots in the sawdust: Writing to learn across the disciplines*. The WAC Clearinghouse; National Council of Teachers of English. https://wac.colostate.edu/books/landmarks/sawdust/ (Originally published in 1985 by National Council of Teachers of English).
Gutas, D. (1998). *Greek thought, Arabic culture: The Graeco-Arabic translation movement in Baghdad and early 'Abbasid society (2nd–4th/8th–10th centuries)*. Routledge.
Hammer, D. & Elby, A. (2002). On the forms of a personal epistemology. In B. Hofer & P. Pintrich (Eds.), *Personal epistemology: The psychology of beliefs about knowledge and knowing,* (pp. 169–190). Routledge.
Harding, S. (1998). *Is science multicultural?: Postcolonialisms, feminisms, and epistemologies*. Indiana University Press.

Hendry, P. M. (2010). Narrative as inquiry. *The Journal of Educational Research, 103*(2), 72–80. https://doi.org/10.1080/00220670903323354.

Hill, D. R. (1991). Mechanical engineering in the medieval near east. *Scientific American, 264*(5), 100–105.

Hill, D. R. (2020). *Studies in medieval Islamic technology: From Philo to al-Jazari–from Alexandria to Diyar Bakr*. Routledge.

Johnson, A. C. (2007). Unintended consequences: How science professors discourage women of color. *Science Education, 91*(5), 80—821. https://doi.org/10.1002/sce.20208.

Kasulis, T. P. (1995). Sushi, science, and spirituality: Modern Japanese philosophy and its views of Western science. *Philosophy East and West, 45*(2), 227–248. https://doi.org/10.2307/1399566.

Lyons, J. (2014). Islam through Western eyes: From the crusades to the war on terrorism. Columbia University Press

Miller, D. I., Nolla, K. M., Eagly, A. H. & Uttal, D. H. (2018). The development of children's gender-science stereotypes: A meta-analysis of 5 decades of US Draw-a-Scientist studies. *Child Development, 89*(6), 1943–1955. https://doi.org/10.1111/cdev.13039.

Nasr, S. H. (1976). *Islamic science: An illustrated study*. World of Islam Festival Publishing.

Nasr, S. H. (1984). The role of the traditional sciences in the encounter of religion and science–An oriental perspective. *Religious Studies, 20*(4), 519–541.

Nasr, S. H. (1988). Islamic science, western science common heritage, diverse destinies. In Z. Sardar (Ed.), *The revenge of Athena: Science, exploitation, and the Third World* (pp. 239–248). Mansell.

Nasr, S. H. (2001). *Science and civilization in Islam*. ABC International Group.

Nasr, S. H. & Iqbal, M. (2003). Islam, science and Muslims: A conversation with Seyyed Hossein Nasr. *Islam & Science, 1*, 5–28. https://cis-ca.org/_media/pdf/2003/1/A_isam.pdf.

Nasr, S. H. (2010). Islam and the problem of modern science. *Islam & Science, 8*(1), 63–75. https://cis-ca.org/_media/pdf/2010/1/TEM_temiatpoms.pdf.

O'Brien, T. (2009). *The things they carried*. Houghton Mifflin Harcourt.

Pearse, S. (1985). Writing to learn: the nurse log classroom. In Gere, A. R. (Ed.). *Roots in the sawdust: Writing to learn across the disciplines,* (pp. 9–30). The WAC Clearinghouse; National Council of Teachers of English. https://wac.colostate.edu/docs/books/sawdust/chapter1.pdf (Originally published in 1985 by National Council of Teachers of English).

Prain, V. & Hand, B. (2016). Coming to know more through and from writing. *Educational Researcher, 45*(7), 430–434. https://doi.org/10.3102/0013189X16672642.

Russ, R. S. (2014). Epistemologies of science vs. epistemologies for science. *Science Education,* 98(3), 388–396. https://doi.org/10.1002/sce.21106.

Saliba, G. (2007). *Islamic science and the making of the European renaissance*. MIT Press.

Singh, R. (2021). Race, privilege, and intersectionality: Navigating inconvenient truths through self-exploration. *Journal of Education for Library and Information Science, 63*(3), 277–300. http://dx.doi.org/10.3138/jelis-2021-0005.

Tibi, S. (2006). Al-Razi and Islamic medicine in the 9th century. *Journal of the Royal Society of Medicine, 99*(4), 206–207.

Todd, S. (2003). *Learning from the other: Levinas, psychoanalysis, and ethical possibilities in education*. State University of New York Press.

# Appendix A: Course Outline and Texts

Week	Class	Questions we'll explore . . .	How to prepare for this class . . .
Week 1	Class 1	Why Arab AND Muslim civilization not Arab OR Muslim civilization?	Video: Islam, Empire of Faith (video on YouTube, 0:00–1:47)
	Class 2		
Week 2	Class 3	Why do some people hate Muslims and Arabs?	Jigsaw Readings: Islam Through Western Eyes: From the Crusades to the War on Terrorism by Jonathan Lyons (chapters 1 &7) Islam Through Western Eyes: From the Crusades to the War on Terrorism by Jonathan Lyons (chapter 3) Covering Islam: How the Media and the Experts Determine How We See the Rest of the World by Edward W. Said (chapter 1, section 1) Video: Islam Through Western Eyes (speech by Jonathan Lyons on YouTube)
Week 3	Class 4	Is there a conflict between knowledge and faith? Where is the oldest university in the world?	Readings: Research online to find out: What are the first Universities? Who is Fatima Al-Fihri? What was women's role is Arab and Muslim education?- follow your curiosity Education in Islamic History (article in Egypt Today, 2017)
	Class 5		Reading: The Basis of the Teaching System and the Educational Institutions by Seyyed Hossein Nasr (chapter 2)
Week 4	Class 6	What did your group find out about the history of Arab and Muslim science?	Reading: Islam Through Western Eyes: From the Crusades to the War on Terrorism by Jonathan Lyons (chapter 4)
	Class 7	What's unique about Arab and Muslim Science?	Group research project on history of Arabic and Muslim science and how cultural beliefs shaped that advancement
Week 5	Class 8	Did all science come from Greek civilization? Who was shining during the dark ages? What did science look like during the Golden Age of Islam?	Videos: The House of Wisdom: How the Arabs Transformed Western Civilization (speech by Jonathan Lyons on Library of Congress website) 1001 Inventions and the Library of Secrets (video on YouTube)
	Class 9		Reading: A young Muslim's guide to the modern world. 1993 by Seyyed Hossein Nasr (chapter 5: Islamic Science)

Week 6	Class 10	What the heck is an epistemology? What does culture have to do with science?	Reading: Epistemology: Internet Encyclopedia of Philosophy (website) Videos: Intro to Epistemology #1: The Nature of Knowledge (video on YouTube, 0:00–3:47) Philosophy of Science: Epistemology Applied (video on YouTube)
	Class 11		Reading: "Is science multicultural?: Challenges, resources, opportunities, uncertainties" by Sandra Harding: (chapter 4: Cultures as a Toolbox for Sciences and Technologies)
Week 7	Class 12	What did you learn about your personal epistemologies of science?	Reading: Cultural Processes in Science Education: Supporting the Navigation of Multiple Epistemologies by Bang & Medin (2010) Video: Recycling is Like a Bandaid on Gangrene (2019 video on The Atlantic)
	Class 13		Individual paper exploring their personal experiences and perceptions of science
Week 8	Class 14	Who else has "other" ways of thinking about science?	Video: The Dalai Lama: Scientist (2019 documentary)
	Class 15		Reading: Sushi, science, and spirituality: Modern Japanese philosophy and its views of Western science by Thomas Kasulis (1995)
Week 9	Class 16	What did your group learn about different epistemologies of science?	Readings: Western science and traditional knowledge- Despite their variations, different forms of knowledge can learn from each other by Fulvio Mazzocchi (2006) Multicultural Science Education: Theory, Practice, and Promise edited by Steinberg and Hines (chapter 10: Defining a Theoretical Framework for Multicultural Science by Samina Hadi-Tabassum)
	Class 17		Group project comparing three different epistemologies of science
Week 10	Class 18	Who's who in Arab and Muslim Epistemology?	Reading: Epistemology in Islamic Philosophy: Islamic Philosophy Online (website) Video: Knowledge Triumphant, The Concept of Epistemology in Islam (documentary on YouTube)

A Curriculum Exploring Arab and Muslim Science | 119

Week 10 continued	Class 19		Readings: Epistemological Foundations of Natural Sciences in Islam by Marziyehsadat Montazeritabar (2019)
			Knowledge Triumphant: The Concept of Knowledge in Medieval Islam by Franz Rosenthal (chapters 3 and 4)
Week 11	Class 20	How do stories shape epistemologies of science?	Readings:
			Braiding Sweetgrass—by Robin Wall Kimmerer, 2013 (chapter 1: Skywoman Falling)
			"The Study Quran." A new translation and commentary: Creation Story from the Quran translation by Nasr, Seyyed Hossein, Caner K. Dagli, Maria Massi Dakake, Joseph EB Lumbard, and Mohammed Rustom (2015): (chapter 2:27–49)
			OPTIONAL: Adam and Eve Story from the Bible
	Class 21		Videos: Hay—The Kid Searching For God (video on YouTube)
			The place of philosophy in religion pt. 3—Hayy bin Yaqzan (video on YouTube)
Week 12	Class 22	How did your group use representation to change perceptions about science?	Readings:
			Ibn Tufayl (A bentofail) and the Origins of Scientific Method by Enrique Cerda-Olmedo (2008)
			Narrative as Inquiry by Petra Munro Hendry (2010)
	Class 23		Group project critiquing and reimagining a media representation of science
Week 13	Class 24	What do you want to learn more about?	Readings: TBD based on student interests
Week 14	Class 25	What does all this mean to us now?	Readings: TBD based on student interests
	Class 26		Reading: The Geopolitics of Knowledge and the Colonial Difference by Walter D. Mignolo (2008)
Week 15	Class 27	What question do you have for Guest Speakers?	Field Trip to the Turkish Islamic Center
	Class 28		Guest speaker: Dr. Seyyed Hossein Nasr
Final paper			Individual paper reflecting on your emerging understanding of multiple epistemologies of science and its impact on your work in their discipline and future careers

Appendix B: Journal Reflections

Why: Writing weekly reflections will help you to put the specific readings into the bigger picture. They will allow you to think about how science is viewed in the world and how that is different and similar in other places and at other times. They will also help you to be self-reflective regarding your own beliefs and experiences, how the reading may challenge your beliefs, and how your perceptions are continually emerging. Additionally, your reflections will enable you to participate well in class discussions and to document the development of your thoughts about key ideas in the course. You should come to class prepared to introduce your ideas in class discussions and to test their significance through scholarly conversation.

What: For all class sessions, you will write journal reflections representing your study and understanding of the readings and videos for the day. Each reflection must include:

- analysis of key concepts, ideas, and perspectives that appear in the course readings and videos.
- exploration of the issues you find most compelling.
- questions that will help you and your classmates develop understanding of the readings.
- relevance to your field of study.

Reflections should not be organized in the form of a thesis and argument. They should not be presented as an essay. Instead, they should represent the diverse range of thoughts, questions, and interpretations that emerge as you study an assigned text in the form of "thinking out loud on paper". This type of journaling is one that many scholars find useful for germinating new and creative thoughts. Reflections should:

- be at least 400 words.
- use 25 words of direct quotation at maximum.
- provide references to sections of the assigned text.

How (you'll be assessed): Instructors will evaluate your reflections in terms of their meeting the criteria outlined above (at least 400 words, no more than 25 words of direct quotation, references to sections of text discussed) and based on thoughtful engagement with the assigned readings. You will not receive evaluative comments on your reflections as they are based on your opinions and developing ideas. However, you may periodically receive comments and questions to further probe your thinking. You should spend time reflecting on these comments but do not need to submit a response. Bring a hard or electronic copy of the reflection to class for reference during the discussion.

Appendix C: Personal Experiences of and Perceptions of Science

Why: The goal of this paper is to make space for you to explore your own history in and beliefs of science. It is vital to reflect on your own understandings in order to be able to form comparisons to the perspectives of others. Through remembering significant experiences you had with science and reflecting on your own perspectives of science, you will examine where your preconceived notions come from, which is a critical step in being able to become more open to learn from other perspectives.

What: This assignment will be in the format of a narrative that demonstrates thoughtful reflection on beliefs and experiences. Consider the following prompts when beginning the narrative:

- What was the first time you recall engaging in science?
- What science activities have you engaged in throughout your life.
- What is science to you?
- Who are scientists?
- What do scientists do?
- How have your experiences (school, friends and family, media, etc.) shaped your beliefs about what science is?

Before writing your narrative, consider starting by doing a few free writes or graphic organizers to get your thoughts down without concern for how to do well on this as a class assignment. In other words, spend time really reflecting inwardly before being concerned about the final outward representation.

The final submission should be between 3 and 5 pages, double-spaced. As this is a self-exploration, reference to other sources is not required. However, if external sources are used, there must be correct APA formatting and a bibliography. The final document should be well-edited. You are encouraged to seek the assistance of a research librarian and of the writing center.

How (you'll be assessed): Papers will be evaluated based on the depth of reflection and analysis. The narrative should describe experiences and events related to your understanding of science, taking into account the context of these experiences and providing an analysis of and an interpretation for the connections between past experiences discussed and perspectives. The narrative should give meaningful consideration to questioning your taken-for-granted assumptions about science by tracing their sources in your personal history.

This paper is worth a total of 10 points: 6 points for depth of reflection, 4 points for clarity, organization, and mechanics.

Appendix D: Emerging Understanding of Perspectives of Science

Why: The goal of this paper is to make space for you to explore your emerging understanding of and perspectives of science. It is vital to reflect on your own understandings in order to be able to continually critique what you are exposed to through classes, media, etc. Reflecting on your own perspectives of science is a critical step in being able to reflect on how you can have a voice in the larger community, which is the second main goal of this assignment.

What: This assignment should be a thoughtful reflection on your shifting beliefs and the experiences, readings, and discussions that caused those shifts both in and outside of this course. Discuss the most transformative aspects of this course explicitly and with citations. You will then reflect on where you will go from here. How will your new perspective shape your future career, studies, relationships, conversations, etc.?

The final submission should be between 3 and 5 pages, double-spaced. Reference to course materials or other materials that you engaged with and found meaningful this semester should be included using correct APA formatting and a bibliography. The final document should be well-edited. You are encouraged to seek the assistance of the writing center.

How (you'll be assessed): Papers will be evaluated based on the depth of reflection, and analysis. The reflection should describe transformations you experienced, engagements that caused those transformations, and future impacts of those new perspectives.

This paper is worth a total of 10 points: 6 points for depth of reflection, 4 points for clarity, organization, and mechanics.

- What changed—Description of transformation
- How it changed—Engagements with course aspects that caused transformation
- What now—Future impacts of new perspectives
- Proper citation and bibliography
- Clarity, organization, and mechanics

Creating Assignments that Put Programmatic Inclusion and Diversity Work into Practice

Justiss Wilder Burry
TARLETON STATE UNIVERSITY

Carolyn Gubala
UNIVERSITY OF CALIFORNIA, DAVIS

Jessica Griffith
JACKSONVILLE STATE UNIVERSITY

Tanya Zarlengo
UNIVERSITY OF SOUTH FLORIDA

Lisa Melonçon
CLEMSON UNIVERSITY

In Technical and Professional Communication (TPC), the service course is an "introductory [course] for nonmajors delivered primarily as a service to other departments and programs on campus" (Melonçon & England, 2011, p. 398), and the course is designed to prepare students to "adap[t] emergent knowledge to specific workplace or community-based contexts" (Scott, 2008, p. 382). The connection between TPC's service course and STEM is long established, with scholarship showing the historical relationship with engineering as early as the 19th century (Kynell, 2000). The first textbook specific to technical and scientific writing was written in 1911 by Samuel Earle and established the rising importance of TPC to technical and scientific fields (Connors, 1982; Cook, 2002). The service course is rooted in late 19th-century courses in writing for engineers (Kynell, 2000). Service courses have advanced since the early emphasis on basic elements of written communication to encompass more nuanced and rhetorical elements of technical and professional writing necessary to succeed in the STEM workplace, which includes issues of diversity, equity, and inclusion (DEI). As a professional practice, TPC uses writing and communication to move audiences to action, and an emphasis on application and practice is fundamental to the work of TPC. This is not to say that TPC has been devoid of theory. Rather, it has been a field where theory moves into practice more smoothly than other humanistic endeavors (e.g.,

DOI: https://doi.org/10.37514/ATD-B.2024.2364.2.08

Melonçon & Schreiber, 2018). TPC is uniquely poised to move the conceptual imperative of justice (e.g., Agboka & Matveeva, 2018; Walton & Agboka, 2021; Walton et al., 2019) to actual practice throughout TPC programs that teach large numbers of STEM students every year. In light of ongoing conversations around justice in TPC, we asked: What happens in a large TPC service course program when it creates a programmatic inclusion vision and then sets out to enact it via diversity work?

We used the TPC service course as the site of our work because it provides "rich locations for program administrators, instructors, and researchers to ask and test central questions about TPC as a field and its role in shaping professional communication practices in both the workplace and the public sphere" (Schreiber et al., 2018a, p. 1). While scholars have taken an interest in the service course (e.g., Boettger, 2010; Read & Michaud, 2018; Schreiber et al., 2018b; Newmark & Bartolotta, 2021), TPC has little research that explicitly focuses on assignments that work toward programmatic inclusion. As noted by Rita Kumar and Brenda Refaei (2021), "STEM is often seen as a more challenging area in which to practice equity and inclusion due to the pragmatic nature of the content and the perceived inflexibility of the curriculum" (p. 113). The possible inflexibility of the curriculum in STEM fields makes required courses like the service course even more important to students' futures, and it makes the necessity for grounding TPC and writing curricula in ways where students can see the necessity of inclusive approaches.

In this chapter, we discuss a way to address this collection's emphasis on "actional steps faculty can enact to make their STEM writing spaces more inclusive and challenge assumptions about disciplinary writing" (see Introduction, this collection). We start by describing our theoretical framework that situates programmatic inclusion within STEM services courses. Next, we move to our educational context, followed by an analysis of student documents based on student learning outcomes (SLOs) and their connection to programmatic inclusion. We end with a discussion of what worked well for this assignment and what could be improved in order to facilitate better implementation of programmatic inclusion.

Theoretical Framework

In Sara Ahmed's groundbreaking book, *On being included* (2012), she writes a cautionary tale of what happens when diversity initiatives are not carried out in practice. Ahmed explained that "when diversity work becomes a matter of writing documents, it can participate in the separation of diversity work from institutional work" (p. 87), and those documents end up being "non-performative," meaning that they stand in place of saying the work needs to be done rather than doing the work. While "what is attended to can be thought of as what is valued" (p. 30), it takes more than documentation because "we have to work on them to make them work" (p. 119). Writing program administrators are taking seriously how issues of "race, accessibility, and assessment" (Voss et al., 2021, p. 14) inform the development of inclusive writing programs and course designs. TPC programs are no exception (Agboka & Dorpenyo, 2021). But how do faculty and program administrators make them work?

From TPC scholarship, we highlight several recent attempts to operationalize the theory of social justice(s) at the course and program level. Chen Chen (2021) and Jennifer Bay (2022) explained the process of developing an undergraduate technical communication introductory course through a socially just pedagogical approach. Cruz Medina and Kenneth Walker (2018) proposed contract grading to disrupt the distributions of power within assessment practices and explained that this framework is "not about mainstreaming shared values" but "making course values explicit" (p. 52). The idea behind "making course values explicit" is also seen in the work of Jennifer Mallette and Amanda Hawks (2020). Using grading contracts, they explained that "instructors can detail what students can expect from instructors and what the instructor expects from students, which can help students see how assessment connects to course outcomes" (Mallette & Hawks, 2020, p. 4). This transparency among instructors and programs can aid in distributing power that would otherwise be hidden.

Connecting assessment to outcomes by aligning course and program outcomes not only provides transparency between student and instructor, but it also allows students to use their previous knowledge and grow over the term. Linda Driskill (2013) argued for the importance of making course outcomes explicit and that "whatever assessment is used must be related to the specific objectives, prior experiences, and long-range plans of the students" (p. 65). Robert Mislevy and Norbert Elliot (2020) discussed the positioning of students and instructors and explained that SLOs as explicit statements of values "advance opportunities to learn for all students" (p. 148). Writing explicit and useful SLOs enables TPC to move toward more equitable assessment practice because the outcomes clearly indicate how students will be assessed, thereby allowing for greater opportunity for all students (Griffith et al., 2024).

In a special issue that foregrounded accessibility, Sushil Oswal (2018) explained the necessity of access, broadly construed to include disabled students, but also as a reminder that focusing on accessibility creates a "rich rhetorical user experience for diverse populations" (Hitt, 2018, p. 62). Pushing this idea further, Lisa Melonçon (2018a) crafted an Ahmed-inspired theory of "orienting access" that asks program administrations and faculty to work toward creating inclusive and diverse learning spaces (p. 46), which expanded previous arguments that called for an "ideology of inclusion" (Oswal & Melonçon, 2017, p. 68). Instantiating Ahmed's concepts of phenomenology, "ideology of inclusion" prioritizes experiences of those who have endured unjust systems and institutions. Holding this ideology means that faculty and administrators who want to perform programmatic diversity work must "acquire critical orientations to institutions in the process of coming up against them" (Ahmed, 2012, p.174). Putting diversity work into practice means articulating our intent for inclusion in the TPC service course program. Programmatic inclusion is an

> antiracist, intentional programmatic perspective that takes as its central aim access and equity by starting with non-harmful considerations for course and curriculum design; creating learning opportunities to achieve equitable outcomes for all students; teaching skills and knowledges that expand students' writing abilities and processes; demonstrating ways to use communication to advocate for or to affect strategic change. (Melonçon, 2024)

The definition does important work in codifying the work of social justice at the program level by making tacit knowledge explicit. Defining programmatic inclusion affords TPC program administrators and faculty to have a clear direction and approach to make their programs inclusive. Too often, the work of program administration is not explicitly codified and documented, but a hallmark of TPC has been in the field's effort to build, maintain, and sustain knowledge management practices for organizations through writing, communicating, and managing information. In this way, creating a definition of programmatic inclusion achieves the same goal that knowledge management practices do in many workplaces and organizations. The definition creates knowledge sharing, encourages interaction and reflection, demonstrates a process for all stakeholders, and makes internal knowledge external and explicit. Jennifer Mallette (this volume) adds to the goals of programmatic inclusion by arguing students need to "understand what they are being asked to do in the class and how it helps them make progress toward course goals" (this collection).

Programmatic inclusion grounds every decision made, including the creation of assignments. Ahmed (2012) forcefully reminds us that diversity work, what we are calling programmatic inclusion, has to be enacted through performance and an attention to the transformative work the policy reflects. Too often, attention is redirected to the policies themselves as evidence of a commitment to diversity, and those lacking in structure for implementation will only perpetuate the problems identified. In the next section, we describe one part of the program's commitment: an assignment that underscores the imperative for an inclusive approach to curricular design that also enacts the learning outcomes of the program.

Educational Context

Our data comes from an English department housed within an R1 university in the Southeast. The TPC service course program within the department serves some

~4,800 students a year in three courses—engineering, allied health sciences, and business. The courses are differentiated by the content brought in by students. For example, in the engineering version of the course, the readings, data, and other examples are drawn from engineering scenarios, and students are encouraged to use their content knowledge when completing assignments. Our focus in this chapter is on the STEM students in engineering and allied health sciences, which account for ~1,700 students a year. The service courses for engineering and allied health sciences majors are not general education courses, but they are required as part of the different major curricula. The rationale for requiring the service course is the emphasis on providing in-depth writing and communication instruction by experts in writing and communication. In limited conversations with stakeholders in engineering and in allied health science, we have been told that the courses are doing what they need them to do for the students. However, we remain interested in advancing conversations so that we can better align the goals with the course, which include issues of DEI enacted through performance and an attention to the transformative work the inclusivity reflects. The program uses a uniform curriculum with a common textbook and four required assignments. The document series, the first project of the term, asks students to engage in real and meaningful inclusivity work outside of the academy. This assignment presents a problem-based scenario (Melonçon, 2018b) that is similar to what they may encounter in the workplace. The students are asked to write three short documents for three different audiences. For most scenarios, students need to write to an external audience as well as to two different internal audiences. Students can choose from multiple scenarios and in each course, one of the available scenarios explicitly addresses an inclusivity problem.

Allowing students to choose from a selection helps students select a meaningful scenario that aligns with the STEM major and specialization. For example, problem-based scenarios in engineering may focus on a computer or civil engineering problem. The document series assignment represents what Michele Eodice and her collaborators (2016) refer to as meaningful writing assignments. In other words, to achieve learning objectives, assignments must do more than interest students; they must meaningfully engage students through their relevance to the students' lives. Meaningfulness is a crucial characteristic, especially as it relates to programmatic inclusion because for students to meaningfully engage, assignments must consider alternate perspectives and experiences. (See Appendix for assignment description, problem-based scenarios, and rubric.)

The TPC program has asked instructors to keep the following questions in mind throughout the term as a key part of their pedagogy:

- How does this document/deliverable affect existing workplace power dynamics, if at all?
- Who does this project leave out?

- How might the final deliverable address, maintain, or facilitate inequitable or unjust practices and power structures in organizations?

Thoughtful and meaningful assignment design presents instructors and administrators with the opportunity to enact program and course goals. Assignments, such as the document series, demonstrate an overall programmatic inclusion framework goal by helping students and instructors put TPC theory into practice.

In what follows, we present the results of our coding of student work that highlight ways that meaningful assignments can move students toward learning outcomes while demonstrating the integration of programmatic inclusion. Explaining the connections between assignments, outcomes, and programmatic inclusion exemplifies a move toward realizing the assignment as the nexus of all the forces in play in the service course program.

Looking at Student Documents Programmatically

The programmatic inclusion scenarios that include an emphasis on DEI were introduced into the document series assignment in fall 2020. Our analysis looks at student finals from spring 2021 and fall 2021 because it allowed instructors to teach the assignment once before we examined the results. The data presented in this section has been exempted and approved for use under University of South Florida Institutional Review Board, #002887.

Table 6.1 shows the total number of student documents for the semesters under examination. The total number of student samples is 1,695, with our focus on the 28 percent (n = 488) of students selecting the DEI-focused scenario. Table 6.1 highlights a difference between allied health science students and engineering students who selected the programmatic inclusion scenario. Less than 15 percent of the Engineering students selected the scenario, whereas a little over 40 percent of the allied health science students chose it.

Table 6.1: Percentage of Students Who Selected the Programmatic Inclusion Scenario (Total N = 1695; DEI n = 488)

Course	Spring 2021	Fall 2021
Allied Health Sciences	44% (n = 204)	41% (n = 189)
Communication for Engineers	11% (n = 47)	14% (n = 48)

To gain insight into student engagement, students were asked to provide a short, written comment to explain their scenario selection. One allied health science student commented, "Cultural and diverse issues are important to me and I found it interesting to fix internal issues such as this." An engineering student remarked, "It struck me the most as I read it. Discrimination is an extremely serious offense for any person or organization to be accused of and I wanted to tackle the most severe and toughest situation." The scenarios require STEM students enrolled in service courses to actively consider meaningful writing outside of the academy and its relevance in workplace settings. While workplace writing may seem far from classroom writing, TPC classroom projects promote connections to "the applicability or relevance of the projects" (Eodice et al., 2016, p. 82) that students may encounter on the job. In other words, the merging of a student's past experiences, acquired skills, and future goals resonates with workplace-centered rhetorical situations of service course writing and student experiences. Students were asked to create three correspondences based on the scenario they selected.

In order to better understand the students' uptake of DEI as it relates to the SLOs of the course, the student examples were coded based on four criteria: empathy, language awareness, power, and point of view (POV).

- Empathy
  - Did the student offer an apology when corresponding to the person who complained? Did the student show an empathic stance (an understanding that the issue was indeed a problem) toward the situation in the documents to their supervisors?
- Language awareness
  - Did the student incorporate appropriate language that shows an awareness of DEI (e.g., including words such as "diversity," "culture/cultural," "inclusion/inclusivity/inclusive," etc.)?
- Power
  - Did the student properly acknowledge the role of organizational power in addressing and solving the problem? For example, establishing the matter was still ongoing, not solved, and/or required stakeholder approval or agreement.
- POV
  - Did the student switch between writing as an individual and as a representative of the organization?

Empathy and language awareness align with the assignment's SLO that asks students to develop an appropriate writing style, while the power and POV align with the SLO for addressing purpose and audience. These four criteria, which became our coding schema, connect the programmatic inclusion framework to SLOs by asking students to work through problem-based scenarios similar to common workplace situations that require students to engage in actions, through writing, that advocate for change to more equitable practices in the workplace.

Beyond aligning with the SLOs, the criteria materialize Ahmed's (2012) diversity work. Empathy and language awareness intentionally engage students in communicating information in an anti-racist and inclusive nature, while power dynamics and POV move students toward practicing critical thinking and problem solving. These ideas connect the vision of programmatic inclusion to the practice of assignment design and SLOs—to engage in a programmatic inclusive framework by doing something about the problem instead of only theorizing. Programmatic inclusion is an extension of the type of action-oriented writing that characterizes TPC and ensures that STEM students have exposure to how change can be enacted within an organization, thereby tying coursework directly to their disciplines and future work. We opted to sample 60 student documents, which is a little more than 10 percent of the student work that focused on the DEI scenarios. Unlike writing analytic models that borrow from quantitative models for a confidence interval or a purely qualitative approach that uses a small sample, we followed our own experiences that suggested results would replicate. We initially coded 30 student samples, then did another 30. When the coding of the second set of 30 aligned with the first set, we felt that this number of student data would achieve a measure of transferability, which suggests that conclusions or processes can be used in other contexts, as well as credibility, which focuses on whether we are accurately describing the thing (Lincoln & Guba, 1985).

After a normalizing session where we coded a couple of student examples together, two of the researchers independently coded the samples using the coding scheme seen in Table 6.2. This simplicity of the coding scheme works well for program evaluation because it aligns with the rubrics used by students and faculty throughout the course to connect SLOs to student drafts and finals.

After both researchers had finished coding the sample independently, they met and discussed disagreement in the codes in order to reach a consensus on the sample (see Clegg et al., 2021; Smagorinsky, 2008). The discussions helped to ensure the codes were applied consistently across the data set; consistency was further verified by an additional researcher (Clegg et al., 2021). We should note that we coded the student sample based on a holistic interpretation of the entire assignment, which is comprised of three short documents. Table 6.3 displays the summary coding results of the STEM students, and Table 6.4 shows the summary of the coding broken down by the two student populations (allied health sciences and engineering).

Table 6.2: Coding Scheme for Analyzing Student Documents

Numerical Code	Definition of Code
1	No evidence or very little evidence of criteria in student writing
2	Some evidence of criteria in student writing; more than 1, but less than 3
3	Substantial and complex evidence of criteria in student writing

Table 6.3: Summary of Coding Schema (N = 60)

	Empathy	Language	Power	POV
Coded 1	17% (n = 10)	3% (n = 2)	17% (n = 10)	18% (n = 11)
Coded 2	22% (n = 13)	43% (n = 26)	82% (n = 49)	78% (n = 47)
Coded 3	62% (n = 37)	53% (n = 32)	2% (n = 1)	3% (n = 2)

Table 6.4: Summary of Coding Schema Split by Course (Allied Health n = 31; Eng n= 29)

	Empathy		Language		Power		POV	
	Health	Eng.	Health	Eng.	Health	Eng.	Health	Eng.
Coded 1	16% (n = 5)	17% (n = 5)	3% (n = 1)	3% (n = 1)	19% (n = 6)	14% (n = 4)	23% (n = 7)	14% (n = 4)
Coded 2	26% (n = 8)	17% (n = 5)	45% (n = 14)	41% (n = 12)	77% (n = 24)	86% (n = 25)	77% (n = 24)	79% (n = 23)
Coded 3	58% (n = 18)	66% (n = 19)	52% (n = 16)	55% (n = 16)	3% (n = 1)	0% (n = 0)	0% (n = 0)	7% (n = 2)

Table 6.4 shows the summary of the coding schema as it is divided per course. When looking closely at the number of students, there are no significant differences between the students in allied health science and engineering. This similarity suggests that once students engage with the material, they are engaging at equivalent levels, no matter their disciplinary background. This information is important for the learning outcomes of the assignment, as well as for framing the results and discussion of the student data.

Results of Coding Categories

As seen in the empathy category, 62 percent (n = 37) of students offered a full apology to their audience. The apology was a key marker to indicate that students understood that the audience deserved some empathy and goodwill. Students' ability to show empathy relates to the purpose and audience learning outcome, as well. A representative example of an apology (coded as a 3) comes from a student who wrote:

> Greetings Ms. Mudnal, I'm responding to your letter pertaining to Coughyfilters' policy on excessive piercings and/or tattoos. Thank you for reaching out to me about this issue and I sincerely apologize that we did not create an environment where you would feel comfortable discussing this policy during the process of your hire.

The student's apology acknowledged the need for empathy, and the student's clear and direct apology shows the student negotiated the dynamic of balancing the needs of the recipient while fulfilling the responsivity of organizational authority. Here is an example coded as a 2:

> As of January 20, 2021, a request was sent out to our account manager at Expedient HR Solution to [sic] have this rule 'excessive piercings and/or tattoos will not be allowed and could result in termination at any time' revised accordingly. We currently await this pending request and we here at Coughyfilters, LLC apologize if you felt any way disrespected by this rule.

While there was an apology issued, this example was coded a 2 because the apology is not forefronted, and the apology also lacks a clear awareness that the company was wrong or at fault. The two examples illustrate the complexity of the assignment but also show why it is important for TPC service courses to integrate assignments focused on inclusivity that ask students to consider an issue from multiple perspectives, including perspectives that indicate a problem exists.

As seen in the data, the majority of students also engaged with language awareness. Our sample shows that 43 percent (n = 26) of students engaged with language awareness concepts at the 2 level, while 53 percent (n = 32) engaged with it fully at the 3 level. Students were addressing and discussing issues around DEI, including aspects of culture, race, identity, and representation. A representative example coded as a 3 from an engineering student demonstrates this language awareness when they wrote: "We at Heartline pride ourselves on having a qualified and diverse staff. We need to ensure that we don't overlook any candidates due to their ethnicity or racial background." Words such as "diverse," "ethnicity," and "racial background" speak to the nature of how most students write away from traditional white, hegemonic standards. The choices students made in their responses to various audiences arguably point to students' awareness of DEI issues via their language choices in their writing. Alternatively, the following is an example coded as a 2:

> I am emailing you to assign a meeting today to discuss the dishonesty of the first round of interviews done by the HR department in the company and further steps to solve the issue. Attached to the email is the letter of the complaint I got from the manager of Diversity Hiring Help about the issue. We must meet as soon as possible as this affects Heartline's reputation hugely.

In this example, the student implicitly talks about diversity and inclusion, but it could be made more explicit with the use of clear language and intent. Our data provides a traceable throughline to show how programmatic inclusion guides SLOs, which in turn can be seen in student final products.

In the power category, 82 percent (n = 49) of students acknowledged issues of power at the coded level of 2, which shows they attempted to solve the problem by communicating with other stakeholders, including superiors, and offered some indication that the solution would take time and be ongoing. We also coded 17 percent (n =10) of student work at 1, while only a single student was coded as a 3 (see example in 'What Could Be Working Better' section). The student deliverables code at a 2 displayed an awareness of power roles within an organization and the need to acknowledge authority—both theirs and others—in their correspondence. Many students wrote similarly to the following example coded as a 2 from an allied health science student who stated that there was a flaw with the hiring practices in the handbook and the student notes there is a rule:

> that talks about excessive piercings and tattoos not being allowed on women; we don't necessarily have to get rid of this rule, but I will like for you to add an exception to the handbook that states that people with certain religious believes should get a pass.

By interrogating power structures and organization practices each time students produce a deliverable, students will learn the impact of their actions and the roles they can play in promoting DEI in their workplace. This stance is especially important for students because it assists in developing their ability to consider the concerns and perspectives of others, which is often seen as something separate in STEM education. Humanities-based approaches to writing challenge STEM students to engage with critical thinking and problem-solving in relation to empathy and power. In contrast, the following is a student example coded as a 1: "The purpose of this memo is to make you aware that the new hire Purnima Mundal, feels that the employee handbook's policy on excessive piercings should be revised to make an exception for cultural and religious observances." This example misses the SLO because it demands their supervisor do something that misses the nuance of the workplace power dynamic. The internal documents that students write as part of this assignment consistently failed to recognize organizational power dynamics. Giving students the opportunity to consider these sorts of power dynamics is key to the purpose and audience SLO, as well as helping them understand the difficulty of effecting change in the workplace.

As seen in the POV category, 78 percent (n = 47) of students switched to "we" or "our" at some point in at least one document, which resulted in their work being coded as 2. However, only two students in our sample wrote an entire document from their company's perspective by using "we" and "our" consistently. A representative example of the 78 percent comes from an allied health science student who wrote:

> I understand that during the first round of interviews, the HR department were in charge. Since I oversaw the second

> session of interviews, this leads me to believe that the HR department may have practiced biased hiring procedures. Due to this, I have sent a memo to the Board of Directors to see if we can investigate the HR department's hiring procedures to make sure that we are practicing equality during these hiring sessions.

With only one "we" as a representative of the company, this example suggests that students are attempting to write both individually and as part of the company in which the scenario requires them to participate; however, the persistent use of "I" indicates that students do not fully understand how they need to represent their organization in documents that are sent to audiences outside of their organization. In contrast, the following is a student sample coded as a 1:

> My name is [Student Name] and I am the office administrator of Coughyfilters, LLC. It has come to my attention that the Expedient HR Solutions Company are the makers of the employee Handbook at my company. According to your handbook, 'excessive piercings and/or tattoos will not be allowed and could result in termination at any time' and one of my employees have brought this specific line to my attention and informed me that they believe their tattoos and piercings are a representation of their cultural heritage.

In this example, there is no awareness of organizational authority and the necessity to shift POV. Our data showed partial acknowledgment of the organizational author with students switching between "I" and "we." This connects to DEI principles by acknowledging the role of the individual within the organization and the importance of responsibly negotiating the impact of the power an individual has when speaking as part of that organization. Shifts in POV signal engagement with critical thinking, problem-solving, and accountability because it asks students to consider their roles and how they are perceived by others in a critical and self-reflective manner. The shift in POV from "I" to "we/our" shows that students can recognize that sometimes they need to communicate as a representative of the organization in order to affect change.

Discussion of Student Data

From the student data, we have come to two broad discussion points that will be of interest to the interdisciplinary audience of this book: what is working well with this assignment and what could be working better.

What Is Working Well

From our analysis, several aspects of this assignment are working well: empathy and language awareness, power and POV, and positive engagement from the allied health sciences students.

Empathy and Language Awareness

As the data illustrates, students did not gain full competency in these areas, but they did show an awareness that, from a programmatic standpoint, should be taken as a positive. For example, students made an effort to apologize while also incorporating language choices that displayed an awareness of DEI concepts. The assignment moves social justice from an abstract idea to more concrete practices and asks students to engage with and respond to problems with diversity and inclusion that occur in business communication. Students addressed and discussed issues around DEI, including aspects of culture, race, identity, representation, and inclusion. The students' ability to navigate issues around DEI demonstrates a practical association from the assignment to workplace practices. When considering the SLO of writing style, our data provides a traceable throughline to show how our programmatic inclusion guides our learning outcomes, which in turn can be seen in student final products.

Power and POV

Our research allowed us to see that students in both engineering and allied health science are starting to negotiate issues of relative power dynamics and POV when positioned as a company employee in a realistic workplace setting. Issues of power represent that students acknowledge hierarchy and authority within organizations and their implications. While this assignment allows students to address inclusivity in the workplace, the service course overall should help students understand that their writing has consequences and effects change. In this case, students need to choose to uphold or dismantle current policies. When students are asked to consider how to challenge policies, they gain experience with the multiple layers and nuance of how communication, and its related power, works at the organizational level. Insights into power and its influence in upholding or dismantling inequalities is a key aspect of the SLO and goals of the assignment.

For example, issues cannot be resolved without input from superior stakeholders, and those stakeholders must be addressed appropriately based on their role within the organization. Practically, authority in these scenarios often means that the issue cannot be solved by the decisions of the author, and the student must address a superior to make a request for change. The following sample student document illustrates these moves, and it was the only document to do it this effectively and receive a code of 3:

Dear Alan Critten

It has come to my attention about the employee handbook that it may not be as inclusive as we have thought. One of our new hires, Mrs. Purnima Mudnal, mentioned about the section that involves women getting too many piercings and or tattoos. Mrs. Mudnal explained to me that the piercings she has is a part of her heritage and felt pressured to sign the book to keep her job. Mrs. Mundal brought this up in hopes that she will not be terminated from her job as it is stated in the handbook.

Since we are a growing business, which will entitle more employees, which would mean that more individuals will have different forms of heritage. I propose that we rewrite the handbook to allow for more freedom of individuality and expression of one's culture provided that it will not get in the way of their work. If we show that we care for our employees and show that we hear them and respect them for their individuality, we will be able to maintain loyal employees. This will make them feel respected and included in the environment and less likely to quit.

Please do consider this as soon as possible. We need to make the work environment as inclusive and less problematic for our employees as possible. Without them, we cannot do our business and grow at a rate that is much befitting our product. I am available by text or email. You can reply to this email if you wish or call me at (813)999–1111. Please do consider what I have suggested, the sooner we can resolve this the better it will be for us.

Sincerely, John Doe

In this example, the student acknowledges the authority of the superior while foregrounding the importance of the situation. The student makes a request and then provides supporting reasons for enacting the request that align with organizational goals. The correspondence concludes with a request for a meeting, acknowledging the need for collaboration and negotiation. This student has appealed for change within the power structures inherent in the organization.

We do, however, concede that only one student received a 3 out of 3 in relation to power, as the majority of students did not acknowledge they could not solve the issue themselves, and many assumed that a positive outcome was a foregone conclusion.

Allied Health Science Student Engagement

Roughly 43 percent of the allied health science students chose to write about the programmatic inclusion scenario despite having three other scenarios as options.

Students who chose these scenarios explained their rationale: "I choose this scenario because it had to do with discrimination and that is a topic that I fully support to end" and "I choose this scenario because it is important." Both of these demonstrate a move between the scenario, a classroom activity, and DEI issues that exist in the world outside the classroom. Helping students make the connection between classroom assignments and situations they will encounter in the world illustrates the impact of programmatic inclusion as it leverages outcomes and assignments in the curriculum to make DEI concepts applied. This connection is aided by an exercise given to students prior to the selection of the document series scenario. In this exercise, students are presented with a scenario in which a co-worker has posted an offensive comment in a company Slack™ (business messaging) channel. Students are asked to write a post to their superiors explaining the situation and how they have handled it. This exercise introduces DEI issues in the workplace and prepares students to deal with the more complex issues they will address in the document series.

While students did not get to full competency of the SLOs with this assignment, the data underscore that the assignment is mostly working as intended. The data illustrates that there were more 2's and fewer 1's across all criteria of the student documents we analyzed (n = 60). This improvement suggests that students are understanding the goals of the assignment and are able to produce a series of documents that show engagement and at least a minimal competency with the concepts. The goal is for students to understand the application of writing as it relates to issues of inclusion following calls to promote inclusion in micro- and macro-social contexts (Riedner et al., in this volume). This is an important distinction and one that circles back to ensuring that what we do in our TPC courses will prepare students for the workplaces they will enter and to perform and engage in their civic lives. We acknowledge that things such as workplace documentation, policies, and the projects that TPC practitioners produce have often contributed to the inequity and exclusion that upholds racist systems. We take seriously our commitment to teach students skills they can use to create more equitable and inclusive organizations.

What Could Be Working Better

The purpose of this section is to highlight what is still not working to successfully reflect the goals of programmatic inclusion. We examine how engineering students were far less likely to engage in DEI scenarios. Students are starting to engage with issues of power and POV, and lastly, we explore how professional development for instructors could potentially help improve the assignment outcomes.

Engaging Engineering Students

The engineering students selected the programmatic inclusion scenario at lower rates than the allied health sciences students. An illustration of the disconnect engineering students experience between DEI issues and their own work is evidenced by how often they select another document series scenario that has DEI overtones but does not foreground them. In this scenario, students are placed in the position of a project manager overseeing wetlands preservation. One of the residents, an important figure in her community, has complained about the noise and mess. In this case, students are working in a community in which a minority population is the majority and must address the concerns to the community's satisfaction. Students selected this scenario in greater numbers than the newer scenarios, highlighting DEI criteria, as it has more of the trappings of an engineering-driven scenario. These students reflect on their reasons for choosing this scenario: "[The wetlands preservation] scenario seemed the most straightforward" and "I chose [the wetlands preservation] scenario because I immediately knew what genres to use for each document." Helping these students to understand that DEI issues will arise in their jobs even when they may not immediately realize it and facilitating connections between the engineering roles they will occupy and the inevitability of white hegemonic values in their workplaces would help them see the relevance of diversity issues in their workplaces.

Aside from the Slack channel exercise discussed previously, the data suggests that engineering students are not seeing the relevance of DEI issues to their jobs. Even among those who selected the programmatic inclusion scenarios, engineering students are not making nuanced connections between their classroom tasks and the work world. Facilitating engagement with DEI issues could be done in the classroom by highlighting that every field and organization will confront issues of diversity and inclusion.

Power in Relation to Audience

We recognize that there are more ways that we can encourage this assignment to work more effectively. In part, this realization is derived from coding the document series assignment materials for the first time in a systematic way with a research team. For example, the fact that only one student received a 3 in the Power and POV categories suggests that this is one area of the assignment that could be improved. Students would benefit from understanding power dynamics and their abuse in organizations. While most students acknowledged that it was necessary to ask for permission to effect change, students often assumed that their superiors would agree without negotiation. In many correspondences, a positive outcome was a foregone conclusion. In preparing to complete the assignment, more in-class discussion of power dynamics and how they manifest in organizations is needed. Adding more direct instruction and engagement could facilitate a more nuanced and realistic approach to workplace communication as students consider how power impacts effect change.

POV as Representative of the Company

An additional area that could be working better is student development of a deeper understanding of the impacts of their personal role on their audience. The majority of students moved from speaking as an individual to a representative of the company—seen through the use of "our" and "we" in at least one document. The exposure to POV signals that students are beginning to understand the complexities of the organizational author and how speaking as the organization complexifies diversity issues.

In order to fully understand purpose and audience as part of this assignment, students should recognize how their role and the way it is signaled in a document impacts their audience and achieves their purpose. The shift in POV speaks to the purpose and audience learning outcome. Personal accountability and critical thought are fundamental to achieving this central outcome and are especially important in a DEI context, in which all aspects of power need to be interrogated.

Professional Development for Instructors

In addition to work in the classroom, which will benefit students, the TPC program used programmatic inclusion as a way to also help sensitize our instructors to issues of diversity and inclusion through professional development opportunities. Aligning with trends in the field, our program is staffed 95 percent by contingent faculty with no TPC background (Mechenbier et al., 2020), and the success of the most well-designed curriculum hinges on how it is taught. Research has shown that contingent faculty desire professional development (Wilson et al., 2020). Because of this, the implementation of programmatic inclusion encompasses how we use professional development to facilitate instructor strategies for approaching DEI principles in the classroom. All the elements of the document series assignment that need improvement could be addressed through professional development, specifically through discussion of power and POV issues, as well as by showing engineering students that DEI issues are relevant to their work. Addressing these issues requires an awareness of organizational culture and applied workplace knowledge.

DEI issues often revolve around issues of power, making the power and POV criteria especially relevant. Students need to understand that power should be acknowledged and integrated in order to attempt to effect change. These power issues are seen in the way that students address their superiors and in the way they adopt organizational authority in reference to their own personas. In professional development, instructors can be sensitized to the role of power in organizational hierarchies and examine the ways in which individuals reinforce or challenge authority through communication. Professional development can empower instructors to leverage the resources included in the curriculum, making DEI relevant and highlighting the necessity for responsibly negotiating DEI issues. Similar to issues of organizational hierarchy, students need to know that DEI issues will arise in every workplace. For the engineering students, this means understanding that their jobs will encompass far more than technical problems. Professional development affords program administrators the opportunity to share disciplinary knowledge with instructors so that they can more effectively draw explicit connections between SLOs and issues of DEI.

Conclusion

We opened this chapter with the recognition that TPC is a field committed to theory-to-practice connections. Our TPC service course program's commitment to programmatic inclusion engenders a programmatic perspective that drives assignment design, pedagogy, and outcomes in ways that give students the opportunities to apply conceptual premises of inclusion, diversity, and equity in practice. Using programmatic inclusion as a consistent guide for programmatic decisions explicitly enacts Ahmed's (2012) theory of diversity work. Assignments that use problem-based scenarios guide discussions around inclusion and belonging and give students the opportunity to confront inequity in the workplace and respond by effecting change.

However, we acknowledge that there is still much work to do. In this case, we wanted students to use an assignment's SLOs as a way to move the conceptual ideals of equity and inclusion into practice. While our analysis found evidence of effective applications of DEI principles, we also found that students are not fully making connections between the importance of language and the documents needed in the workplace to change embedded and implicit issues of inequity.

Our goal has been to explicate an assignment that applies principles of DEI in a way that is replicable. Assignments such as the document series, created through the vision of programmatic inclusion, lead to using curricular elements such as outcomes and problem-based scenarios as opportunities for advancement toward learning opportunities that address equity and justice. The assignment outlined can be adapted in any service course or other writing courses, such as an introduction to TPC, editing, proposals or instructions, and capstone courses, which are all common courses in TPC degree programs. This example allows programs to put theory into practice by giving students experience with the types of diversity issues they will face in the workplace.

Acknowledgments

We gratefully acknowledge the support of the University of South Florida and Eric Putz with USF Writes for use of the data presented.

References

Agboka, G. Y. & Dorpenyo, I. K. (2021). Curricular efforts in technical communication after the social justice turn. *Journal of Business and Technical Communication, 36*(1), 38–70. https://doi.org/10.1177/10506519211044195.

Agboka, G. Y. & Matveeva, N. (Eds.). (2018). *Citizenship and advocacy in technical communication: Scholarly and pedagogical perspectives*. Routledge. https://doi.org/10.4324/9780203711422.

Ahmed, S. (2012). *On being included: Racism and diversity in institutional life*. Duke University Press. https://doi.org/10.2307/j.ctv1131d2g.

Bay, J. (2022). Fostering diversity, equity, and inclusion in the technical and professional communication service course. *IEEE Transactions on Professional Communication, 65*(1), 213–225. https://doi.org/10.1109/TPC.2021.3137708.

Boettger, R. K. (2010). Rubric use in technical communication: Exploring the process of creating valid and reliable assessment tools. *IEEE Transactions on Professional Communication, 53*(1), 4–17. https://doi.org/10.1109/TPC.2009.2038733.

Chen, C. (2021). Trial and error: Designing an introductory course to technical communication. In M. J. Klein (Ed.), *Effective teaching of technical communication: Theory, practice, and application* (pp. 111–129). The WAC Clearinghouse; University Press of Colorado. https://doi.org/10.37514/TPC-B.2020.1121.2.06.

Clegg, G., Lauer, J., Phelps, J. & Melonçon, L. (2021). Programmatic outcomes in undergraduate technical and professional communication programs. *Technical Communication Quarterly, 30*(1), 19–33. https://doi.org/10.1080/10572252.2020.1774662.

Connors, R. J. (1982). The rise of technical writing instruction in America. *Journal of Technical Writing and Communication, 12*(4), 329–352. https://doi.org/10.1177/004728168201200406.

Cook, K. C. (2002). Layered literacies: A theoretical frame for technical communication pedagogy. *Technical Communication Quarterly, 11*(1), 5–29. https://doi.org/10.1207/s15427625tcq1101_1.

Driskill, L. (2013). Designing a visual argument course in an era of accelerating technological change. In E. R. Brumberger & K. Northcut (Eds.), *Designing texts: Teaching visual communication* (pp. 49–69). Baywood.

Earle, S. C. (1911). *The history and practice of technical writing*. The Macmillan Company.

Eodice, M., Geller, A. E. & Lerner, N. (2016). *The meaningful writing project: Learning, teaching, and writing in higher education*. Utah State University Press.

Griffith, J., Zarlengo, T. & Melonçon, L. (2024). A field wide snapshot of student learning outcomes in the technical and professional communication service course. *Journal of Technical Writing and Communication, 54*(1), 46–68. https://doi.org/10.1177/00472816221134535.

Hitt, A. (2018). Foregrounding accessibility through (inclusive) universal design in professional communication curricula. *Business and Professional Communication Quarterly, 81*(1), 52–65. https://doi.org/10.1177/2329490617739884.

Kumar, R. & Refaei, B. (Eds.). (2021). *Equity and inclusion in higher education*. University of Cincinnati Press.

Kynell, T. (2000). *Writing in the milieu of utility: The move to technical communication in American engineering programs, 1850–1950* (2nd ed.). Ablex Publishing.

Lincoln, Y. S. & Guba, E. G. (1985). *Naturalistic inquiry*. Sage.

Mallette, J. C. & Hawks, A. (2020). Building student agency through contract grading in technical communication. *The Journal of Writing Assessment, 13*(2), 5. https://escholarship.org/uc/item/4v65z263.

Mechenbier, M., Wilson, L. & Melonçon, L. (2020). Results and findings from the survey. *Academic Labor: Research and Artistry, 4*(1), 27–64. https://digitalcommons.humboldt.edu/alra/vol4/iss1/4.

Medina, C. & Walker, K. (2018). Validating the consequences of a social justice pedagogy: Explicit values in course-based grading contracts. In A. M. Haas & M. F. Eble (Eds.), *Key theoretical frameworks: Teaching technical communication in the twenty-first century* (pp. 46–67). Utah State University Press.

Melonçon, L. (2018a). Orienting access in our business and professional communication classrooms. *Business and Professional Communication Quarterly, 81*(1), 34–51. https://doi.org/10.1177/2329490617739885.

Melonçon, L. (2018b). Critical postscript on the future of the service course in technical and professional communication. *Programmatic Perspectives, 10*(1), 202–230.

Melonçon, L. (2024). *Programmatic imaginations through a curricular history of technical and professional communication*. Manuscript submitted for publication.

Melonçon, L. & England, P. (2011). The current status of contingent faculty in technical and professional communication. *College English, 73*(4), 396–408.

Melonçon, L. & Schreiber, J. (2018). Advocating for sustainability: A report on and critique of the undergraduate capstone course. *Technical Communication Quarterly, 27*(4), 322–335. https://doi.org/10.1080/10572252.2018.1515407.

Mislevy, R. J. & Elliott, N. (Eds.). (2020). *Ethics, psychometrics, and writing assessment: A conceptual model*. Utah State University Press.

Newmark, J. & Bartolotta, J. (2021). Creating the "through-line" by engaging industry certification standards in SLO redesign for a core curriculum technical writing course. In M. J. Klein (Ed.), *Effective teaching of technical communication theory, practice, and application* (pp.147–165). The WAC Clearinghouse; University Press of Colorado. https://doi.org/10.37514/TPC-B.2020.1121.2.08.

Oswal, S. K. (2018). Can workplaces, classrooms and pedagogies be disabling? *Business and Professional Communication Quarterly, 81*(1), 3–19. https://doi.org/10.1177/2329490618765434.

Oswal, S. K. & Melonçon, L. (2017). Saying no to the checklist: Shifting from an ideology of normalcy to an ideology of inclusion in online writing instruction. *WPA: Writing Program Administration, 40*(3), 61–77.

Read, S. & Michaud, M. (2018). Hidden in plain sight: Findings from a survey on the multi-major professional writing course. *Technical Communication Quarterly, 27*(3), 227–248. https://doi.org/10.1080/10572252.2018.1479590.

Schreiber, J., Carrion, M. & Lauer, J. (2018a). Guest editors' introduction: Revisiting the service course to map out the future of the field. *Programmatic Perspectives, 10*(1), 1–11.

Schreiber, J., Carrion, M. & Lauer, J. (2018b). Afterword: Service courses as an extension of technical and professional communication disciplinary identity. *Programmatic Perspectives, 10*(1), 231–242.

Scott, J. B. (2008). The practice of usability: Teaching user engagement through service-learning. *Technical Communication Quarterly, 17*(4), 381–412. https://doi.org/10.1080/10572250802324929.

Smagorinsky, P. (2008). The method section as conceptual epicenter in constructing social science research reports. *Written Communication, 25*(3), 389–411. https://doi.org/10.1177/0741088308317815.

Voss, J., Sweeney, M. A. & Serviss, T. (2021). A heuristic to promote inclusive and equitable teaching in writing programs. *WPA: Writing program administration, 44*(2), 13.

Walton, R. & Agboka, G. (Eds.). (2021). *Equipping technical communicators for social justice work: Theories, methodologies, and pedagogies.* Utah State University Press.

Walton, R., Moore, K. & Jones, N. (2019). *Technical communication after the social justice turn: Building coalitions for action.* Taylor & Francis.

Wilson, L., Mechenbier, M. & Melonçon, L. (2020). Data takeaways. *Academic Labor: Research and Artistry, 4*(1), 65–87. http://digitalcommons.humboldt.edu/alra/vol4/iss1/5.

Appendix: Document Series Project Description

This project asks students to consider how letters, memos, and emails function rhetorically in various scenarios.

Learning Objectives:

- Practice writing various forms of business correspondence and documents (i.e., email, letters, memos)
- Address purpose and audience in business correspondence
- Practice selecting the appropriate correspondence genre (i.e., email, letters, memos) for a specific rhetorical situation
- Develop a professional writing style, paying particular attention to concision (i.e., avoiding wordiness), paragraph construction, and tone

Allied Health Sciences Diversity Scenario:

You are the manager of the Graphics Department at Heartline, Inc., a medium-sized company with three offices and 300 employees that sells a mobile app that monitors customers' heart function. Your department has recently begun hiring to fill up to eight positions from entry-level to middle-manager. As a department manager, you have been sitting in on the second round of interviews. The first round of interviews, consisting of phone interviews, is solely completed by the HR department. The second round of interviews consists of Zoom or Teams online meetings with several members of your organization, including yourself, your boss, an employee specialized in the position, and someone from HR.

A few weeks into interviewing, you receive a letter from someone named Xaviare Roberts. She is connecting with you from a non-profit organization called Diversity Hiring Help. Ms. Roberts informs you that they have had more than a dozen qualified applicants apply to available positions, but not a single person has been contacted by your HR department. Ms. Roberts explains that each applicant has worked closely with a hiring consultant to perfect their resume and cover letter for your firm's specific job listing. Additionally, each applicant meets, or exceeds, the required qualifications in your online job posting. Yet, still, not a single person from her organization was contacted for an interview.

Ms. Roberts explains her company's mission is to help people of color find jobs. She suggests that none of her applicants were contacted because they do not possess Caucasian-sounding names.

While the first round of interviews is determined by HR, you know that Heartline values diversity in the workplace. As a manager, it is your responsibility to encourage equitable hiring practices. Ms. Roberts' allegation merits investigation and revision of hiring practices.

Deliverables

Based on the scenario above, your deliverables will be the following:

- Document to Ms. Roberts at Diversity Hiring Help
- Document to the Board of Directors at Heartline
- Document to Heartline's HR department

Communication for Engineers Diversity Scenario:

As the office administrator of Coughyfilters, LLC, a small mask-making company (less than 50 employees in one office), you are in charge of staffing and training. Your primary job is to oversee daily operations, new employee training, and on-going employee development. Your firm has gotten busier, which has required more hires in a short period of time, so it is imperative that all staff are trained and ready to begin work. Your boss, president and founder, Alan Critten, has approved additional hires for the increase in business. Your newest hire, Purnima Mudnal, who works in Marketing, started work at the beginning of the month.

Two weeks after Ms. Mudnal went through training, you receive a letter from Ms. Mudnal. She explains that, on her first day, she was asked to read through and sign the employee handbook. Upon reading the handbook, she was surprised to read that "excessive piercings and/or tattoos on women will not be allowed and could result in termination at any time." Ms. Mudnal states that her numerous ear and nose piercings are a cultural representation of her heritage. She felt pressured to sign the handbook in order to keep the job, but she feels strongly that the handbook should be revised.

As you investigate the matter further, you discover that the handbook was written by an outside human-resources consulting firm called Expedient HR Solutions. At Expedient, you work with your assigned account manager, Ms. Linda Fleming.

In order to make revisions to the handbook, you will need to get approval from your boss, Mr. Critten. You also will need to communicate with Ms. Fleming to explain the need for revisions and what revisions are necessary. Finally, you will have to respond to Purnima Mudnal.

Deliverables

Based on the scenario above, your deliverables will be the following:

- Document to your boss, Mr. Critten
- Document to Linda Fleming at Expedient HR Solutions
- Document to Purnima Mudnal

Section 2. Challenging Orientations to Instruction and Assessment

In the second section of the collection, we turn our attention explicitly to pedagogy and assessment. All disciplines have their own ways of communicating—from the genres they select to the syntactical structures and citation practices they use. These communicative practices serve very important, practical purposes and reflect the work the community does. Similarly, all disciplinary communities have ways of viewing the world that shape the ways in which those communicative and practical aspects are performed. What counts as evidence, for example? What kinds of questions are worth exploring, and which are not as relevant? What are our end goals in disciplinary writing courses or disciplinary courses that incorporate writing? What kind of writing is fair and appropriate to assign? These are questions we must consider when planning and enacting our courses, but they are not always questions with easy answers. These systems of doing help communities function and contribute to the world in specific and necessary ways. Yet, the hidden assumptions that often come with this work contribute to the marginalization of individuals in STEM.

The chapters in this section ask us to think actively about the ways in which disciplinary practices and norms are reified in the language choices we make and the types of questions we consider, as well as the practical applications of social justice orientations to our work. They ask us to move beyond teaching formulaic approaches to communicating as a member of a disciplinary community and into spaces where individuals have agency to critique and question those practices. Finally, they ask us to consider our own roles in designing inclusive spaces and to be more intentional and conscious about the impacts of assignments and classroom interactivity. Here, we are more actively engaging with instructor worldviews and orientations to education. While there are still clear take-aways and applications, the chapters are much more philosophical in content.

This section continues the theme of creating space with reflections by Madison Brown and Madeline Dougherty. Brown's reflection on the power of process-oriented assessment on her identity and sense of belonging in a physics course asks us to challenge our conceptions of the types of knowledge we can assess in our courses. Similarly, Dougherty highlights how the type of feedback we provide students can have significant impacts on their sense of belonging within educational and disciplinary spaces (in Dougherty's case, a STEM vocational program). These vignettes are followed by chapters discussing the ways in which we can create space for students to learn and grow as members of their discipline while still leaving space for negotiating disciplinary orientations with their own identities.

The chapters in this section explore topics such as the transfer of critical reasoning and judgment in engineering spaces (Riedner et al.), frameworks and assignments that exemplify best practices in writing across the curriculum and inclusive teaching (Mallette), creating scaffolded, meaningful writing assignments that engage students in activities that lead to public-facing artifacts (Seraphin), challenging traditional grading frameworks in a biochemistry undergraduate research course to illustrate inclusive strategies (Newell-Caito), and incorporating public-facing genres into a STEM student teaching program that call for an activist lens as it relates to disability (Johnson et al.). The section concludes with a demonstration of the power of liberatory frameworks within neuroethics courses (Fink). In this final chapter, Fink shows how to apply Freire's concept of *conscientização* to a pedagogical orientation applicable to STEM educational spaces.

Student Vignette

Madison Brown
UNIVERSITY OF MAINE

When reviewing our General Physics II midterms, one question had stumped everyone. Dr. Sirvon shared that no one scored full points. He proceeded to perform the "correct" solution on the whiteboard, but before moving on, he paused to share the "success" of one respondent. Given certain items (a wire and nail), the question asked the student to construct a magnetic field. It was a short answer test question, no calculations required.

    Dr. Sivron always said he wanted to see our sweat on the test, wanted us to *figuratively* beat our heads against our desks a little at first, and always, always, always draw a free body diagram. He encouraged us to think both critically and creatively. I felt a sense of belonging in my major when my professor opted for a process-oriented versus product-oriented grading approach. It wasn't about whether I was right, how good my maths were, or if I could substitute variables and follow instructions; it was a measure of my understanding and comprehension of the theory and principles of electricity and magnetism.

    The student with such an inventive response, one that lacked practicality but oozed with ingenuity, was mine. I was being anonymously honored in front of my whole class of all-male peers, lauded for my creativity, and given half points on a question I didn't answer as expected. In that midterm review, I was no longer the only girl in the class, but I was the only student to earn points on a test question that had stumped everyone, even me. I was able to bask in my "incorrect" but inventive response because only Dr. Sivron and I knew who submitted that answer. I had used a science-driven process to arrive at the objectively wrong answer. Everything my formal education suggested up to this point equated incorrect responses with no credit, zero points.

    Dr. Sivron challenged my conceptions of education and learning. I thought I had to be "right" to learn. I thought 100 percent meant faultless effort, and the only way for me to have pride in my work was perfection. He taught me I was wrong because my short answer had, in fact, produced a magnetic field, albeit a weak one. Seven years later, I still remember the precious gift Dr. Sivron gave me on that small liberal arts campus during our midterm review. He gave me a chance to believe in myself, to call myself a physicist, and to belong.

Student Vignette

Madeline Dougherty
UNIVERSITY OF MAINE

In 2008, I began an 18-month STEM-centered vocational school. Female students were not rare, but we were not common either. Throughout the school, students were given constant written feedback in the form of grade sheets. My grades were consistent but below average, and the comments on my grade sheets tended to focus on areas where I could improve. The instructors offered useful, constructive criticism but almost no praise, even when I thought something had gone well. As a result, I began to believe that I was a terrible student; I had stacks of grade sheets to confirm my beliefs.

The more I listened to other students brag about their current aptitude and imagine their future post-graduation success, the more I felt that I did not belong in this community. Those who were most successful seemed to be large, loud, type-A personality men, not me. As the school continued, I learned to work harder and study smarter, but my grades continued to be below average. In fact, some of my worst grades came at the end of the course in areas where I knew I would be spending the majority of my professional career. However, despite my lack of confidence, I graduated with my peers and continued into work experience.

During the first two years of work experience, we continued to receive frequent written evaluations on our abilities. The grades themselves were pass/fail, so everyone received more or less the same score, but the comments continued as before: "Needs to improve . . . ," "Unable to . . . ," "Failed to. . . ." Again, instructors gave respectful, constructive criticism but very few positive comments. After nearly two years, I approached the evaluation for a major qualification. Based on the written feedback I had received, I was so convinced of my ineptitude that I considered withdrawing from the qualification process. After one particularly difficult day, I began to mentally prepare myself to talk to my boss about declining the final evaluation. Luckily, my boss got to me first. Looking back, I think he had some idea of what was going through my head because he pulled me into his office, looked me straight in the eye, and said, "You know you're really good at this, right?" I was flabbergasted. I had no idea. I knew I was working incredibly hard, but I felt that was the effort required to barely keep up with my peers. I didn't know I had gone from below average to excelling because *no one ever told me.* Even as I improved, my graded comments continued to be the same lines of "Fix this . . . ," "Work on that . . ." After that conversation with my boss, I started to feel as if I belonged in my position in ways that I never had before. It only took one person telling me *Yes, you are good enough* for me to see myself as an equal member of the workforce.

Years later, I returned to the school as an instructor. More experienced instructors told me, "Be the instructor you wished you'd had." I adapted my teaching strategy in many ways to abide by that advice, but one major technique was to tell students when they were doing well. I never blew smoke; I was always honest, but I was very careful to give credit where credit was due. I praised students verbally, and I wrote it down on their grade sheets. Of course, I provided the same constructive feedback that I had received, but if a student excelled, I made sure they knew it and had a reminder to read amongst all the generally negative written feedback. When I started training new instructors, I passed the technique along as a best practice. I taught new instructors that part of teaching is building confidence and belonging. I told them one of the easiest ways to do that is to tell the students when they are doing well and *write it down* so they can go back to it when their confidence starts to slip.

Promoting Inclusion Through Participation in and Construction of Engineering Judgments

Rachel C. Riedner and Royce A. Francis
The George Washington University

Marie C. Paretti
Virginia Tech

There has been significant research scholarship that reports on differential treatment of STEM students and faculty around race, gender, ethnicity, and other social categories (Tonso, 2006; Foor et al., 2007; McGee & Martin, 2011; Secules et al., 2021). As one example, Mary Blair-Loy and Erin A. Clef (2022) provide an important study of culture around scientific merit, which devalues the contributions of faculty women and people of color. As Blair-Loy and Clef point out, it is important to respond to differential treatment as well as research and document this treatment. Aligned with Blair-Loy and Clef's emphasis on responding, our chapter focuses on creating inclusive classroom practices in engineering classrooms. This chapter discusses the goal of promoting inclusion by supporting engineering students to recognize their own capacities, each other's capacities, and the social and discursive contexts in which they learn and work—a concept that we call engineering judgment.

The chapter takes engineering judgment as a starting point for discussions of recognition and inclusion in engineering classrooms. In previous work, we have discussed engineering judgment as a holistic, participatory capacity that integrates the technical and social context of engineering work, the cultural and discursive production of professional identities, and the cognitive processes underpinning naturalistic decision making (Francis et al., 2022). This previous work situates engineering judgment as a learning process through which students come to recognize a range of patterns and social practices as they accumulate decision-making experience over the course of their career trajectories. However, the capacity to learn judgment, as we expand upon below, rests upon students being included and recognized in engineering classrooms.

This chapter brings recognition into the conversation of engineering judgment. It argues that participation in engineering judgment practice requires a learning process and pedagogical structure where a student attains both recognition from others and self-recognition that they are a legitimate participant in engineering work. In learning structures that support the development of engineering judgment capacity, students are recognized by their peers, faculty, and professionals as engineers when they exercise engineering judgment and when they are able to successfully communicate their judgment to multiple audiences.

We add to previous discussions the idea that such learning requires recognition by others. In an engineering context, discussion of recognition by self and others raises questions about who is included as contributing to learning, and more specifically, how recognition is situated within micro contexts (i.e., classrooms and teamwork) and macro contexts (i.e., larger social structures and histories in which learning takes place). Recognition as an aspect of engineering judgment and an aspect of engineering education more broadly, therefore, draws attention to inclusion and belonging and, concurrently, to marginalization and exclusion in the contexts of engineering education.

To extend this discussion of recognition, this chapter puts engineering judgment in conversation with scholarship on belonging and inclusion, as well as marginalization and exclusion, that are currently circulating in multiple disciplinary spaces, including rhetoric and composition and engineering education. Building upon data collected from student interviews, the chapter concludes by pointing out the need for classroom strategies that acknowledge, foreground, and integrate practices that enable recognition and inclusive learning of engineering judgment.

Judgment and Recognition

Engineering judgment is an important concept because it addresses how students are taught capacities to participate in professional life and to identify as engineers. Previous research has argued that engineering judgment as a capacity, an individual skill, or self-understanding can be taught and learned through embedded writing assignments where the process of developing reports, presentations, and posters about ongoing projects creates contexts that require students to exercise and justify a range of decisions (Francis et al., 2020; Francis et al., 2021; Paretti et al., 2019).

Elsewhere, we have developed the concept of engineering judgment by drawing from frameworks of academic literacies (e.g., Lea, 1998; Lea & Street, 1998), discourse identities (Gee, 2000), and naturalistic decision making (Mosier et al., 2018). Taken together, these theoretical frameworks describe how students develop fluency for participating in the discourse of their discipline and thus create a sense of belonging in the discipline. This approach to engineering judgment involves not only understanding the communicative language of the discipline but also addresses how students learn to express themselves through communicative forms appropriate to a task's context, purpose, and audience expectation (Carter, 2012;

Mathison, 2019; Russell, 2002; Thaiss & Zawacki, 2006). This communicative fluency is taught in writing in the disciplines (or WID) curriculum where, as Susan McLeod (2012) argues, faculty "teach students to observe disciplinary patterns in the way [that their language] is structured, helping students understand the various rhetorical moves that are accepted within particular discourse communities" (p. 59). Teaching the communicative practices and tasks of a disciplinary community teaches communicative conventions of a discipline, enabling students to identify and be recognized as part of the disciplinary community (see Allie et al., 2009, for a discussion of this process in engineering specifically).

Emphasizing the importance of explicit instruction in communicative conventions of disciplines, scholars in engineering education have pointed to the need for writing instruction that introduces students to the genres, recognizable rhetorical moves that achieve particular outcomes within disciplinary discourse (Berkenkotter et al., 1988; Miller, 1984; Russell, 1997). Scholars have argued that engineering disciplines with a focus on writing that include a nuanced concept of genre, audience, purpose, conventions, and attention to professional engineering contexts and traditions is a means of creating entrance into professional spaces and fields and for students to learn to identify as engineering professionals (Artemeva, 2007; Conrad, 2017; Dannels, 2000; Paretti, 2008). This scholarship has focused on how to make the teaching of writing in engineering contexts more nuanced by focusing on rhetorical awareness, where students learn to develop writing practices that respond to different audiences (Artemeva, 2005, 2007, 2009; Dannels, 2000, 2002, 2003; Paretti, 2006; Winsor, 1996). Rhetorical awareness suggests attention to genre conventions of the discipline, its purpose, goals, audiences, and other areas that may be implicit in written communicative practice but are essential for academic and professional success. The pedagogical goal is for students to learn and practice engineering judgment capacities through writing and for them to be recognized and self-recognize as having these capacities.

These observations that students develop engineering judgment and communicative capacities in which they are recognized as engineers suggest that writing skills (i.e., understanding and use of genre, audience, purpose) are a means through which students are recognized and recognize themselves as belonging to the engineering community of practice. When students learn forms of communication of their engineering discipline and use these forms of communication to make and communicate judgments, they see themselves and are seen by others as part of a disciplinary or professional community (Wenger, 1998). This recognition acknowledges student ability to apply and communicate specialized knowledge and analytic techniques to interpret information in ways that lead to meaningful engineering judgments. In other words, to be successful, students must be recognized by others (faculty and peers) and, thus, come to recognize themselves as individuals capable of exercising engineering judgments. Judgment, in this line of reasoning, is the capacity for understanding and responding to situations, adapting thinking, making decisions, and communicating decisions. However, to exercise engineering judgment, students must demonstrate proficiency in these capacities in ways that are recognized by others, including their faculty and peers or other engineering professionals.

To summarize, in an engineering teaching context, developing and learning engineering judgment requires:

- Analytical skills: developing capacities to understand and respond to situations, to adapt thinking, and to make decisions;
- Rhetorical and communicative skills: understanding and use of genre, audience, purpose, etc.;
- Social context that enables these capacities to be developed creates opportunities for students to fully participate in learning and creates opportunities for students to be recognized as full participants.

This chapter develops the third area of engineering judgment: recognition. It draws upon scholarship in engineering education, rhetoric and composition, and other fields that are interested in processes of inclusion and exclusion, where gender, race, and other attributes designate some as inside a community while others are designated as outside of the community (Riedner, 2015; Young, 2003).

This expanded discussion of recognition places learning and teaching of engineering judgment in larger contexts than just classroom-level pedagogies or team building. In an engineering context, the valuation and worth of student work (grading, feedback, and other modes of evaluation), participation on teams, and contribution to labor of teams (including writing and analysis) all result in recognition of students as included (or excluded) members of the community. This process of recognition is thus situated within local sites of engineering contexts and broader educational and institutional contexts. Recognition occurs within social interactions among and between faculty and students, among individual students and student teams. Recognition also takes place within larger institutional contexts and histories, professional standards, and the wider social interactions and historical situations that provide the context in which learning and teaching take place. Recognition is situated within histories, structures, and discourses through which "individuals are socially assigned and ascribed" (McCall et al., 2020, p. 81). Thus, the teaching of engineering judgment—a learning process through which students come to recognize a range of patterns and social practices as they accumulate decision-making experience over the course of their career trajectories—must account for the social and discursive contexts in which students are included, excluded, or marginalized, and subsequently evaluated and valued.

To put it another way, recognition takes place in micro social contexts (classroom interactions, grading and other forms of evaluation, feedback from instructors and peers, support for awards and honors, letters of recommendation, etc.) that resonate with macro educational structures, systems, traditions, and histories (Claris & Riley, 2012; Inoue, 2015). In order to teach engineering judgment so that students achieve recognition of their participation in disciplinary communities and for students to see themselves as members of their disciplinary communities, engineering educators must understand both the micro and macro social and written and oral communicative contexts in which classroom teaching and teamwork take place. Even more specifically, engineering educators need to understand how micro and macro contexts impact recognition of students themselves and their contributions to teamwork. To expand a discussion of micro and macro contexts in which engineering judgment is taught and learned and in which students are recognized as capable of engineering judgment, we turn to scholars of rhetoric and composition who argue that classroom-level practices and strategies are not removed from these larger systemic, structural, and historical contexts but in fact are deeply embedded within them (Inoue, 2015; Walsh, 1991). Our effort here is to understand and expand an understanding of how and where engineering judgment is taught by engaging with this scholarship and to understand the possibilities and constraints of recognition.

Social and Discursive Contexts of Teaching

The current moment, as Deborah Brandt (2015) argues, is a period of mass literacy where "the rise of mass writing has accompanied the emergence of the so-called knowledge or information economy" (p. 3). An information economy creates a context in which workers, nations, regions, industries, and globally minded universities, in some instances with directives from governments or professional organizations, shift their curriculum to facilitate acquisition of literacies to create curricula that will enable students to discern, use, apply, and communicate information—in other words, exercise judgments.

Brant's work allows us to approach teaching broadly as shaped by historical situations (p. 7), the particular political and national economies that necessitate the development and teaching of particular kinds of literacies and judgments. This contextual approach to teaching is echoed by Brian Street (2017), who points out that academic literacies take place within "social context and with cultural norms and discourses" (p. 24). More pointedly, as Street's work suggests, academic literacies are developed in contexts of multiple forms of power that are immersed in political economy and social worlds at the local, national, and global level (see also Burry et al., this collection). Power—in all its forms, institutional, historical, discursive, social forms organized around race, gender, and other social categories —is always present in educational settings, including classrooms, curricula, interactions, and scholarship (Inoue, 2015; Walsh, 1991). Social relations, multiple forms of power, and economic forces shape the experiences that students bring to education and take from their education, the capacities for discernment they develop, the writing that they do, and the recognition and self-recognition they develop.

This understanding of teaching and classroom practice as situated within larger contexts resonates with scholarship in writing studies that emphasizes that educational practices and educational outcomes are deeply imbued in real, lived experiences. These experiences include work, class identity, racial, gender, disability, or other social formations, and are connected to the historical and the structural to the personal and the lived (Mohanty, 2003). As James P. Gee (2000) notes, constructions of [student] identity are always embedded in the socially and historically constructed community or cultural narratives that, to a large extent, shape the identities (discourse or otherwise) that are available to individuals and that individuals refer to and negotiate with. Focusing on the social and institutional features that constitute the context within which students learn, Karen Tonso's work on engineering identity looks at how social and institutional features, and in particular, language, create cultural spaces that yield particular expectations and pressures. These existing cultural forms define sets of norms and expectations that individuals engage with as they negotiate and construct their identities (Tonso, 2006, pp. 273–274).

Importantly, a substantial body of work in engineering education over the past decade or more has repeatedly demonstrated the ways in which these existing social practices exclude and marginalize students who do not fit what Alice L. Pawley (2019) refers to as the "ideal engineering student": "White, male, between the ages of 18–22, lives on campus and lacks major obligations such as full-time employment or family care" (p. 24). Tonso's ethnographic work of engineering student design in the early 2000s highlights the ways in which women (as well as men who do not fit key stereotypes) were both discursively and practically excluded from conceptions of what it means to be an engineer (Tonso, 2006, 2007). In her study of the identities used to describe engineering students at one public, engineering-focused university in the US, she found that collectively, the available set of terms and the images they invoked "gave unequivocal messages that women are generally not recognized *as engineers*" (Tonso, 2006, p. 292). Cynthia E. Foor, Susan E. Walden, and Deborah A. Trytten's (2007) seminal study of "Inez," a first-generation, multi-racial, low socio-economic status female engineering student uses critical cultural theory to demonstrate the ways in which students who are outside the dominant culture (white, middle-class, heterosexual, male) are othered and excluded from the culture of engineering programs. Using theories of intersectionality, Erin A. Cech and Tom J. Waidzunas (2009) highlighted the ways in which engineering culture is heteronormative, positioning homosexuality as incompatible with technical competence in their qualitative study of engineering students who identified as gay, lesbian, or bisexual. Ebony O. McGee and Danny B. Martin (2011), drawing on stereotype threat and critical race theory, detail the repeated exclusions experienced by Black undergraduate students in mathematics and engineering at four universities in the Midwestern US as they confronted the implicit and often explicit stereotypes that suggest Black students cannot succeed in STEM fields despite years of work to develop and promote a more inclusive engineering culture. Recent work by Stephen Secules et al. (2021) on experiences of "professional shame" among engineering students demonstrates the ways in which students experience the social worlds of engineering differently based on their demographics, with women and racially diverse students demonstrating more awareness of the gendered and raced construction of norms and expectations in engineering. Moreover, these experiences extend into graduate education as well; quantitative research on engineering identity by Matthew Bahnson and colleagues (2012) found that white and male engineering graduate students experienced statistically significantly higher recognition of their engineering identity than female graduate students and graduate students of color. Across numerous quantitative and qualitative studies over time at a wide range of institutions, researchers continue to find that the socially constructed culture of engineering programs continually reproduces implicit biases, cultural norms and expectations, interpersonal interactions, uneven access to resources, and more that marginalize and exclude students who do not match the implicit white, male, middle-class, single, heterosexual norm.

Classroom learning and evaluation is addressed by Asao Inoue (2015), who discusses classroom ecologies, or material conditions and discursive contexts in which complex interactions take place that are influenced by local events and histories (pp. 77–86). Writers, as Inoue emphasizes, "learn to write in "real social contexts," with real people in mind as their audience, from real people's words about their words and worlds, from material action and exchange in material environments" (p. 91). Multiple, intersecting, and intersectional forces shape the institutional places and instructor approaches to the teaching of writing. As a formal curriculum—one that is authorized by institutional committees and by other authorizing bodies at universities, supported implicitly and explicitly by corporations, and sanctioned by nation-states—the lived experiences of writing are situated within complex contexts that link students and faculty to institutions and places; engineering exists in complex social and historical contexts. As Inoue argues, "environments," that is, economic, political, and historical contexts along with social beliefs and practices, all complex and intersecting forces, "affect people . . . as we dwell and labor because we dwell and labor in those places" (p. 79).

To expand this discussion, students learn, write, participate in teamwork, learn engineering judgment, and are assessed for their learning in socially constructed cultures of engineering programs. In terms of evaluation of student performance and valuation of student contribution, Inoue argues that assessments of student writing and classroom performance by instructors are located within racialized (and we add gendered) systems. Inoue observes that instructor assessment of student writing performance is neither free of macro social ideas of race (and gender and other social designations) nor are they free of judgments about the appropriate or valued forms of creation and communication of knowledge. Assessment, he says, "ha[s] uneven effects on various groups of people . . . [and] privilege some students over others" (p. 19). Thus, assessing student writing and learning engineering judgment both take place within systems that are shot through with racialized and gendered meanings that can create marginalization and exclusion.

Judgment of Student Writing

Understanding the evaluation of student learning and student performance as a social activity that is part of larger, powerful structures that are present in micro-teaching contexts and classrooms raises questions about how students whose identities or whose contributions to teamwork do not fit the normative stereotype of engineering (this point will be elaborated below). Scholars in composition studies draw attention to obstacles that students who are underrepresented in STEM, minoritized students whose voices in STEM have been pushed to the margins, first-generation students, and other students face in writing classrooms that are not set up to recognize their knowledge, experience, or other mitigating factors that impact classroom participation.[1] This attention to how personal experiences of students, and their development of identities, is echoed in scholarship that considers the experiences of disabled students in engineering curricula. Cassandra McCall and colleagues (2020) suggest that students' ability to acquire professional identity can be impacted by disability. As they argue, "little work has examined the way students with disabilities experience, interpret, and engage the field to become professional engineers" (McCall et al., 2020, p.80).

Inoue and other writing studies scholars describe a felt sense of failure produced by teaching systems, pedagogical practices, and assessment of student writing that are not attentive to the knowledge and learning of minority students. To understand the broader context in which marginalization and exclusion take place in educational contexts, Inoue, therefore, looks to "broader patterns" (p. 21) and

---

1   The position of students can vary depending upon institutional and other social contexts. Full participation requires recognition, and a minoritized individual is more at risk of not attaining that recognition. This is different from under-representation, which may relate to the number or proportion of individuals sharing an identity in a given context (e.g., African American, queer, male, etc.). For the purpose of discussing recognition, minoritization may be more relevant. For example, we are aware that a student who is minoritized at one institution may not be minoritized at another. As a result, it is important to attend to the particular institutional contexts and histories where teamwork takes place.

"historical exigencies" (p. 64) that influence the assessment of student writing. His work investigates how assessment of student writing that does not understand student experience can have a negative impact on individual students and calls upon instructors to be "attentive to structural racism, the institutional kind . . . that makes many students of color like me when I was younger believe that their failures in school were purely due to their own lacking in ability, desire, or work ethic" (p. 4). Inoue asks teachers of writing to consider the evaluation and assessment of student materials, asking "how does a teacher not only do no harm through [their] writing assessments but promote social justice and equality" (p. 3).

Current conversations in composition studies suggest that teaching, including teaching in engineering contexts, must consider how instructor feedback can have a marginalizing impact on students whose experiences do not fit with social norms. How instructors respond to, assess, and communicate assessment of student writing can result in marginalization and exclusions that are linked to broader patterns, powerful structures, and embedded institutional practices.

This discussion impacts and develops how we view engineering judgment as a learned skill where students develop capacities to understand and respond to situations, to adapt thinking, and to make decisions and a learning process through which students come to recognize a range of patterns and social practices as they accumulate decision-making experience over the course of their career trajectories. To undertake a learning process where students learn engineering judgment necessitates consideration of how instructors recognize and evaluate student performance and how this recognition and evaluation can, as Inoue points out, impact student learning. At the micro level, how instructors evaluate student products and performance, provide feedback, guide (or fail to guide) teamwork, and understand and evaluate student contributions to teamwork can have a significant impact on student's development of engineering judgment. As we go on to discuss in the next section, the need for a focus on the social and institutional contexts in which judgment is learned is suggested by data we gathered from student interviews. This data indicates that processes of marginalization and exclusion are active in engineering teaching contexts.

Case Study in Student Engineering Judgment Experiences

In the larger research project that we are undertaking (IRB# NCR192007), we explore how students participate in the construction and communication of engineering judgments through their writing projects (Francis et al., 2022). Although a full discussion of this project is beyond the scope of this chapter, student interviews from this wider project suggest a need for discussion of processes of marginalization and exclusion that interfere with the acquisition of engineering judgment capacity. Our data come from students in a systems engineering senior project cohort of 2020–2021 at the first and second author's institution. The senior project course (i.e., capstone course) holds a critical place in the undergraduate systems engineering curriculum, as it is the course that provides the most extensive integration of professional practice with mastery of foundational systems engineering science and concepts. Moreover, the systems engineering capstone course emphasizes teamwork and professional communication. Although the systems engineering students will have worked on several team projects by the time they have reached the capstone course, the senior project is unique in that it allows students full autonomy over their project selection, problem formulation, and course(s) of action. Thus, teams must work together to enact judgments and choices related to the type of project they'd like to construct, the problems they will focus on throughout that project in response to their key stakeholders' concerns, and the types of solutions they'd like to deliver.

Therefore, the data we collected provide in-depth insight into the construction and communication of engineering judgments by undergraduate students. Data were collected from 11 semi-structured interviews with six students enrolled in the systems engineering senior project. All of the students have received prior instruction in WID courses that focus on the application of risk, uncertainty, and statistical decision theory to engineering problems and have had prior experiences completing substantial semester-long projects in engineering teams. These projects have required the student participants to apply engineering judgment to problems with significant uncertainties and conflicting objectives.

Our analysis of this data has allowed us to explore the choices students express in their writing about their judgments, as well as the processes used to construct both the judgments and the written document. These data suggest several important subthemes instructors must be aware of when designing assignments, course objectives, or classroom experiences. For example, one important subtheme has emerged from the data collection that indicates possible processes of marginalization at work in the formation of teams and the evaluation of student contributions to teams by instructors. At least one student reported occasions where marginalization impacted team construction and how recognition influenced the steps taken when team members needed to resolve conflict or otherwise work through unspoken or implicit processes of marginalization to complete their work. For example:

> The teams—it was mostly—I liked working with [name redacted], so we decided that we were going to do something together. [Name redacted] was last man standing at some point, so we told him to join. And then there was [name redacted], who I think he joined late or something so he needed a team, and we had him come on board. So there was that. That's how the team came about.

This excerpt shows that some teams are the result of marginalized students being forced by circumstance to work together. The reasons these students were unable to find teams are not clearly elucidated in the interviews. However, this brief thought shared by one of the participants points to a greater need for understanding how student project teams are formed and may indicate a lack of guidance from instructors about how students are included in teams.

Additionally, our data suggests that conflict resolution is another need from faculty that is potentially under-described in the corpus. Consider the following:

> At that point, I [pushed for] my team, I'm like I'm willing to change the whole thing myself. Just let me do it, because I feel like now that clarified a lot of things that maybe we were not getting, and I kind of—at the very end I saw where the issue was. My team was reluctant so there was a lot of dynamics where like, no, we don't want to change anything. My issue was we weren't changing anything and that wasn't taking us anywhere. Now that we've found the thing that gives us the best chance at understanding what it is that we should do, and we should actually do it, even if it means that there is a change [that's kind of my mentality is], I will work day in and day out to get it done. But they were like no, we don't want to change it. I understand, they didn't want to change everything so radically with only one submission left. So . . . Our paper was very patchy. I basically—I tried to incorporate the latest feedback that we got [in the sections that I wrote]. They were not on board, so half the paper was on one topic. The other one was all over the place. So, yeah, I totally understand why we didn't get the grade that we wanted.

Although conflict was not widely discussed in these interviews, this excerpt clearly shows that team dynamics affected judgments about the problem being formulated, the analyses being constructed, and the interpretation of those results that could be constructed by the team. In this excerpt, the student felt that the team should be more willing to make changes to their project scope and deliverables, even down to the last submission (e.g., "I understand, they didn't want to change everything so radically with only one submission left."). The student reported, "They were not on board, so half the paper was on one topic. The other one was all over the place."

As many scholars have argued in recent years, the marginalization and exclusion of students who do not fit the normative stereotype of engineering (i.e., white, male, cis-gendered, heterosexual) is a function of many facets of engineering culture that serve to continuously reproduce and validate some identities over others. Tonso's work on engineering identity production highlights the ways in which the cultural production of engineering identity often excludes women and some men, and work by Donna Riley, Amy E. Slaton, and Alice L. Pawley (2014), McCall et al. (2020),

Cech and Waidzunas (2011), McGee and Martin (2011), and others have similarly highlighted cultural exclusions along race, (dis)ability, and sexual orientation. In such environments, students need both guidance and support in exploring their own sense of identity, including both their personal understanding of self and their view of how they are viewed by others. There is a need for instructors and teaching assistants to play an explicit role in helping students understand how these cultural production dynamics influence engineering teamwork and knowledge production.

Consequently, it is increasingly important to help instructors and teaching assistants determine how identity production should intersect with team formation. Moreover, instructors and teaching assistants can help students to understand how identity production dynamics influence decision making within teams. This guidance is not external to the goals of teamwork; it is, in fact, fundamental to it due to its centrality in the construction of and participation in engineering judgment. If students are to develop the participatory capacity of engineering judgment, they must be recognized as legitimate contributors to their teams, and they must be fully included in teamwork. Because social and power dynamics can limit the recognition of some students' contributions to teamwork and can interfere with the learning of engineering judgments on the basis of the perception of identity, pedagogies of inclusion are central to this learning.

Inclusive Teaching

Our review of the data generated by our student interviews suggests some possible avenues of development of inclusive classroom practice that can support student learning. The second excerpt from a student interview demonstrates that teamwork often involves decision making and complex engineering judgments that require the collaborative participation of multiple team members. Engineering educators who aim to foster engineering judgment skills may consider guiding students throughout the teamwork process to explore intra-team dynamics *while* identifying some of the complex judgments teams will be required to make in order to complete their work. Our data suggest that these judgments include but are not limited to: understanding audience or framing important problems; selecting appropriate analytical methods or work processes; synthesizing and interpreting work products, including addressing unexpected research findings or scope changes; consulting clients, subject matter experts, or external resources; and, determining how, when, and to whom to communicate their findings or work products. These are complex tasks that necessitate a collaborative and inclusive approach to teamwork, which must be guided and cultivated.

Engineering judgment and intra-team dynamics are implicated in processes of recognition introduced earlier in this chapter and observed by other investigators such as Tonso (2006). Recognition takes place within larger institutional contexts and histories, professional standards, and the wider social interactions, discourses, and historical situations that provide the context in which learning and teaching take place. Our data suggests that recognition occurs within and between teams as students build their own perceptions of who they and their peers are within the local social and cultural contexts they inhabit. These perceptions of each other can influence, as the second student quotation suggests, how the students choose whom to work with and how they delegate the roles and tasks that team members are responsible for within teams. These perceptions and decisions based upon these perceptions can influence the delegation of work and respect given to different contributions that are necessary for successful teamwork. Without explicit acknowledgment of social contexts that privilege certain groups over others, teamwork can contribute to practices of marginalization. Thus, the teaching of engineering judgment—and relatedly, guidance given to students about teamwork and the evaluation of individual student contributions—must account for the social and discursive contexts in which students are included, excluded, or marginalized, and subsequently evaluated and valued.

As our data suggests, engineering educators must be aware of how students recognize each other's professional skills and capacities and how this recognition is integrated into team dynamics and decisions. Pedagogical approaches that explicitly and carefully guide students to consider aspects of group formation, decomposition of work processes and synthesis of work products, and exploration of cultural, social, or political factors that influence and partially determine student work are key to promoting inclusion in the engineering classroom. This guidance is crucial for student learning because, as Scott Weedon (2019) observes and as our data suggests, engineering work is mediated through embodied and enacted communication practices. These communication practices have the potential to either be sites of recognition and inclusion or marginalization and exclusion. To account for social and discursive contexts and to promote inclusive practices, we put forward questions that provide a conceptual framework for teamwork design:

1. How might engineering educators design transparent pedagogical practices that address the micro (i.e., university culture, classroom dynamics) and macro (i.e., racial and gender dynamics) social contexts in which students develop engineering judgment capacities?
2. How might engineering educators design transparent pedagogical practices and assignments that enable students to recognize, address, and integrate differences among team members, recognize historical practices of marginalization, and develop a teamwork culture that cultivates full participation and recognition of the contributions of all team members (see Mallette, this collection)?

Instructor Assessment

How an instructor's view of student performance may be disconnected from actual student skills and disconnected from the dynamics of teamwork is another factor in creating more inclusive recognition. Tonso (2006) shows that student skills and capacities may be different from professor expectations or what an instructor recognizes or is able to see in her comparison of two student teammates, "Martin" and "Marianne" (pp. 293–297). First, Tonso notes that although Marianne possessed "technical skills that exceeded those of most senior students" (p. 293), Marianne's "being considered a bona fide engineer in the team did not carry over into her being considered that way" across a range of other situations in other courses and in on-campus recruitment by prospective employers (p. 294). Marianne's part-time job as a research assistant gave her real-world insights that made her better prepared than her teammates for design work and made her an indispensable part of her team whose work could not proceed without her input or authentication. Similarly, "Martin" was not known widely outside of his team (the same team as Marianne's) "as a 'star' student engineer *because he was not visible to faculty and administration*" (p. 295, italics added). Tonso notes that Martin did not participate in certain aspects of identity production that could have earned him greater recognition by declining to "exploit and control others, act as if he were superior to women in normative heterosexual relations, or beat his own drum" (p. 295). Instead, he "generously shared his work so teammates [whose other responsibilities interfered with project work] would have something to say during presentations to faculty and client" (p. 295). Tonso notes that Martin embodied "counter-hegemonic leadership" and "prototypically feminine practices during teamwork" (p. 296), including empowering and valuing teammates' voices and putting engineering work quality above classroom-required products. Importantly, Martin's leadership style contrasted with that modeled by at least one professor described by Tonso as recommending "a divide-and-conquer model where the leader cracked the whip and told teammates what to do" or with other more recognized students who were "doing very little themselves, telling others what to do, and later taking credit for that work" (p. 296).

This discussion from Tonso's work indicates three factors relevant to our discussion of how instructor perspective can impact student recognition. First, most student team dynamics are invisible to the professors and, in some cases, the clients who must evaluate the products of and the individuals constituting student teams. Next, the students who comprise student teams are evaluated both by professors and other students against recognized gender, racial, and other identities. Finally, faculty and clients who occupy positions of institutional authority recognize and legitimize a subset of the possible student identities available to each of the students. This has a range of implications for our discussion. On one hand, faculty and instructors are important agents in the curation and reproduction of recognized campus engineer identities because they incentivize and legitimize certain roles or types. In this case, both Marianne and Martin were not the types of students widely recognized as occupying the highest levels of the hierarchy described by Tonso, but both students developed a wider range of skills and capacities that are critical to engineering practice, such as teamwork than their peers who may have received more personal recognition from professors, clients—and possibly, prospective employers—than their accomplishments warranted. In concordance with Tonso's observations, our own data suggest that inter-dynamics of teamwork are more complex than an instructor may be aware of (or perhaps interested in). In assessing teams, the information that instructors have about individual student contributions may not recognize reality on the ground, particularly when that reality is intertwined with micro and macro social contexts and communicative practices that exclude certain student identities from processes of institutional recognition. Instructors need to understand that their own social context, their position of authority, and their insight into student dynamics may not align with student capacities or teamwork dynamics while providing a strong stimulus to the reification of processes of engineer identity recognition and legitimization.

Engineering educators might consider how to design transparent assignments and experiences that enable students to intentionally and reflexively engage in the processes of forming work teams, making decisions as a team, distributing work, or resolving team conflicts. As Jennifer Mallette argues in this collection, transparent course and assignment design is the practice of clearly communicating gloss, tasks, and evaluation criteria with the goal of inclusion for all students. These experiences or assignments could involve foregrounding recommendations about team decision making when data, tools, techniques, or findings conflict with *a priori* expectations. These experiences or assignments should also foreground how decisions are made by the team and how intra-team conflicts should be resolved. Many students do not receive explicit instruction in team dynamics, and such dynamics are among the key changes in the transition from school to work. For example, Ben Lutz and Marie C. Paretti (2021) point out that relationship building is critical to engineering work, where learning processes are "(mostly) informal, unstructured, sporadic, and motivated by production of goods or services." (p. 134). Their findings suggest that while students often wrestle with cultural and institutional factors during their schooling, assignment designs or experiences that explicitly highlight social processes at the organizational, workgroup, and interpersonal levels have the potential both for improved professional preparation and classroom inclusion.

There is extensive research from a number of fields that provides pedagogical guidance on how to set up productive teams that can come to collective decisions. For example, in their review of the literature on engineering and computer science project teams, Maura Borrego et al. (2013) suggest that team-based assignments are effective for training in team-based skills such as communication and coordination and have the potential to involve interdependence among team members. The authors include several recommendations for instructors, including i) establishing activities for goal-setting and establishing team interaction rules, ii) project scaffolding, iii) guidelines for dealing with conflict, iv) guidelines for forming smaller teams (including trial periods and rules for switching members), v) exercises for developing mutual understanding and respect, and vi) utilizing grading schemes that motivate participation in team projects (Borrego et al., 2013, p. 497).

These findings suggest several important actions that can be taken by instructors who seek to create classrooms and course activities that foster inclusion. Assignments and classroom activities can be designed to address and provide guidance with areas that require engineering judgment, such as how to integrate feedback, how to make collective decisions, how to include all team members in decision making, how teams address unexpected results, dealing with uncertainty and ambiguity, and iteratively moving toward a solution as much as they are designed to assess the ability to understand and apply knowledge.

Instructors can begin by guiding students to develop a system of mutual accountability. This system of mutual accountability should be inclusive in assessing strengths and weaknesses of members, be aware of gendered perceptions of certain types of skills (such as writing tasks often assigned to women), and be inclusive of how different voices are acknowledged and heard. Race can also weigh in as teams can be a place where students experience microaggressions. If there is no way to get past biases and past experience, if there is no guidance on how to equitably distribute labor, then certain team members won't be considered for certain types of tasks, tasks may not be aligned with students' capacities, or students may not receive recognition for the work they in fact produce.

If engineering judgment involves learning to work through complexity and act, this capacity includes how teams of engineers work through complexity and act as a group in and through written language. The marginalization reflected by the first student quotation suggests that these key skills may not be learned by some students who are not included in teamwork or who are included as an afterthought. These skills may not be recognized by some students who, intentionally or unintentionally, exclude others from full participation in teamwork. As a result, guidance with inclusive participation and inclusive communication should not be an afterthought or left to chance but an explicit aspect of pre-professional pedagogical practice—including guidance on team formation, intra-team communication, teamwork decomposition and distribution, and other important judgment processes affected by the dynamics of recognition and inclusion.

To address these dynamics, assignments and classroom activities can address and provide guidance concerning the construction and communication of engineering judgments to audiences, including faculty evaluators and peer co-workers, but with an eye to the wider range of audiences that students will interact with in the professional world and that can recognize student's capacity to enact engineering judgment. Although many students wish to create more inclusive learning and work environments, they need guidance on how to recognize their own capacities and experiences, how to recognize each other's capacities and experiences, and how to recognize the social and discursive contexts in which they learn and work. Students need to be guided to cultivate inclusivity and need resources to be capable and inclusive partners. Assignments can be designed to make students aware of the experiences of others and how those might influence engineering work products and judgment. Instructors can foreground the following questions as they design teamwork assignments: How can we guide the ways in which students manage their projects? How can we design assignments that draw out specific work processes and team contributions? How do we assess contributions to project formation and uncertainty management? How might we ask team members to assess their own skills before assigning tasks? How might assignments ask students to evaluate their own growth and learning?

Acknowledgments

This research is supported by the National Science Foundation (NSF) under Grant Numbers 1927035 and 1927096. Any opinions, findings, conclusions, or recommendations expressed in this material are those of the authors and do not necessarily reflect the views of the National Science Foundation.

References

Allie, S., Armien, M. N., Burgoyne, N., Case, J. M., Collier-Reed, B. I., Craig, T. S., Deacon, A., Fraser, D. M., Geyer, Z., Jacobs, C., Jawitz, J., Kloot, B., Kotta, L., Langdon, G., le Roux, K., Marshall, D., Mogashana, D., Shaw, C., Sheridan, G. & Wolmarans, N. (2009). Learning as acquiring a discursive identity through participation in a community: Improving student learning in engineering education. *European Journal of Engineering Education*, *34*(4), 359–367. https://doi.org/10.1080/10288457.2010.10740678.
Artemeva, N. (2005). A time to speak, a time to act: A rhetorical genre analysis of a novice engineer's calculated risk taking. *Journal of Business and Technical Communication*, *19*(4), 389–421. https://doi.org/10.1177/1050651905278309.
Artemeva, N. (2007, August). Becoming an engineering communicator: Novices learning engineering genres. In *4th International Symposium on Genre Studies* (pp. 15–18).
Artemeva, N. (2009). Stories of becoming: A study of novice engineers learning genres of their profession. In C. Bazerman, A. Bonini & D. Figueiredo (Eds.), *Genre in a changing world*, pp. 158–178. The WAC Clearinghouse; Parlor Press. https://www.doi.org/10.37514/PER-B.2009.2324.2.08.

Bahnson, M., Perkins, H., Psugawa, M., Satterfield, D., Parker, M., Cass, C. & Kirn, A. (2012). Inequity in graduate engineering identity: Disciplinary differences and opportunity structures. *Journal of Engineering Education, 110*(4), 949–976. https://doi.org/10.1002/jee.20427.

Blair-Loy, M. & Clef, E. (2022). *Misconceiving merit: Paradoxes of excellence and devotion in academic science and engineering.* University of Chicago Press.

Brandt, D. (2015). *The rise of writing: Redefining mass literacy.* Cambridge University Press.

Berkenkotter, C., Huckin, T. N. & Ackerman, J. (1988). Conventions, conversations, and the writer: Case study of a student in a rhetoric Ph.D. program. *Research in the Teaching of English, 22*(1), 9–44. https://www.jstor.org/stable/40171130.

Borrego, M., Karlin, J., McNair, L.D. & Beddoes, K. (2013). Team effectiveness theory from industrial and organizational psychology applied to engineering student project teams: A research review. *Journal of Engineering Education, 102,* 472–512. https://doi.org/10.1002/jee.20023.

Carter, M. (2012). Ways of knowing, doing, and writing in the disciplines. In T. Myers Zawacki & P.M. Rogers (Eds.), *Writing across the curriculum: A critical sourcebook* (pp. 212–238). Bedford/St. Martin's.

Cech, E. & Waidzunas, T. (2011, June 14–17). *Engineers who happen to be gay: Lesbian, gay, and bisexual students' experiences in engineering.* [Conference presentation.] 2009 American Society for Engineering Education (ASEE) Annual Conference & Exposition, Austin, Texas. https://doi.org/10.18260/1-2--5583.

Claris, L. & Riley, D. (2012). Situation critical: Critical theory and critical thinking in engineering education. *Engineering Studies 4*(2), 101–120. https://doi.org/10.1080/19378629.2011.649920.

Conrad, S. (2017). A comparison of practitioner and student writing in civil engineering. *Journal of Engineering Education, 106*(2), 191–217. https://doi.org/10.1002/jee.20161.

Dannels, D. P. (2000). Learning to be professional: Technical classroom discourse, practice, and professional identity construction. *Journal of Business and Technical Communication, 14*(1), 5–37. https://doi.org/10.1177/105065190001400101.

Dannels, D. (2002). Communication across the curriculum and in the disciplines: Speaking in engineering. *Communication Education, 51*(3), 254–268. https://doi.org/10.1080/03634520216513.

Dannels, D. P. (2003). Teaching and learning design presentations in engineering: Contradictions between academic and workplace activity systems. *Journal of Business and Technical Communication, 17*(2), 139–169. https://doi.org/10.1177/1050651902250946.

Foor, C. E., Walden, S. E. & Trytten, D. A. (2007). "I wish that I belonged more in this whole engineering group:" Achieving individual diversity. *Journal of Engineering Education, 96*(2), 103–115. https://doi.org/10.1002/j.2168-9830.2007.tb00921.x.

Francis, R., Paretti, M. & Riedner, R. (2020, October). *Exploring the role of engineering judgment in engineer identity formation through student technical reports.* [Conference presentation.] 2020 IEEE Frontiers in Education Conference (FIE).

Francis, R. A., Paretti, M. C. & Riedner, R. (2021, July). *Engineering judgment and decision making in undergraduate student writing.* [Conference presentation.] 2021 ASEE Virtual Annual Conference.

Francis, R., Paretti, M., Riedner, R. (2022). *The WRI2TES project: Writing research initiating identity transformation in engineering students*. [Conference presentation.] The Proceedings of the American Society of Engineering Education 2022.

Gee, J. P. (2000). Chapter 3: Identity as an analytic lens for research in education. *Review of Research in Education, 25*(1), 99–125. https://doi.org/10.3102/0091732X02500 1099.

Inoue, A. B. (2015) *Antiracist writing assessment ecologies: Teaching and assessing writing for a socially just future*. The WAC Clearinghouse; University Press of Colorado. https://doi.org/10.37514/PER-B.2015.0698.

Lea, M. (1998). Academic literacies and learning in higher education: Constructing knowledge through texts and experience. *Studies in the Education of Adults, 30*(2), 156–171. https://doi.org/10.1080/02660830.1998.11730680.

Lea, M. R. & Street, B. V. (1998). Student writing in higher education: An academic literacies approach. *Studies in Higher Education, 23*(2), 157–172. https://doi.org/10.1080/03075079812331380364.

Lutz, B. & Paretti, M. (2021). Exploring the social and cultural dimensions of learning for recent engineering graduates during the school-to-work transition. *Engineering Studies, 13*(2), 132–157. https://doi.org/10.1080/19378629.2021.1957901.

McCall, C., Shew, A., Simmons, D. Paretti, M, McNair, L. (2020). Exploring student disability and professional identity: Navigating sociocultural expectations in U.S. undergraduate civil engineering programs, *Australian Journal of Engineering Education, 25*(1), 79–89. https://doi.org/10.1080/22054952.2020.1720434.

McGee, E. & Martin, D. (2011) "You would not believe what i have to go through to prove my intellectual value!" Stereotype management among academically successful Black mathematicians and engineers. *American Educational Research Journal 48*(6), 1347–1389. https://doi.org/10.3102/0002831211423972.

McLeod, S. (2012). The pedagogy of writing across the curriculum. In T. M. Zawacki & P. M. Rogers (Eds.), *Writing across the curriculum: A critical sourcebook,* (pp. 53–68). Bedford/St. Martin's.

Mathison, M. (Ed.). (2019). *Sojourning in disciplinary cultures: A case study of teaching writing in engineering*. Utah State University Press.

Miller, C. R. (1984). Rhetorical community: The cultural basis of genre. In A. Freedman & P. Medway (Eds.), *Genre and the new rhetoric*, pp. 67–78. Routledge.

Mohanty, C. (2003) *Feminism without borders: Decolonizing theory, practicing solidarity*. Duke University Press.

Mosier, K., Fischer, U., Hoffman, R. R. & Klein, G. (2018). Expert professional judgments and "naturalistic decision making." In K. A. Ericsson, R. R. Hoffman, A. Kozbelt & A. M. Williams (Eds.), *The Cambridge handbook of expertise and expert performance* (pp. 453–475). https://doi.org/10.1017/9781316480748.025.

Paretti, M. C. (2006). Audience awareness: Leveraging problem-based learning to teach workplace communication practices. *IEEE transactions on professional communication, 49*(2), 189–198. https://doi.org/10.1109/TPC.2006.875083.

Paretti, M. C. (2008). Teaching communication in capstone design: The role of the instructor in situated learning. *Journal of Engineering Education, 97*(4), 491–503. https://doi.org/10.1002/j.2168-9830.2008.tb00995.x.

Paretti, M. C., Eriksson, A. & Gustafsson, M. (2019). Faculty and student perceptions of the impacts of communication in the disciplines (CID) on students' development as engineers. *IEEE Transactions on Professional Communication, 62*(1), 27–42. https://doi.org/10.1109/TPC.2019.2893393.

Pawley, A. (2019). Learning from small numbers: Studying ruling relations that gender and race the structure of U.S engineering education, *Journal of Engineering Education, 108*(1), 13–31. https://doi.org/10.1002/jee.20247.

Riedner, R. (2015). *Writing neoliberal values: Rhetorical connectivities and globalized capitalism*. Palgrave Macmillan

Riley, D., Slaton, A. & Pawley, A. L. (2014). Social justice and inclusion: Women and minorities in engineering. In A. Johri & B. M. Olds (Eds.), *Cambridge handbook of engineering education research* (pp. 335–356). Cambridge University Press. https://doi.org/10.1017/CBO9781139013451.022.

Russell, D. R. (1997). Rethinking genre in school and society: An activity theory analysis. *Written communication, 14*(4), 504–554. https://doi.org/10.1177/0741088397014004004.

Russell, D. R. (2002). *Writing in the academic disciplines: A curricular history*. Southern Illinois University Press.

Secules, S., McCall, C., Mejia, J. A., Beebe, C., Masters, A., Sánchez-Peña, M. & Svyantek, M. (2021). Positionality practices and dimensions of impact on equity research: A collaborative inquiry and call to the community. *Journal of Engineering Education. 110*(1), 19–43. https://doi.org/10.1002/jee.20377.

Street, B. (2017). Can library use enhance intercultural education? In J. Pihl, K.S. van der Kooij & T.C. Carlsten (Eds.), *Teacher and librarian partnerships in literary education in the 21st Century* (pp. 23–32). Sense Publishers.

Thaiss, C. J. & Zawacki, T. M. (2006). *Engaged writers and dynamic disciplines: Research on the academic writing life*. Boynton/Cook.

Tonso, K. L. (2006). Student engineers and engineer identity: campus engineer identities as figured world. *Cultural Studies of Science Education, 1*, 273–307. https://doi.org/10.1007/s11422-005-9009-2.

Tonso, K. L. (2007). *On the outskirts of engineering: Learning identity, gender, and power via engineering practice* (Vol. 6). Brill.

Weedon, S. (2019). The role of rhetoric in engineering judgment. *IEEE Transactions on Professional Communication, 62*(2), 165–177. https://doi.org/10.1109/TPC.2019.2900824.

Walsh, C. (1991). *Pedagogy and the struggle for voice: Issues of language, power, and schooling for Puerto Ricans*. Bergin and Harvey.

Wenger, E. (1998). *Communities of practice: Learning, meaning, and identity*. Cambridge University Press.

Winsor, D. A. (2013). *Writing like an engineer: A rhetorical education*. Routledge.

Young, I. (2003). From guilt to solidarity: Sweatshops and political responsibility. *Dissent 3*, 39–44. https://www.dissentmagazine.org/article/from-guilt-to-solidarity/.

Engineering an Inclusive Integrated Writing Course

Jennifer C. Mallette
BOISE STATE UNIVERSITY

Students know they need to communicate effectively to be good engineers, and engineering programs are required by the Accreditation Board for Engineering and Technology (ABET) to provide opportunities for graduates to develop effective writing and speaking skills (ABET, 2019; Williams, 2002). As engineering communication research demonstrates, integrating writing into engineering courses is crucial for student success (e.g., Ford, 2012; Ford & Riley, 2003; Paretti, 2008; Reave, 2004). Furthermore, situated learning offers the most effective approach for introducing and building students' disciplinary knowledge and expertise, as well as creating the conditions for the potential transfer of writing knowledge from classroom to workplace (Ford, 2004; 2012; Ford et al., 2021; Paretti, 2008; Walker, 2000). How students receive communication support, however, can vary widely from university to university (Reave, 2004; Ford & Riley, 2003) and even across engineering programs within a single university (e.g., Ford, 2012; 2018; Mallette & Ackler, 2019).

One issue is that writing and communication-based assignments may be incorporated into engineering courses without specific and explicit writing instruction (Paretti, 2008; Reave, 2004), a challenge that instructors attempt to address through various integration models (e.g., Ford, 2012; Ford & Riley, 2003). Because so many of the norms and conventions of the discipline are left unsaid, students may struggle to navigate what instructors require (Paretti, 2008), and employers find that new graduates are often unprepared to communicate in the workplace (Ford et al., 2021). As the editors argue in the introduction to this section, "the hidden assumptions that often come with this work contribute to the marginalization of individuals in STEM" (this collection). These tacit requirements and expectations can serve to widen gaps between students with and without access to stronger preparation in writing, better mentoring, or effective peer educational networks. Thus, engineering assignments may further exacerbate inequities among students who are less prepared or less able to ask for and receive mentorship, those who are multilingual writers (and thus learning conventions of written English alongside disciplinary-specific demands), or those who might otherwise struggle to acquire writing knowledge that isn't sufficiently or explicitly outlined. For underrepresented students (e.g.,

DOI: https://doi.org/10.37514/ATD-B.2024.2364.2.12

women or racially minoritized students) who already find themselves overcoming barriers, struggling on these assignments may reinforce messages that they are unable to succeed in engineering or do not belong in engineering courses or professional settings.

However, if engineering communication assignments and expectations potentially exacerbate inequities, then integrated writing courses could be used to remove or reduce barriers and make this writing knowledge explicit for all learners. Discipline-specific technical communication classes can also be intentionally designed with inclusion as a core value, as addressed by Justiss Burry et al. in this volume. Furthermore, as Rachel Riedner, Royce Francis, and Marie Paretti (this collection) argue, writing offers "a means through which students are recognized, and recognize themselves, as belonging to the engineering community of practice" (this collection). These courses can also serve to disrupt ideas about who belongs, what success means, and how rigor is enacted, exposing factors that contribute to structural inequities impacting student success. This chapter examines one such course, a one-credit online writing course for electrical engineering and computer systems majors. This course was not only designed to teach students engineering-specific writing skills but also to support their success through labor-based contract grading (Inoue, 2019), flexible policies (Boucher, 2016; Cheney, 2020; Santelli et al., 2020), and effective course design that uses transparent assignment frameworks (Fink, 2003; CAST, 2023; Reynolds & Kearns, 2017; Wiggins & McTighe, 2005; Winkelmes et al., 2016).

Institutional and Programmatic Context

In spring 2021, Boise State's electrical and computer engineering (ECE) program had 339 enrolled undergraduate students. Like many engineering programs across the United States (ASEE, 2020), this program is predominately white and male: just 56 (or approximately 17 percent) enrolled students identify as female (J. Browning, personal communication, 23 February 2021). The ethnic/racial background of undergraduate students in ECE as compared to the total Boise State is summarized in Table 8.1. No students reported Native Hawaiian or other Pacific Islander ethnicity, nor Indigenous or Native American ethnicity. These numbers are roughly representative of the population of Boise State as a whole, with a higher representation of students with Asian ethnic backgrounds in ECE. Nonresident international students comprise 1 percent of the total student population (Boise State University, 2021); two international undergraduate students were enrolled as of fall 2021 (J. Browning, personal communication, 11 November 2021).

Table 8.1. ECE Student Demographics by Ethnicity Compared to Boise State 2020–2021 Totals

Enrollment by Ethnicity	ECE Enrolled	ECE % Overall	Boise State Enrolled	Boise State % Overall
American Indian/Alaska Native	0	0%	77	<1%
Asian	46	14.0%	641	3%
Black/African American	5	1.5%	399	2%
Hispanic/Latino of any race	40	12.2%	3,047	13%
Native Hawaiian/Other Pacific Islander	0	0%	98	<1%
No Race/Ethnicity Reported	7	2.1%	720	3%
Two or More Races	14	4.3%	1,130	5%
White	227	69.2%	17,679	73%

Prior to fall 2020, ECE students and students in several other engineering disciplines were required to take Introduction to Technical Communication as a social science elective within the general education curriculum at Boise State. Programs required this introductory course with the goal that students would gain more writing skills while also fulfilling general education requirements. In ECE, the course was also a prerequisite for Electrical Engineering Practice, a junior-level professional skills course that was formerly a communication in the disciplines class. This junior-level course focuses on ethics, communication, and other professional skills, and it also serves as a course that supports the senior project course sequence. However, starting in fall 2020, the introductory technical communication course was no longer listed as a general education option. Engineering programs could not add an additional three-credit writing course without reducing the number of required technical credits or exceeding the 120-credit limit set by the state board. However, faculty in several engineering programs, including those in ECE, worried that their students would be unprepared to write in upper-level courses such as Senior Design Project, let alone when they entered the workplace.

While the program could not find space for a full three-credit course, they were able to add one credit for a writing course at the sophomore level, which they asked me to design and teach. The result was a one-credit, co-requisite course with the required sophomore-level Circuit Analysis and Design lab, which typically has twelve short lab reports completed by two-person teams. I designed this course using my expertise and experience with engineering communication and based on my ongoing collaboration with ECE. For ECE's undergraduate students, the one-credit writing course would offer a chance to be introduced to technical communication within an engineering context, thus bridging their writing education from the first-year writing course sequence. The course would also allow students to receive more writing support before entering the junior professional skills course and provide a more scaffolded writing education throughout the entire ECE curriculum.

Given the class launched in fall 2020 in the midst of the pandemic, the course was designed as an online experience, which created greater flexibility for students to complete work within the various demands on their time and scheduling constraints, even when instruction returned to more in-person modes. In addition, the course served as an opportunity for the program to demonstrate that they were meeting ABET communication outcomes. The writing course supports and works with the content from Circuits Analysis and Design, allowing students to submit writing assignments to both courses. Integration is the goal each semester, but we continue to manage challenges. For example, the instructor and curriculum in the ECE course can change without the writing instructor's knowledge, and expectations about the reports are not always communicated to students in a unified way. Despite these challenges, the students taking both courses experience them as more connected to their engineering education than when they took the three-credit writing Introduction to Technical Communication course. Finally, the instructor of the junior-level professional skills course makes efforts to align course content and approaches with this class and was involved in conversations on how to scaffold writing across the ECE curriculum.

Intents and Goals of the Engineering Writing Course

With these goals and programmatic context in mind, I applied a backward course design approach (Fink, 2003; Reynolds & Kearns, 2017; Wiggins & McTighe, 2005). Instead of designing a course to move through a textbook chapter by chapter, instructors use backward course design to begin by defining learning objectives and then creating scaffolding activities, formative assessments, and summative assessments aimed at helping students to achieve those learning goals (Fink, 2003). This design approach requires more work to understand what students should leave a class being able to do or what they should know. The instructor then designs the daily activities, readings, and assessments to support student progress toward those goals. When done effectively, students understand what they are being asked to do in the class and how it helps them make progress toward course goals. And when paired with course documents that clearly communicate expectations and the purposes behind assigned readings and assessments, backward course design ensures that an instructor has specific reasons for the work a student must complete. In this course, backward course design helped me decide which major projects (or summative assessments) would align with the outcomes and support student learning; for example, I decided to leave out an oral presentation assignment that would not align with the outcomes.

I began this process by reviewing the College of Engineering's mission and the ECE program outcomes. To frame the one-credit course specifically, I also generated program goals for a full engineering writing program to outline the orientation and focus for this course and others like it, should other engineering programs opt in. ECE program outcomes include an emphasis on technical skills, ethical decision making, lifelong learning, and strong professional skills (Boise State University, 2023). In addition to these program objectives, the ABET-driven student outcomes focus on 1) solving problems, 2) engineering design, 3) communication, 4) ethical and professional responsibility, 5) effective teamwork/collaboration, 6) experimentation, data analysis, and drawing conclusions, and 7) lifelong learning and professional development (Boise State University, 2023). The two outcomes specific to a communication class are effective communication and teamwork, though students would be communicating about data drawn from their lab and would also be exploring how writing aligns with their professional goals.

Thus, I developed the course outcomes based on where the course would fall in the students' education, how it could support program outcomes, and what was feasible in one credit (see Table 8.2 for specific course outcomes). After developing the outcomes, I planned out specific in-class formative assessments and major projects as summative assessments, as listed in Table 8.2.

Table 8.2. Backward Course Design Outline for ECE Engineering Communication Course

Course Outcomes	Exercises	Major Project(s)
1. Investigate and apply the conventions and genres of engineering communication, with a focus on their specific discipline	• examine practitioner examples • project updates/status reports • report sections • style analysis	Engineering reports + revision
2. Connect communication skills development with career goals	• reflections • revisions • professionalization plan	Resume
3. Communicate research findings to a technical audience	• project updates/status reports • report sections • creating engineering visuals	Engineering reports + revision
4. Identify the range in audiences and situations that will affect how they communicate and articulate differences in approaches	• reflections • teamwork exercises • writing for multiple audiences • style comparison	Resume + engineering reports

Backward course design was also critical for designing an effective online class, where students most frequently engage with content in the course's learning management system. Because this writing course was designed to be taught online, I created discrete weekly modules that would take students 2–4 hours per week to complete, adhering to the time recommendations for online courses based on university policy. The course focused on rhetorical awareness, and I used genre theory to frame the main assignments and contextualize writing skills. For example, students were asked to consider how the lab reports they created in the co-requisite class were more focused on meeting instructor requirements and demonstrating their learning. After completing a lab report for the class, they were then asked to write an engineering report that would be more aligned with what professional engineers would create. While similar to a lab report, the engineering report required students to use rhetorical awareness and engage with genre theory to understand how the two documents differed and how the differences in audience, context, and genre expectations affected how they wrote. As what might be students' first situated writing experience, the course itself was also structured so that students would understand that they would build on what they learned in subsequent engineering courses. Students need sustained, integrated writing throughout their full engineering education, and this course offered a place to begin those efforts.

Inclusive Practices Used

Since the course is heavily focused on writing and serves as a co-requisite for Circuit Design and Analysis and a prerequisite for Electrical Engineering Practice, I designed it to be a supportive, inclusive experience. Essentially, I set out to design a transparent, clearly outlined online course, which in itself is an inclusive practice (CAST, 2023; Design Justice Network, 2018); I also incorporated practices explicitly aimed at inclusion, such as labor-based contract grading and flexible policies. Ultimately, a course tied to other required courses could potentially function as a gatekeeper course (Jaschik, 2009), preventing students from continuing in their education. This writing course should instead provide a dedicated space for students to learn how to write like an electrical engineer, supporting their technical education in the corequisite lab. Thus, my designing and planning process was influenced by my desire to ensure the class supported student success instead of creating another barrier to degree completion. Furthermore, by focusing on student support structures, I attempted to disrupt white, hegemonic frameworks, particularly around narrow concepts of rigor, success, and responsibility (Brooks & McGurk, 2021). This disruption continues to be an ongoing process as I learn more about how these structures are enacted in my class and reflect on what students need.

Equitable assessment became a key way I attempted to disrupt the white, hegemonic frameworks that pervade university settings, particularly in engineering. In traditional grading frameworks, students may be sorted into groups (the A students, the C students), or faculty may compare students against each other or against a vaguely defined concept of success with roots in inequitable structures (Brooks & McGurk, 2021; Inoue, 2019). To push against these ideas about success and achievement, I opted to use contract grading or labor-based assessment (Inoue, 2019) as the basis of the course design to be used regardless of course instructor. In other words, instead of retrofitting a class designed around traditional grading frameworks or leaving it up to individual instructors teaching the course to opt into contract grading, I aligned the entire course structure with contract grading. The contract focuses on completing specific labor and assessments: a student who is actively engaged and meets expectations on major assignments would be able to earn an A. Asao B. Inoue (2019) describes labor-based contract grading as having three dimensions: how students labor, how much, and what it means. To demonstrate these three dimensions, students complete work, reflect in various ways, and keep labor logs. In my course, the contract outlined the work expected from them for each grade, and students completed weekly reflections where they shared how much time they spent on work for the week and what was meaningful from that week's work. Similar to Inoue's outlining of the elements of the contract, Table 8.3 below shows the categories for engagement and overall criteria for each grade, with the emphasis on meeting expectations on major projects, completing most homework assignments and reflections, and being involved in peer review.[1] The contract indicates that major assignments must meet expectations, which is further defined in the contract itself and is one way the approach may differ from Inoue's (2019) approach (see Appendix A for one iteration of the full contract). The goal, however, is aligned with Inoue's (2019) arguments that we define labor, communicate expectations clearly for students, and provide reasons for each task students will complete.

Because the labor-based assessment approach was central to the overall course design, this integration also meant that the instructor who co-taught with me in spring 2021 had a clear model for understanding contract grading since she had not used it before. Given that one engineering outcome is a focus on lifelong learning, contract grading also aligns learning outcomes with assessment, encouraging students to focus on their progress rather than a predetermined product with strict rules governing success. This inclusive assessment strategy thus encourages students to understand themselves as in process, as writers who will continue to learn more about effective communication beyond this class.

---

1   I want to recognize the work of Dr. dawn shepherd, who shared the tables she uses to summarize labor expectations for students. I have adapted the table for my courses and specific contexts, and I have altered categories/expectations, but the base design is hers.

Table 8.3. Minimum Expectations for Each Letter Grade

	A	B	C	D	F
Participation Expectations					
Weekly reflections	13+	11–12	9–10	8	7 or fewer
Revision workshop draft posts	3	2	2	1	0
Revision workshop peer responses (2 or more per workshop)	6	4–6	3	2	0
Consultation with outside reader (Writing Center, Career Services, mentor)	1+	0–1	0	0	0
Writing Conference with Dr. Mallette	2+	1+	1	0	0
Homework and Projects Completed					
Submitted major projects	4	4	4	3	2 or fewer
Completed weekly homework	90%+	80–89%	70–79%	60–69%	Less than 60%
Projects Meeting or Exceeding Expectations					
Exceeding expectations	0–4	—	—	—	—
Meeting expectations	3 or more	3 or more	2 or more	2	0
Not meeting expectations	0	1 or fewer	2 or fewer	2	3+

I began using this approach in the three-credit Introduction to Technical Communication course housed in the English Department (see Mallette & Hawks, 2020) and now housed in the Department of Writing Studies starting in 2022. However, contract grading requires some experimentation and adaptation to make it most effective for a given context and to ensure it is indeed an inclusive practice. In this one-credit course, I revised the contract several times based on student feedback and input from the co-instructor in spring 2021, and I continue to reflect and revise based on student experiences whenever I teach the class. For instance, after the first semester, I added language about the level of expectations because students were not submitting work that demonstrated they could meet the outcomes (e.g., they did not revise the lab report to reflect an understanding of professional engineering reports, essentially submitting the same report for the subsequent assignment). However, I lost the focus on process by requiring students to exceed expectations on one assignment to earn an A. To support inclusion and student success, I adjusted the criteria again to enable students who fully meet expectations and are active in the course (but never, perhaps, exceed expectations on major projects) to still earn an A. This adjustment allowed me to continue to shift away from assessing students based on narrow standards of success and instead focus on assessing their ability to meet the outcomes.

Because contract grading is a new assessment approach for many students, I asked them to review the contract at the start of the course and to complete several check-ins as a self-assessment tool (see Appendix B). In these reviews and check-ins, they could share any questions or concerns they might have, ensuring that they understood how they were being assessed. In addition to using a contract, I wanted to involve students to move toward a more democratic classroom and disrupt the idea that the instructor is the absolute authority over student learning. At the start of each semester, I offer students a chance to review the contract's terms and negotiate. In the first semester, several students thought that requiring a Writing Center consultation to earn a B was overly burdensome, but they agreed that it was a good requirement to earn an A if A means exceeding expectations, so I altered that requirement. This review also functioned to help students acclimate to an assessment approach that may be completely new to them. In the review, students asked questions and sought clarification on aspects of the contract, which helped me communicate elements more clearly and led to other adjustments.

Flexibility and Late Work

One adjustment was if I would accept or penalize late work. Originally, the contract outlined a set number of allowed late homework assignments, though students had a grace period in which to submit work with no questions. In the response period, some students asked if work would count as late if it was submitted in the grace period, so I clarified that it would be counted as on time if submitted within that period. However, late work policies have been criticized as an exclusionary tactic because penalties are more likely to undermine student success, particularly among neurodivergent students as well as students with family and work responsibilities (Boucher, 2016; Santelli et al., 2020). These policies may also be confusing and inconsistent across a student's classes or in relationship to university-wide late work policies, or their understanding of policy may differ from the instructor's intent to be more lenient than they appear in the syllabus (Santelli et al., 2020), meaning some students may not understand that they can request extensions. Furthermore, syllabus language and tone, as well as penalties, can imply that an instructor is inflexible or unaccommodating, even if the instructor may intend to create an inclusive, supportive educational space (Cheney, 2019; 2020).

Given the pandemic-induced shift to remote learning and increased attention on the pressures students faced in that period, more faculty have advocated for removing late policies and creating more flexibility in classes (Ezarik, 2021; McMurtrie, 2021; Kent State, n.d.; Schacter et al., 2021) or creating approaches and policies that lead to what Matthew Cheney (2019) calls a "cruelty-free syllabus." Thus, to remove a barrier, in the second iteration of the course, I decided to accept all late work completed. The impact on inclusion was immediate: several students who stopped submitting work mid-semester for a variety of reasons were able to submit enough to demonstrate they could meet the course's learning objectives and earn at least a C. For the students who expected to fail, the flexible submission policy meant they were able to pass—and ultimately demonstrate they could meet the learning objectives.

When combined with labor-based assessment, flexible late work policies disrupt the exclusionary norms that govern classroom interactions. Namely, students and faculty perceive submitting work on time as evidence of an individual's responsibility, and they believe late penalties are fair because they reward students who submit work by the deadline (Santelli et al., 2020). Some faculty fear allowing students to submit work at any point would be considered unfair to the students who submitted work on time (Bosch, 2020; Harrington, 2019). However, in my view, these ideas of fairness are too often part of capitalistic ideologies that dictate productivity (and preparation for the working world) as the ultimate goal of education while ignoring the varied conditions students face. In addition, I argue that students should not be compared to one another since students have different needs and abilities. From my experience, students who complete work on time actually gain an advantage; allowing a few students to submit work late will not affect the experiences or achievements of those who submit work on time. Allowing flexibility in submitting work creates space for students to prioritize their needs without sacrificing academic success. Finally, a classroom is not a workplace, and while what we teach can apply to professional settings, it is my belief that the classroom should be a space where students can be supported if they make mistakes or need additional support without undermining their success.

Student Opportunities to Revise and Reflect

With this goal of supporting success, the course structure not only allowed revisions to projects but encouraged them. Some students, for instance, were motivated to revise projects to reach the "meets" or "exceeds expectation" category for their work, partly to meet the terms of the contract but also because they were motivated to improve their written products. For example, many students needed an effective resume for their first major assignment, which they could then use to apply to internships and other opportunities. The revision flexibility also supported students who needed more time to write and revise, so if they didn't meet expectations on the first submission of an assignment, they could revise and resubmit.

Finally, the course also embedded regular, ongoing reflection as an inclusive practice. The reflection served a learning purpose: students had to articulate what elements from the week they engaged with and how they might apply it outside the class. However, the reflection also functioned as a check to ensure students were not spending too much time on the class (2–4 hours a week)—or to prompt them to spend more time. Students could also share what was confusing or what they had questions about, which allowed me to adjust or connect with students periodically if they seemed to be struggling. These reflections ultimately gave them space to think and to ask questions, which was not only useful to them as learners but also allowed them to see the instructor as responsive to feedback and supportive of questions, creating more instructor presence in the online space. At times, students would use the reflections as an opportunity to share about the challenges in their lives, which would prompt an email to check in with them and to alert them to the ways they could take advantage of some of the course's flexibility if needed. These methods thus helped counter the ways online classes can make students feel isolated from their peers as well as their instructor (Stavredes, 2011).

Transparent Assignment Frameworks

All these practices around assessment, flexible submission, revision, and reflection fit within transparent assignment frameworks, a practice aimed toward creating equitable and inclusive classroom spaces. Transparent course and assignment design—or Transparency in Learning and Teaching in Higher Education (TILT)—is the practice of clearly communicating goals, tasks, and evaluation criteria to students (Winkelmes et al., 2016). Transparent design thus helps instructors ensure students understand the goals for all assignments (from low-stakes formative assessments to high-stakes summative ones), what specifically they are being asked to do, and what success looks like on a given task. The TILT assignment template requires instructors to provide the purpose of each assignment, the task or tasks students need to complete to produce the assignment, and the criteria by which they will be evaluated (Winkelmes, 2013). Faculty may already use some elements of transparent design in their courses and assignment descriptions, but they may not articulate these practices as inclusive and equitable ones. I learned about the transparent assignment framework in a semester-long faculty learning community focused on designing courses for student success hosted by Boise State's Center for Teaching and Learning. This professional development experience helped me better understand what elements of course design are inclusionary, so I began using transparent assignments as an intentionally inclusive practice. Multiple students remarked on how clear and easy to navigate the course was, which demonstrates that this approach removed yet another barrier to their learning, particularly in a fully online course.

Ultimately, my argument here is that practices that focus on clear communication, organized materials, and fully planned and effectively structured course design can be a tool for inclusion. This argument is at the heart of design justice approaches and Universal Design for Learning (UDL) (CAST, 2023; Design Justice Network, 2018). However, sometimes, this aspect of design can remain hidden as a strategy for inclusion, as teachers cast it as an effective or evidence-based practice rather than an explicitly equitable one. Thus, in addition to the practices that we frame as explicitly inclusive and equitable, we should consider how carefully planning and designing a course so that students have a seamless user experience is also a tool for inclusion. For instance, in the first semester I taught the course, I had a student with visual accessibility needs. The student needed to be able to work ahead in the class, so I made sure weekly modules were available at least a month ahead of time, and I used accessibility tools to ensure that the screen reader worked effectively with all documents. In addition, I presented information in multiple ways, as recommended by UDL approaches (CAST, 2023): I created videos, text to accompany the videos, the slides from the videos as separate files, course texts that I created, and opportunities to meet with me regularly. The student remarked that the course was one of the more accessible classes they had taken at Boise State. Furthermore, all students continue to benefit from these approaches. While these efforts required significant planning, these materials can continue to be used in future iterations of the course and revised/revisited periodically.

Student Responses to Course Approaches

Based on student responses in reflections and evaluations, my design and approaches succeeded in creating an inclusive writing course. On final course reflections, students responded that the course was thoughtful and accommodating and that the content was the most applicable out of all the writing courses they had taken. On their weekly reflections, students commented on the structure and flexibility of the class, as well as how organized and navigable the materials were. Some students indicated relief that they had a class that reduced their barriers to learning, particularly in an environment where they were forced to take more remote/online courses than they would normally. The reflections also allowed them to share their learning and thinking as they progressed through the class. In these reflections, they indicated that the class allowed them to connect writing knowledge to their specific engineering discipline, perhaps for the first time. That alone made the class invaluable because it was situated within the academic and professional spaces they occupied. In the final reflection, they called out the genres they felt more familiar with that would apply to electrical engineering contexts and how they might apply their learning in their professional lives. They also talked about learning about technical style as well as strategies for successful teamwork, content they could see as immediately applicable to their needs as students and future professionals.

One need was developing teamwork skills. In the first semester, students had to apply what they learned about teamwork more abstractly, but in the second semester, we asked them to work in their lab teams in our class to complete the final project. Given that engineering students may be asked to participate in teams without adequate support—instructors may provide little practical instruction and structure in favor of theoretical content (Adams, 2003)—the focus on teamwork and conflict management as a set of skills was new to most students (see Riedner et al. in this volume for a discussion of teamwork, inclusion, and engineering judgment). One student applied those skills to get back on track with his partner when their collaboration had started to deteriorate, and that team was able to work together effectively and productively for the rest of the semester. Another student, who tended to take over projects because he worried his teammates would slack off, decided to give his teammate a chance; he discovered that his teammate was able to contribute actively. Other students remarked on the templates they could use to assign roles, schedule tasks, and make progress toward their final goal, and these lessons were impactful to many of the students who recognized that they would frequently be working collaboratively.

Overall, students noted that they had beneficial experiences and felt supported in their learning. The class was applicable to their discipline, and the situated learning meant that they could better understand what it meant to be an engineering communicator, which they believed would help them be successful as students and professionals. They made plans to take their resumes and apply to internships, and they understood that report writing would be a significant part of their future—and felt that they would be able to craft those reports successfully. In the final reflection, many students expressed gratitude for the chance to take such a useful course that was also enjoyable, a course designed to lessen burdens for student learning and engagement. Ultimately, students had a positive experience because the design of the course facilitated their success through transparent design, flexibility, and equitable practices.

Currently, we have some evidence that the course may have had an impact on student experiences. The instructor who taught the junior-level Electrical Engineering Practice at the time of writing observed a modest increase in student scores in that course, though she noted that students still struggle with using some of the writing concepts covered in the one-credit course (E. McKinney, personal communication, Feb. 28, 2023). It may be that students are improving as communicators, but they are not yet fully transferring the knowledge and skills into other electrical engineering writing contexts. In addition, the instructor for the Circuit Analysis and Design co-requisite lab has changed several times, which has disrupted some of the integration as new instructors make changes. Future research should collect more specific data to assess the impact of the course and determine other avenues to support student writing across the curriculum. However, this one-credit class did spark a move through the department to focus on writing more throughout the curriculum. For example, in fall 2021, I received a small grant from our Center for Teaching and Learning to lead professional development with a small group of faculty around writing and inclusive pedagogies. As part of that work, we also attempted to map where and how writing was occurring throughout the degree. I also met with and supported the senior project instructor to reconfigure writing assignments. Overall, the interest in writing across the entirety of the curriculum indicates the potential for broader impacts for electrical engineering graduates by creating a stronger culture of writing instruction.

Lessons Learned from the Unexpected

As noted above, students found the course focused and organized, in part because of effective design and clear communication. This experience was also partly indicative of the one-credit nature; I could only require about four hours of work each week, so each module was focused on a manageable amount of content. The disadvantage of the shorter time needed for the class, however, is that it was easy for students to put off the work until the last minute. A few students would often set the goal to do their work well before the deadlines for the next week, only to lament that they had to do it at the last minute yet again. These comments helped me understand the ways that they would use time allotted for the writing class for other purposes, consciously making choices to give less time and energy to this course. I supported these decisions, even if it meant the students may earn a lower grade. This honoring of student choice disrupts ideas that educators know what is best for students and that students have little autonomy. It also encourages students to make the choices they need to care for themselves and to choose how to prioritize their time.

Another unexpected element was how students responded to my overt statement that contract grading was an antiracist teaching strategy. One Latinx student particularly pushed against the contract, challenging how I had framed it. He indicated that he didn't want to receive what he perceived as special treatment for his background and identity while also pushing against the framing of the contract as supporting BIPOC students specifically. However, he made me realize how I implied that I saw BIPOC students as deficient (and thus in need of special treatment). For the second iteration of the class, I added a reference to Inoue's book and clarified that the "emphasis is on effort and progress" in an effort to clarify how all students would benefit from the approach because it makes space for a range of experiences, expertise, backgrounds, and abilities (see Appendix A). I will continue to revisit how I communicate these goals with students, given the political climate in Idaho and my continued efforts to avoid deficit thinking.

A final unexpected element was the ways students were able to recover from missing a significant portion of the work. In spring 2021, a few students had various personal crises that interfered with their ability to participate in the class. For example, toward the end of the semester, one student let me know he had disappeared because of his mental health. When I charted a path for him to earn a C in the class, he leaped at the chance, which was made possible by our late work flexibility and the structured modules in the class. He worked through the modules that would be most useful and submitted work that met expectations; if the class had used more traditional late-work policies and grading criteria, it's likely he would have failed. In addition, despite fears that this type of flexibility would create undue burdens for instructors, I have found that this flexibility did not substantially add to my workload, and it provided a path to success for the few who needed it since most students turned their work in on time or near the original deadline. Ultimately, this flexibility allowed students to demonstrate their ability to meet the course's learning outcomes, and their progress toward degree completion was not derailed.

Reflections and Recommendations

Students appreciated that this class gave them a space to learn what it means to communicate in an engineering setting. Many students also saw this one-credit class as a supportive space with usable content and materials that reduced their fears about online courses. Students were also empowered to make choices that served them and their learning. These students also allowed me to understand the benefits of a carefully, fully planned course with a usable, accessible, and useful course site. In a time when students were taking more online or remote classes than they ever expected, a well-designed course was a respite from other courses where faculty may have been less experienced with effective online/remote delivery or were less transparent with their assignments. Repeatedly, students commented on how the class was easy to navigate, and they rarely struggled to find information to complete tasks. They were able to benefit from my experience with online teaching and my technical communication expertise, which I used to create useful, usable course materials to support their learning.

Their reactions and comments highlighted how effective and inclusive teaching isn't always just about the content; if instructors can take the time to plan and use effective design principles to craft their materials, then students will benefit. Thus, one inclusive teaching practice is to make course materials *accessible* in terms of supporting screen readers and other accessibility tools and *usable* in terms of creating documents using design principles (such as contrast, repetition, alignment, and proximity) and consistent navigation aids such as headings, as well as structuring the course sites to be easy to navigate within the constraints of a learning management system. For instance, each week's module had an overview page that summarized the content and main tasks for the week and then had a separate page for the week's readings and another for the week's assignments descriptions/submission links. The same formatting and navigation structure was used each week, along with clear headings, bullets, and tables to make specific information easy to find. Furthermore, having the course fully planned out and the course modules available at least a few weeks ahead sent students the message that the course had clearly defined goals and outcomes. This planning reduces anxiety and allows students to anticipate upcoming assignments—or, if they want to work ahead, gives them the chance to do so. Together, these experiences underscored how inclusive practices are augmented by clarity, transparency, and consistency in materials and content.

Ultimately, what I take away from this course design and instruction is that students require multiple avenues to success and that they should be allowed to define what "success" means to them in their own contexts within various constraints. By using flexible policies, labor-based contract grading, and transparent assignments, I was able to provide a structure where students could map their way to learning as best suited their goals, constraints, and abilities. This experience was made possible with careful backward course planning as well as the use of effective document design and communication strategies that are the focus of the technical communication field. These strategies augmented my desire to create an inclusive, supportive class for students.

Recommendations for Course Design and Teaching

Sometimes, faculty think that inclusive teaching requires the most innovative strategies that take a lot of time to implement, sentiments echoed in professional development. In addition, practices like contract grading can challenge both faculty and students. However, sometimes the small elements—choices that indicate care and support, that don't necessarily take us much time or energy, and that may seem generally good practices—can add up to a class that is inclusive. A class that is designed to support all students must disrupt ideas of success and rigor that are part of white, hegemonic, and capitalistic structures because success cannot be framed as only possible for a subset of the student population. Thus, I offer the following recommendations for instructors:

Start by defining what rigor is in your courses. As Jamiella Brooks and Julie McGurk (2021) stressed in a recent workshop, a careful definition of rigor that is detached from deficit mindsets and examined critically leads to purposeful teaching. With this in mind, what does rigor look like in your classroom and discipline? Who might be more likely to succeed based on that definition of rigor, and how can you shift that definition to include all students? How can you make that definition and expectations clear and transparent to all students?

Involve students in the planning and assessment process. Invite student perspectives throughout the class, ask them to reflect frequently, request feedback on various elements of the class, and be willing to shift, revise, or otherwise adapt the course based on their feedback and experiences.

Reframe good teaching and effective communication practices as inclusive practices. The practices we use to teach and to clearly communicate are inclusive because they are responsive to all students' needs. Thus, you can use tools like the transparent assignment framework (Winkelmes, 2013; 2016), accessibility tools, and UDL frameworks (CAST, 2023) to communicate tasks and expectations clearly to students.

Revisit various course policies, such as penalties around late work. These policies often are detached from the course's learning goals and approaches and can serve to burden already struggling students, such as neurodiverse students, students struggling with mental health, or students who already see themselves as outsiders in STEM spaces.

Rethink assessment and evaluation within the context of inclusion. Traditional grading often participates in white, hegemonic frameworks, even if the instructor resists these structures. In addition, traditional grading often means assessment approaches are unaligned with course outcomes and student needs. Alternative assessment approaches—such as labor-based contract grading (Inoue, 2019), specifications grading (Nilson, 2014), or other forms of ungrading (Blum, 2017)—offer the potential to better align assessment with course goals and to support student learning (see also Newell-Caito, this volume).

Thus, I conclude with an invitation. We must disrupt the frameworks that too narrowly define success and imply that certain students do not belong in these spaces, particularly given concerns around participation and retention in STEM. To engage in this disruption, we must be reflective practitioners who continue to learn and change our approaches based on how they impact our students. As we reflect on how our practices might unintentionally support the ideologies that are in opposition to our own values, we can then find ways to disrupt them in our classrooms. What I share here is just one point in my own process of unlearning; my own goal is to use reflection to adapt or completely revise what I do in the classroom. This class is likely to change as I continue to critically examine what practices contribute to inequitable structures and what works to support student success. I invite each of you to join me in this process of reflection and revision as we work to open up our classroom spaces to support all students.

References

ABET Engineering Accreditation Commission (2019). *Criteria for accrediting engineering programs.* https://tinyurl.com/ypp5t3sk.

Adams, S.G. (2003). Building successful student teams in the engineering classroom. *Journal of STEM Education, 4*(3), 1–6. https://www.jstem.org/jstem/index.php/JSTEM/article/view/1096.

American Society for Engineering Education (2020). *Engineering & engineering technology by the numbers 2019.* https://ira.asee.org/wp-content/uploads/2021/02/Engineering-by-the-Numbers-FINAL-2021.pd.

Blum, S.D. (2017, November 14). Ungrading. *Inside Higher Ed.* https://tinyurl.com/mwn3ew8u.

Boise State University (2023). *Electrical and computer engineering.* https://www.boisestate.edu/coen-ece/.

Boise State University (2021). *Facts and figures.* https://www.boisestate.edu/about/facts/.

Bosch, B. (2020) Adjusting the late policy: Using smaller intervals for grading deductions. *College Teaching, 68*(2), 103–104. https://doi.org/10.1080/87567555.2020.1753644.

Boucher, E. (2016, August 22). *It's time to ditch our deadlines.* The Chronicle of Higher Education. https://www.chronicle.com/article/its-time-to-ditch-our-deadlines.

Brooks, J. & McGurk, J. (2021, November 9). *Rigor as inclusive practice: Improving equitable outcomes in teaching.* [Conference presentation]. POD Network Conference 2021. https://tinyurl.com/458kd446.

CAST (2023). *About universal design for learning.* https://tinyurl.com/4hypwadt.

Cheney, M. (2019, February 16). *Cruelty-free syllabi.* Finite Eyes. https://finiteeyes.net/pedagogy/cruelty-free-syllabi/.

Cheney, M. (2020). (Against) the syllabus as instrument of abuse. *Syllabus, 9*(1), 1–2.

Design Justice Network (2018). *Design justice network principles.* https://designjustice.org/read-the-principles.

Ezarik, M. (2021, June 21). *How COVID-19 damaged student success.* Inside Higher Ed. https://www.insidehighered.com/news/2021/06/21/what-worked-and-what-didn%E2%80%99t-college-students-learning-through-covid-19.

Fink, L. D. (2003). *A self-directed guide to designing courses for significant learning.* https://tinyurl.com/39a6p7m3.

Ford, J. D. (2004). Knowledge transfer across disciplines: Tracking rhetorical strategies from a technical communication classroom to an engineering classroom. *IEEE Transactions on Professional Communication. 47*(4), 301–315. https://doi.org/10.1109/TPC.2004.840486.

Ford, J. D. (2012, Fall). Integrating communication into engineering curricula: An interdisciplinary approach to facilitating transfer at New Mexico Institute of Mining and Technology. *Composition Forum, 26.* https://tinyurl.com/3pze9hjp.

Ford, J. D. (2018). Going rogue: How I became a communication specialist in an engineering department. *Technical Communication Quarterly, 27*(4), 336–342. https://doi.org/10.1080/10572252.2018.1518511.

Ford, J. D., Paretti, M., Kotys-Schwartz, D., Howe, S. & Ott, R. (2021). New Engineers' transfer of communication activities from school to work. *IEEE Transactions on Professional Communication, 64*(2), 105–120. https://doi.org/10.1109/TPC.2021.3065854.

Ford, J. D. & Riley, L. A. (2003). Integrating communication and engineering education: A look at curricula, courses, and support systems. *Journal of Engineering Education, 92*, 325–323. https://doi.org/10.1002/j.2168-9830.2003.tb00776.x.

Harrington, C. (2019, May 8). *Examining the why behind your late or missed work policies.* NOBA Blog. https://tinyurl.com/mwxsrf68.

Inoue, A. B. (2019). *Labor-based grading contracts: Building equity and inclusion in the compassionate writing classroom* (2nd ed.). The WAC Clearinghouse; University Press of Colorado. https://doi.org/10.37514/PER-B.2022.1824.

Jaschik, S. (2009, December 2). *Long road to "gatekeeper" courses.* Inside Higher Ed. https://www.insidehighered.com/news/2009/12/03/long-road-gatekeeper-courses.

Kent State University Office of the Provost (2020). *What it means to be reasonable, flexible, and equitable when students are required to quarantine or isolate due to the COVID-19 pandemic.* https://tinyurl.com/36juf3wa.

Mallette, J. & Ackler, H. (2019, June 16–19). *Using reflection to facilitate writing knowledge transfer in upper-level materials science courses.* [Conference presentation]. American Society for Engineering Education (ASEE) Annual Conference and Exposition Proceedings, Tampa, FL, United States. https://doi.org/10.18260/1-2--33516.

Mallette, J. C. & Hawks, A. (2020). Building student agency through contract grading in technical communication, *Journal of Writing Assessment, 13*(2). https://escholarship.org/uc/item/4v65z263.

McMurtrie, B. (2021, March 17). *Good grades, stressed students.* The Chronicle of Higher Education. https://www.chronicle.com/article/good-grades-stressed-students.

Nilsen, L. B. (2014). *Specifications grading: Restoring rigor, motivating students, and saving faculty time.* Stylus Publishing. https://doi.org/10.4324/9781003447061.

Paretti, M. C. (2008, October). Teaching communication in the capstone design: The role of the instructor in situated learning. *Journal of Engineering Education, 97*, 491–503. https://doi.org/10.1002/j.2168-9830.2008.tb00995.x.

Reave, L. (2004). Technical communication instruction in engineering schools: A survey of top-ranked U.S. and Canadian Programs. *Journal of Business and Technical Communication, 18*(4) 452–490. https://doi.org/10.1177/1050651904267068.

Reynolds, H. L. & Kearns, K. D. (2017). A planning tool for incorporating backward design, active learning, and authentic assessment in the college classroom. *College Teaching, 65*(1), 17–27. https://doi.org/10.1080/87567555.2016.1222575.

Santelli, B., Robertson, S. N., Larson, E. K. & Humphrey, S. (2020). Procrastination and delayed assignment submissions: Student and faculty perceptions of late point policy and grace within an online learning environment. *Online Learning Journal, 24*(3), 35–49. https://doi.org/10.24059/olj.v24i3.2302.

Schacter, H. L., Brown, S. G., Daugherty, A. M., Brummelte, S. & Grekin, E. (2021, December 1). Creating a compassionate classroom. *Inside Higher Ed.* https://tinyurl.com/3z7834u8.

Stavredes, T. (2011). *Effective online teaching: Foundations and strategies for student success.* Jossey-Bass.

Walker, K (2000). Integrating writing instruction into engineering courses: A writing center model, *Journal of Engineering Education, 89*(3), 369–375. https://doi.org/10.1002/j.2168-9830.2000.tb00538.x.

Wiggins, G. & McTighe, J. (2005). *Understanding by design* (2nd ed.). Association for Supervision and Curriculum Development.

Williams, J. (2002). Technical communication, engineering, and ABET's Engineering Criteria 2000: What lies ahead?, *Technical Communication, 49*(1), 89–95.

Winkelmes, M. (2013). Transparent assignment template. *TILT Higher Ed*. https://www.tilthighered.com/assets/pdffiles/Transparent%20Assignment%20Templates.pdf.

Winkelmes, M., Bernacki, M., Butler, J., Zochowski, M., Golanics, J. & Weavil, K. H. (2016). A teaching intervention that increases underserved college students' success. *Peer Review 18*(1/2), 31–36. https://tinyurl.com/3udbmcsr.

Appendix A: Grading Contract

The following language is placed on the course syllabus under the heading "Evaluation":

This course uses **contract grading**. Contract grading has been demonstrated to support student learning and offers an antiracist tool for evaluating writing.[2] Contract grading also emphasizes labor/effort, progress instead of products, and continuous improvement. Since writers can come from varied backgrounds with vastly different levels of preparation and different writing experiences, contracts also allow you to build on the skills you have currently and set your own goals for learning. While quality does factor in, particularly for the A grade, the emphasis is on effort and progress.

Table 1 outlines the minimum expectations for each letter grade. In order to earn a B, for instance, you must complete each requirement within the B column. Even if you sometimes complete the requirements for the A column, your final grade will still be a B. We anticipate that most students will earn either an A or a B in this class.

A change to the contract per negotiation from Fall 2020

The Writing Center visit requirement is now an "Outside Reader" requirement and is only required to earn an A in the class. To meet this requirement, you will take your writing to anyone outside of the class. This person can be at the Writing Center or Career Services, or you can have your work reviewed

---

2 See Asao Inoue's *Labor-Based Grading Contracts: Building Equity and Inclusion in the Compassionate Writing Classroom*, available through the WAC Clearinghouse.

by someone outside of the class, such as an upperclassman in ECE, a faculty member, such as a professor or adviser, or another mentor, such as someone in engineering you work within your workplace or at an internship. When you meet this requirement, have the outside reader send one or both course instructors an email saying they met with you, or you can forward emails you have with them about your writing.

Table 1. Minimum Expectations for Each Letter Grade

	A	B	C	D	F
Participation Expectations					
Weekly reflections	13+	11–12	9–10	8	7 or fewer
Revision workshop draft posts	3	2	2	1	0
Revision workshop peer responses (2 or more per workshop)	6	4–6	3	2	0
Consultation with outside reader (Writing Center, Career Services, mentor)	1+	0–1	0	0	0
Writing Conference with Dr. Mallette	2+	1+	1	0	0
Homework and Projects Completed					
Submitted major projects	4	4	4	3	2 or fewer
Late projects	1	2	3	4	4
Completed weekly homework	90%+	80–89%	70–79%	60–69%	Less than 60%
Projects Meeting or Exceeding Expectations					
Exceeding expectations	0–4	—	—	—	—
Meeting expectations	3 or more	3 or more	2 or more	2	0
Not meeting expectations	0	1 or fewer	2 or fewer	2	3+

What about pluses or minuses?

If a student generally meets all the requirements for a specific grade but misses in one column, the student may be able to earn a minus letter grade for the next tier, even if, technically, they would be in the lower tier. This approach means that the class offers more flexibility and enables students to be successful in whatever way they can, regardless of things that might pop up in the semester. For example, a student who manages to meet expectations on all major projects and completes all other requirements for the B but only does 77% of the homework may still be eligible for a B-.

What does it mean to meet expectations?

In general, meeting expectations will mean that the assignment attempts to include all required components using the parameters provided, even if they aren't fully effective. Essentially, you will meet expectations if your attempt (on both homework and major projects) clearly makes an effort to follow guidelines and demonstrate learning, even if you have areas to improve on. An example of not meeting expectations is taking your draft for A2 (the lab report) and not significantly revising it and then resubmitting it for A3 (the engineering report) or neglecting to use the appropriate report template.

If you submit a homework assignment or project that doesn't meet expectations, you'll be able to revise and resubmit. Again, the goal is for you to learn, so if you can demonstrate how you're attempting to meet course outcomes, you'll be meeting expectations and the grading contract. If you are asked to revise and resubmit a project, and the revision meets expectations, you will still be able to earn an A in the class based on the contract.

What does it mean to exceed expectations?

While meeting expectations is focused on giving it a good attempt and demonstrating effort toward meeting the course outcomes, exceeding expectations is characterized by being particularly effective, impactful, and/or successful. What this usually means is that you've revised a draft a few times and met with one of the course instructors or with other writing support to get feedback to make your attempts more effective overall.

Do I need to exceed expectations on assignments to make an A?

If you are wanting to make an A in the class, you can aim to have all your assignments exceed expectations or just have them all meet expectations—that way if you show your progress through the semester and your final project meets or exceeds expectations, then you'll still earn an A and will have demonstrated your learning. What you will need to do is at least meet expectations on all major projects and complete the additional work required for an A, including visiting the Writing Center or another form of writing support.

How Do I Know Where I Stand and Track My Progress?

You can use the table above to assess your current standing in the class, and we'll periodically ask you to assess your grade/progress and make sure you understand how to stay on track for the grade you wish to work toward. To help you track your progress in the course, you'll be given **a self-assessment tool** that you and the course instructors can both see and access. We'll periodically ask you to update your self-assessment and to reflect on your progress in the class. You'll also complete **weekly reflections** where you reflect on your learning for the week and set goals for the next week.

Participation & Engagement

To receive full credit for activities, reflections, assignments, you should submit your work by the designated due date, **typically Friday of each week** unless otherwise noted.

Communicate with us if you're struggling to complete your work—we'll work with you to find a solution!

Late work

As with fall 2020, we have no idea what might happen this semester, and we all have the possibility of getting ill, experiencing scheduling changes, taking on caretaking responsibilities, and other challenges. For instance, Dr. Mallette currently has two young children at home, so we know how hard it can be to focus on work while also taking care of other responsibilities. Thus, deadlines are flexible, **so think of them more as a "best-by" date.**

Getting in work on time will be most beneficial to you for your learning and progress in the class, but you have space to submit work late as needed.

Weekly work will be accepted late (particularly individual assignments), though it will benefit you if you turn them in on time to support your learning and progress toward the major projects. If you need more time to complete weekly work or feel that you're falling too far behind, reach out to talk to us so we can figure out options.

Major projects will be accepted up to 48 hours late with no questions asked. We'll also accept projects up to 1 week late as long as you let us know that you need more time. If you need more than 1 week for major projects, you'll need to talk to us to create a plan for when you will be able to submit those projects. The contract builds in flexibility for late projects.

The key here is to communicate with me if something will impede you completing your work—we'll work with you to find a solution! When you reach out, you don't need to give us full details about what is causing you to submit work late unless you really want to or need help finding resources. And we will never, ever ask you to provide documentation for illness or anything else (and honestly, it violates HIPPA, so none of your other teachers should either). Our goal is for you to be successful, and as long as you're able to complete work and meet the course objectives, then turning work in late is OK.

Appendix B: Self-Assessment Tool
ENGR 207 Contract Check-In

Use this tool to track your progress in the class. You can maintain this as a Word document that you update for each of the check-ins throughout the semester (Week 4, Week 8, Week 13, and Final Course Reflection). You can also create a Google Document and share it with me so that we can both comment.

Below is Table 1, which outlines the minimum expectations for each letter grade.

Table 1. Minimum Expectations for Each Letter Grade

	A	B	C	D	F
Participation Expectations					
Weekly reflections	13+	11–12	9–10	8	7 or fewer
Revision workshop draft posts	3	3	2	1	0
Revision workshop peer responses	6	6	4	2	0
Consultation with outside reader (Writing Center, Career Services, mentor)	1+	0–1	0	0	0
Writing Conference with Dr. Mallette	2+	1+	1	0	0
Homework and Projects Completed					
Submitted major projects	4	4	4	3	2 or fewer
Late projects	1	2	3	4	4
Completed weekly homework	90%+	80–89%	70–79%	60–69%	Less than 60%
Projects Meeting or Exceeding Expectations					
Exceeding expectations	0–4	—	—	—	—
Meeting expectations	3 or more	3 or more	2 or more	2	0
Not meeting expectations	0	1 or fewer	2 or fewer	2	3+

In the following sections, you'll be filling out what you have completed so far to document your progress in the class.

Participation Expectations Table

Fill out the following table based on completed activities.

Participation Activity	Number Completed
Weekly reflections	
Revision workshop drafts posted	
Revision workshop responses	
Writing Center visit	
Writing conferences	

Homework and Projects Completed Table

Fill out the following table based on work submitted, late work, or missed projects.

Activity	Number for Each Item
Submitted major projects	
Late projects	
Completed weekly homework	

Meeting or Exceeding Expectations

Expectation	Number in Each Evaluation Category
Exceeding expectations	
Meeting expectations	
Not meeting expectations	

Current Standing

Based on your work completed as detailed in the tables, what letter grade are you currently meeting the expectations for? Look at your performance so far in the class (versus comparing against the final total—in other words, what is your standing currently?) Is this the performance in line with where you want to be? Type your answers below.

Goals

Reflect briefly on what your goals are for the next phase of the class. How will you continue moving toward those goals? What do you need to do to stay on track with your work or to get back on track? Type them below.

Questions or Concerns

What questions or concerns you'd like me to know about? Type them below.

Putting Science in Black and White: Intensive Technical Writing Through Non-disposable Assignments as a Path for Decolonizing STEM

Sally B. Seraphin
Trinity College

Humanity demands science, technology, engineering, and math (STEM) that can nimbly respond to its global health, economic, and environmental challenges. Unfortunately, as argued by Alo Basu (2021a, 2021b), the lack of gender and, especially, racial diversity in STEM disciplines threatens progress by the continued reliance on structural mechanisms for hoarding opportunity, which ultimately stifle innovation. Based on data from the United States Bureau of Labor Statistics (2020, 2022), employment in STEM disciplines offers greater earning potential than non-science or technology related occupations. Considering such a financial incentive, the fact that graduation rates in STEM disciplines (Riegle-Crumb et al., 2019) are lowest for women and people of color suggests they are blocked from access. Indeed, while they represent 28.7 percent of the population, underrepresented minorities (American Indian, Alaska Native, Black or African American, and Hispanic or Latino) obtained only 14.2 percent of doctoral degrees in science or engineering between 2019 and 2020 (NCSES, 2020). Further, while women earned nearly half of doctorates in that time frame, they constituted only one-third of doctorates in physical or earth sciences and merely one-quarter of doctorates in engineering, math, or computer sciences. At the same time, the diversification of these fields offers several advantages for both historically marginalized people and the general population. For example, in healthcare, which is dominated by white cis-male medical models, there is a particularly urgent need to address health disparities by the inclusion of diverse female and racial perspectives. Additionally, prevailing evidence of the "edge effect" —where creative solutions are likely to emerge from multicultural collaborations—suggests that only a diverse body of STEM practitioners can yield the necessary innovation to address pressing global challenges such as climate change.

    A brighter future requires our deliberate and relentless cultivation of inclusion in STEM, beginning with education (Basu, 2021a, 2021b). How do we interrupt the process of STEM attrition, enable more minorities to flourish in that arena, and achieve the richly diverse perspectives needed for future innovation? In this chapter, I argue that writing through non-disposable assignments (NDAs) can be an effective means for chipping away at the inequities that block diversification in STEM. I, with Shannon Stock (Seraphin & Stock, 2020) and others (Seraphin et al., 2019), propose that in contrast to assignments that are discarded at the end of each semester (e.g., quizzes, individually-prepared term papers, or lab reports), NDAs (assignments that are prepared through peer-collaboration or produce publicly disseminated learning objects) have the potential to enhance student learning and retention outcomes in the culturally responsive classroom while sustainably generating useful applications or new tools that can benefit society. Like the meaningful writing projects highlighted in the work of Michelle Eodice, Anne E. Geller, and Neal Lerner (2017), NDAs also have the potential to be transformative for student learning and engagement.

In this chapter, I begin by defining NDAs, first in the context of the Open Pedagogy movement and then using a six-leveled, three-dimensional (6x3D, nonagonal) framework through which they can be considered. Once we have observed that learning objects represent the tangible outcome of NDAs, writing will be presented as the ultimate learning object. Next, I describe how the writing powers of STEM students can be shaped to meet course grading specifications through a three-stage process, using examples and student feedback from my own teaching of a recent neuroscience course. Finally, I address the question of "Why teach writing through NDAs as a means for diversifying STEM?" by invoking my own marginalized perspective as a teacher-scholar navigating the intersectional identities of a Black woman, immigrant, and mother reentering the workforce.

Non-Disposable Assignments: A Tool for Breaking Barriers Through Open Educational Praxis

Information enthusiasts may agree that knowledge should be freely shared for the benefit of all humanity. Indeed, Maha Bali, Catherine Cronin, and Rajiv S. Jhangiani (2020) argue for a social justice perspective on open education. Open educational practice is characterized by an application of instructional methods and the integration of teaching materials that are broadly distributed and commonly shared. These often free and reusable teaching resources and techniques represent "learning objects" (Retalis, 2003), constituting what is generally referred to as an Open Educational Resource (OER). In this spirit, an expanding culture of openness governs the creation and use of vital educational tools that are OER. Pedagogical practices advancing the objective of openness include those that either generate OER or facilitate the transfer of acquired knowledge between students, outside the academy, and even globally. The OER used in STEM courses may range from large course apparatuses and designs (i.e., learning management course templates, course syllabi, textbooks) and moderately sized course content (instructional materials, assessments) to granular course components (individual course elements such as slides, illustrations, simulations). These OER traditionally originate with field experts in academia or publishing but are also amenable to student creation, modification, and reuse. Student-generated instructional materials, developed through "renewable" or "non-disposable assignments," represent some of the best examples of culturally rich and effective learning objects available for blended learning (Alvarez, 2013; Falconer & Littlejohn, 2007). Furthermore, I suggest that the potential of OER depends on the pedagogical practice of using NDAs, which can sustainably generate the large number of learning objects of diverse origin that are needed for future open education. After many years of using group writing NDAs in anthropology, biology, neuroscience, and psychology courses, I can identify several assignments that both fulfill the objectives of open education while providing useful writing practice. These assignments vary in the gravitas or temporal and spatial reach of learning objects or deliverables. They also represent possible entry points for instructors wishing to experiment with this approach.

As formative assessments that shape individual practice, NDAs can be conceptualized through a 6x3D or nonagonal framework with learning products spanning six levels (i.e., Peers, Class, College, Community, National, International) across three key dimensions (Time, Space, and Gravity) (Seraphin et al., 2020). This framework is illustrated in Figure 9.1, which has been adapted from Seraphin et al. (2020). On the X-axis of *Time*, NDAs are marked by openness because they self-perpetuate through direct adoption, customization, and reuse. Since OER are easily modified to suit current learning objectives, they exhibit shelf lives surpassing the ordinary limitations of copyright and traditional publication-expiration cycles. For example, a learning object or teaching resource that was created and shared by a colleague last year could be customized by another for deployment in a new course and even further modified for future reuse as teaching needs or standards change. On the Y-axis of *Space*, OER also circumvents the physical and social structural boundaries that normally confine information within closely guarded spaces. Learning transfers across and transcends the usual margins separating those inside/outside the classroom, institution, community, and nation. For example, a learning object or teaching resource that was created and shared by students in one class can be adopted, modified, and reused in informal as well as formal educational circles—eliminating the longstanding identity- or affinity-based barriers of privilege. This includes barriers such as the English language supremacy identified by Elizabeth Blomstedt and bias against non-Western epistemological science challenged by Alicia Bitler and Ebtissam Oraby (both in this collection).

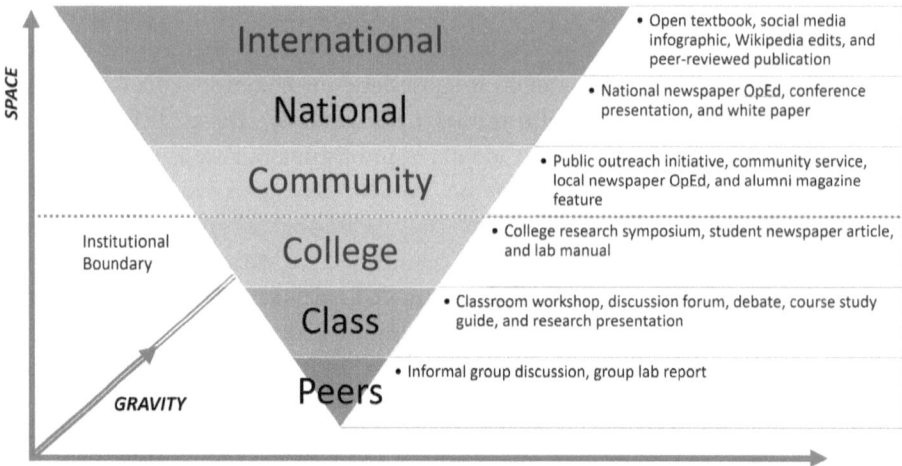

Figure 9.1: The space-time-gravity continuum for non-disposable assignments (NDAs). Adapted from Seraphin et al., 2018.

NDAs yield results in the form of learning objects that can be of tremendous value to students, their communities, and society. Thus, their *Gravity* can be viewed as proceeding along an imaginary Z axis whereby the results of open pedagogy have varying gravitas or significance, as determined by the degree of impact on the individual creator or a shared knowledge base. Depending on the information conveyed and the stakes involved, the learning object, for instance, a scientific meme about climate change, can simply educate or even serve to mobilize activism around causes such as the climate crisis and environmental justice. In this way, students develop important literacy skills while generating texts that reflect their unique ideas and diverse perspectives.

The most common NDA used in STEM courses unfolds at the level of Peers, where student–student teaching occurs through informal discussion, planning, and collaboration on learning objects, such as lab reports, shared among group members and with the instructor. Despite having the smallest temporal and spatial reach, the "Peer Level" NDA is foundational because it emphasizes peer-collaboration, elevating student perspectives and decentering the instructor. Being largely informal and contained between the peer-peer-instructor triad, this may provide a safe space for underrepresented students to experience the freedom of articulating their viewpoints and practice skills necessary for eventual success, with NDAs offering broader reach. It is important to note that the critical distinction making a lab report an NDA is this group requirement—which removes it from the realm of typical disposable assignments relegated to the classic student–instructor dyad.

At the "Class Level," NDA deliverables emerging from asynchronous discussion forums, synchronous learning activities such as workshops, debates, study guide development, and research presentations have an intermediate impact within the college. At the "College Level," NDA writing could generate learning objects for a public research symposium, student newspaper article, or reusable laboratory manuals and protocols. "College Level" NDAs have the greatest impact within that imagined or real physical boundary between the institution, its affiliates, and outsiders. While they have spatial reach across departments, student cohorts and may be preserved, the learning objects may remain confined to the academy.

Community-based learning or service-learning opportunities are increasingly demanded for college students. Through partnerships between the institution and community stakeholders, "faculty [are] able to take the classrooms out into the city and bring the city into their courses," according to Davarian Baldwin (2021, p. 68). Such new initiatives enrich student learning objectives by imparting a sense of meaning and the added purpose of serving the public good. Community and service learning also function to better engage the surrounding people and neighborhoods that are often adversely impacted by the so-called "Ivory Tower," which has an "elitist tradition of enclosure" (Baldwin, 2021). "Community Level" NDAs help to bridge the town–gown divide by generating learning objects that support public information or construct new channels of communication between entities ordinarily separated by college walls. By writing with, to, and for the benefit of their surrounding community, STEM students directly challenge the elitist tradition of enclosure. While working in close collaboration with community partners, students can generate research reports to facilitate their organization's mission. For example, through NDAs, STEM students can develop and disseminate scientific learning modules for use in public schools or craft op-eds to inspire public engagement around health and environmental problems. For example, students in my social neuroscience course recently partnered with community youth to build understanding on the developmental neurobiological impact of peer-bullying via learning objects they created. Their work was, in turn, celebrated in a college alumni magazine article by Andrew Concatelli (2023). In highlighting their science advocacy and community involvement, STEM students can appeal to their alumni and trustees on the mutual benefits from inter-collaboration (as opposed to coexistence) and begin to erase the legacy of suspicion between 'town and gown.' In a predatory trend, higher educational institutions have partnered with cities in building "technology communities" or "knowledge districts" that ultimately generate college revenue at the expense of surrounding neighborhoods under the guise of urban revitalization (Baldwin, 2021). Restoratively, STEM students can use NDAs to contribute to the communities they serve by generating learning objects through a fair process of exchange.

The broader the geographic reach of STEM student writing accomplished through NDAs, the more altruistic or intrinsically motivated is the endeavor as reciprocal demands from an identifiable stakeholder become impossible. By this ultimate service to humanity at large, STEM students can generate objects for learning that open higher education up to anyone with access to the internet nationally and internationally. Examples of "National Level" NDAs include a white paper describing a policy issue that can inform government deciders, a professional academic society or undergraduate research conference presentation, or an op-ed in a national newspaper.

At the "International Level," written NDAs can take the form of an open textbook, social media infographic, a peer-reviewed publication, or editing Wikipedia for accuracy. Depending on the mode of dissemination, the learning objects created at these final levels have excellent reach and greater permanence. For example, infographics made to inform the public on an issue can live on the internet forever once distributed through social media (e.g., Facebook, X/Twitter, Instagram, Pinterest, Tumblr), but their global accessibility depends on the route and frequency of redistribution. Contrastingly, Wikipedia edits are immediately accessible everywhere in the world but may easily be revised or erased by others. Finally, peer-reviewed articles in open-access journals have infinite reach as scientific learning objects. They also allow students to demonstrate disciplinary literacy and actively position themselves as authoritative practitioners in the field, which Rachel Riedner and colleagues (this collection) explain as necessary for promoting inclusion through professional identity formation.

Writing as the Ultimate Learning Object

Writing supports student learning across STEM curricula. First, writing is thinking. Necessarily, the process of writing involves planning, drafting, reading, and revision. As such, it requires thinking about the subject of writing as well as thinking about one's ideas on the subject in a non-linear and recursive manner (Hacker et al., 2009). That writing is thinking is also supported by the fact that metacognitive knowledge increases with writing skill, and both can be enhanced by pedagogical approaches emphasizing direct instruction on the metacognitive aspects of writing (Harris et al., 2010). Self-regulation, which involves goal setting, self-evaluation, and self-accommodation or help-seeking, is another key component of skilled writing (Harris et al., 2010). In this vein, Self-Regulated Strategy Development (SRSD) is an empirically supported pedagogical approach to facilitate the acquisition of effective writing strategies by providing students with knowledge of various writing tactics, supporting their self-management, and enhancing their motivation throughout the development process (Harris et al., 2010).

According to Karen R. Harris, Tanya Santangelo, and Steve Graham (2010), SRSD includes six instructional stages. In the first stage, the student develops and activates awareness of what is needed for good writing in a particular genre. This may be accomplished through exposure to examples of that literature with an eye on the declarative, procedural, and conditional elements therein. In the second stage, subjective aspects of writing, such as personal attitudes and beliefs about writing and the purpose and potential benefits of specific writing strategies to be learned, are discussed. In the third stage, the instructor models, for instance in collaboration with students, the constructive process of composition. A fourth phase involves memorization of certain mnemonics related to writing. More importantly, the self-regulation of student writing is supported by various means in stage five, culminating in their independent self-regulation and performance of writing tasks in stage six. A parallel of this is the "integrated knowledge" model described by Kara Taczak and Liane Robertson (2017), where students combine the acquired awareness of writers and the writing process with understanding gained from experiences outside of the classroom. Importantly, this model recognizes the value in the diverse perspectives that students bring to learning and writing.

Second, adding to the premise that "writing is thinking," this further represents a means for apprenticeship. Through writing practice, students can adopt a disciplinary framework, acquiring skills for technical communication with other scientists. Despite the instructor's propensity to recruit and indoctrinate their pupils into her own discipline, it is important that the writing strategies we teach serve students in many settings. Thus, special attention should be paid to science writing for different purposes and through different modalities. Students of STEM should be able to transfer their acquired writing skills to achieve effective communication or translation of science through audio-visual presentations and various genres or modalities of writing (e.g., technical reports, white papers, op-eds, social media). Third, contrary to learning strategies like rote memorization, writing is a form of tool used in the behavioral ecological sense because the technique is observed, imitated, practiced, and reworked with increasing mastery. According to Ian McGilchrist (2019), the neurobiological phenomena that underpin our ability to grasp facts are akin to those that coordinate our ability to grasp the pen for communication through language. Thus, although typically conceived as a skill, writing is fundamentally a tool. As a tool, writing also facilitates the transmission of knowledge between individuals and groups, as well as across generations and time. It is in this final way that writing represents the ultimate learning object.

To curate high-quality, knowledge-based learning objects, the instructor must first consider factors influencing students' motivation for working on NDA "products." Instructors should give full advanced disclosure of ultimate use(s) for student work with an option to contribute shared work anonymously (opt out of public exposure). By extension, the instructor could consider offering "traditional" disposable assignment options (e.g., essays) of equivalent weight for students who are not inclined toward public service or engagement. The instructor must also recognize a need for extensive scaffolding (support), develop a means for the internal vetting (i.e., quality control) of student-sourced learning objects, and adoption of grade-based incentives to facilitate the production of high-quality materials. For example, requiring multiple drafts separated by peer-assessment and revision allows student work to keep pace with the course and improved by positive feedback while they exercise self-regulated learning. An additional benefit is double-blind peer assessment (in large courses where some degree of anonymity can be maintained), which may enhance knowledge gain and metacognition by exposing students to new information or ways of thinking.

One early adopter of non-disposable assignments, Rajiv Jhangiani (2015), observed significant creativity in what students brought to bear on their projects. For similar results, instructors should give latitude or flexibility to accommodate students' creativity and heterogeneity of resulting learning objects/products. For example, in his initial attempts to incorporate NDAs in undergraduate courses, Jhangiani (2015, 2017) began by encouraging projects aligned with program and course learning objectives. Requiring prior approval for student project ideas or offering students a limited range of projects that suit preexisting learning objectives may inevitably lead to empirically grounded solutions in the outcome of student work. Thus, principles of backward course design can be used as a preventative technique for failing NDAs. To increase the probability that high-quality learning objects will emerge from student NDAs, the instructor should model the creation process and show examples of NDAs achieved through best practices. Finally, it cannot hurt to review guidelines for open licensed publication (for true OER) or release (for assignments that are shared outside of the course, but not by definition OER, such as "letters-to-the-editor") of student work. The evaluation of learning objects created by students through NDAs requires the development of hitherto non-existent, empirically based standards for their classification and associated metadata. This metadata would facilitate future reuse or adaptation of learning objects. In the meantime, one can develop personalized methods for rating (external quality control of) student-sourced materials, keeping in mind that consistency in student outcomes and convergent solutions will emerge from empirically grounded work.

Example Non-disposable Assignments Featuring Intensive and Multipurposed Writing in a Scaffolded Environment

In the spirit of open pedagogy and with the aim to equalize access to the STEM professions, I have implemented intensive scientific writing through NDAs. To illustrate how this could be incorporated into STEM courses, I offer specific examples from my Brain and Behavior course. Student writing serves multiple functions throughout the semester in this writing intensive course, which is required for second-year psychology and neuroscience majors. What follows is a detailed description of some NDAs as well as an overview of the method by which I have incorporated a focus on writing in this and other STEM courses. The major assignment phases are illustrated in Figure 9.2.

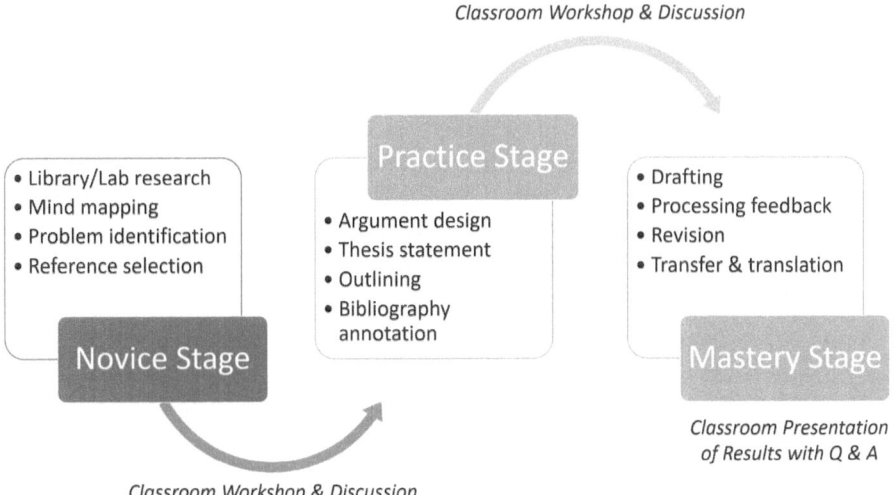

Figure 9.2: The three stages of STEM writing progression

Written NDA I: Writing as a Path to Establishing Healthy Classroom Norms

From the outset, themes of collaborative and inclusive writing are invoked when we begin the semester with an activity where students help to write a portion of the course syllabus. On the first day of lecture, each student is asked to describe, in one sentence entered on a shared electronic document, the attributes or values that they would like to have epitomized in our classroom culture. This information is summarized in a word cloud (an infographic that gives visual prominence to higher-frequency words), which becomes embedded in the "Class Norms" section of our course syllabus. Individual participation makes up 5 percent of the course grade, and the participation rubric includes a section related to the student's adherence to and support of the collective norms established through this activity. This process of going from crowd-sourced information to a single infographic that serves as a semester-long learning object also foreshadows the ongoing production process for NDAs.

STEM Writing NDA II: Active Reading through Writing

Student writing should be used to support reading as a critical skill component of any college education (Klucevsek & Brungard, 2016). I address this through interconnected individual and group learning activities. We use a digital textbook that has a built-in notebook and journal, which invite students to summarize and reflect on their course readings through daily writing. Whether similarly accomplished with a physical textbook and note-taking, the act of paraphrasing course reading and personally relating important concepts through writing represents "active reading." This exercises individual metacognition and provides benefits for comprehension and retention of the information therein (Fisk & Hurst, 2003; Hirvela & Du, 2013). With "translanguaging," where multilingual students apply all their linguistic knowledge toward making sense of an assigned reading (Hungwe, 2019), this paraphrasing may equalize reading skills in traditionally marginalized students. Thus far, the notebook and journaling are ungraded and merely offered as an opportunity for study skill enhancement. This could easily be formalized as an individually graded component to maximize the incentives for improving literacy in STEM majors. In a connected learning activity culminating in shared learning objects, students can enhance their scientific literacy by collating what they and classmates identify as important information from the textbook while note-taking or journaling. Using a shared electronic document that I provide, groups of students are required to contribute written content to "fill in the blanks" on course exam guides containing only an initial list of keywords or phrases. Being crowd-sourced, this written NDA conserves individual studying effort by making light work of an otherwise labor-intensive task. In keeping with what Kristin M. Klucevsek and Allison B. Brungard (2016) described as the need for STEM domain-specific literacy, this written NDA may also level the learning playing field by exposing important information that may have been missed by students with less experience reading or deciphering discipline-specific text. By using their own words to fill in the study guide, students also model skills for scientific translation, effectively peer-teaching. The class comes to realize first-hand a benefit of the NDA.

STEM Writing NDA III: Moodle Discussion Forums

Students practice communicating their own perspective or analysis through regular asynchronous discussion forums maintained on Moodle, a course learning management system. These required Moodle forums comprise 10 percent of the final grade and involve a two-step process whereby students initially respond to a posted discussion prompt (e.g., a case study, video, news article related to that week's lecture topic). Next, they must comment on the responses of one or two peers, depending on the length of the multimedia prompt, for full credit. While I monitor the thread for adherence to class norms of conduct and may periodically inject additional resources for their consideration, I regard this as a predominantly student space devoted to their discovery through peer-peer interaction. Beyond allowing them an opportunity to practice short-form science writing as they hammer out controversies related to brain and behavior, the forums also represent a "Class

Level" NDA because of the compulsory and visible inter-peer exchange of perspectives (here, the learning objects).

STEM Writing NDA IV: Semester Research Project

Two fundamental components linking all my courses are group research projects and intensive writing. Besides simulating the practical aspects of everyday science, these afford an opportunity for students to build upon or transfer previously acquired skills and integrate their curricular knowledge, which is key for inclusive STEM learning (Basu et al., 2017). In laboratory courses, groups of students produce highly technical writing in the form of lab reports. In lecture courses, students have more flexibility to communicate their ideas using a scientific framework through writing analytical or persuasive research papers. In both cases, the group element qualifies this form of writing as a "Class Level" NDA because students together plan, create, organize, refine, and combine individual subcomponents of the ensuing learning object, which may then be revised for resubmission. In a semester-long NDA comprising 12 percent of the course grade, Brain and Behavior students practice a sequential approach to written communication by completing several ungraded, low-stakes assignments or learning activities that build up to two final group research deliverables: a paper (9 percent) and a presentation (3 percent). Not long after the syllabus review and introductory lectures, the semester research projects are launched with a class conversation about which of the course topics covered particularly interest them. Students are invited to enter three areas of research interest into a shared Google document. I then identify relevant topics that will not be covered in detail and would complement the course before students are invited to sign up to research these topics in groups of three to five. The semester-long research projects then proceed through three successive stages: Novice, Practice, and Mastery. Each lasts approximately four weeks and includes ample opportunity for instructional guidance within as well as between stages.

To begin the "Novice Stage" of the STEM Writing NDA, students receive specialized instruction from a Science and Electronic Resources Librarian about best practices for conducting a literature review, tools for managing bibliographic data, and the American Psychological Association (APA) Style (see Figure 9.2 earlier). At this stage, it may be useful to map the chosen research topic. This can be accomplished by simply brainstorming with paper and pencil or using a sophisticated library resource such as CQ Researcher or Credo Reference: Academic Core, which graphically displays related concepts, issues, events, and pro/con information. With a mind map in hand, students can better choose their search terms and decide which rabbit holes to pursue as they probe scholarly literature databases for reference information. After determining their topic parameters, students are encouraged to identify a problem or question to guide their research. One week after the library workshop, groups submit an ungraded topic declaration form including a preliminary bibliography, paper title, presentation title, and three to four scholarly resources per person. The "Novice Stage" concludes with a whole class discussion about successes and problems encountered during the initial library research process. In this way, the separate experiences of each group become an example or "learning object" from which others can benefit.

In preparation for the "Practice Stage," a classroom workshop on the nuts and bolts of writing is held. First, students receive a list of resources outlining the basics of English grammar, how to paraphrase, when to use quotations, and how to avoid plagiarism. After discussing argument design around a thesis statement and the selective deployment of resources to explain a principle or present evidence, we review different strategies for creating outlines (e.g., chronological, topical) and annotated bibliographies. Students then have approximately three weeks to prepare a graded paper Outline and Annotated Bibliography assignment. The outline must identify which students will be responsible for each part (i.e., which questions or subtopics) of the overall paper. Annotations must include a sentence explanation about how each resource will be used to advance their argument. This last requirement forces students to be mindful about how sources help their product. It encourages them to be more selective and even consider substituting or supplementing their source material at this stage. Since the writing-thinking-rewriting process is part of what is being assessed with NDAs, there are minimal benefits for students using artificial intelligence (AI). Along with an interim grade (0–3 out of 3 points), the student groups receive very detailed feedback about their thesis statement, outline, and reference choices. We then devote class time to discuss overall trends observed in these early submissions so that all students can benefit from my observations. To emphasize the importance of a central theme, we also "workshop" their thesis statements. Notably, Jhangiani (2015) observed peer assessment of a quiz positively influences subsequent test scores in an introductory psychology course. In this vein, groups formally announce their semester-long research projects by sharing their prepared thesis statement. As a class, we discuss and troubleshoot the statements, clarifying the group writing goals in the process. Once each group addresses my feedback on their Outline and Annotated Bibliography, a completion grade of 3/3 typically replaces the interim grade.

As part of the Practice Stage, students begin the process of writing their first draft. In preparation for this, they receive a brief workshop on how to construct a paragraph from components of the approved outline. At this stage, students may be able to identify parts of the outline that lend themselves to topical, explanatory, and transitional sentences. They are also encouraged to rearrange elements of the outline for improved argument structure or paper flow. Each member is required to contribute 500–750 words, not including the bibliography, to an APA formatted group paper due at the end of this stage—approximately three weeks after the Outline and Annotated Bibliography. As an additional support, I encourage individual students or entire groups to meet with me as questions emerge while preparing this first draft.

The "Mastery Stage" begins with the submission of a first draft of group research papers for formal assessment. In addition to a grade, the student groups receive a detailed mark-up of their paper with individual-specific and group-level feedback. In my response to each group, I also include some overall remarks based on all the essays submitted in the class. This circumvents students committing new errors with subsequent drafts. Student groups are also encouraged to submit a copy of their paper and assignment instructions for review by the campus Writing Center staff. Ideally, once both forms of feedback have been received, we devote a small amount of class time to review and discuss next steps. After processing, collating, and organizing the feedback received, the student groups create and share their plan for revision over a series of weeks. I then work closely with individual students or groups needing extra support while implementing the necessary changes for achieving full credit on the final submission. The culmination of their semester-long STEM Writing NDA is a graded final draft that is due at the conclusion of the semester.

STEM Writing NDA V: Public Presentation of Semester Research Project

Continuing the "Mastery Stage" of the STEM Writing NDA, student groups prepare a 15–20 minute final presentation. This presentation is intended to introduce new information into the course content. I instruct student groups to craft the presentation around their thesis. Using a data-centered argument, they are advised to tactically deploy resources in a manner designed to persuade the class of their perspective. A draft presentation is due two weeks in advance of the final presentation deadline, and we discuss necessary changes. Finally, the last week of class is devoted solely to a symposium on their semester-long projects.

STEM Writing NDA: Student Feedback

Throughout the course, students had several small assignments to complete along the way. For example, topic choice, outline, preliminary bibliography, annotated bibliography, and first draft were all required before submitting a final draft and research presentation. At the course conclusion, 25 out of 30 students (83.33 percent) participated in an optional 4-question Moodle survey where they were asked to rate the statements in three questions according to the Likert scale (a. Strongly agree, b. Somewhat agree, c. Uncertain, d. Somewhat disagree, and

e. Strongly disagree) with the ability to select multiple options. This was followed with a fourth short response question: "How has it been carrying out a semester-long research project? (Share anything you'd like me to know about your science communication journey)." These results are described in Table 9.1.

Table 9.1. Responses from Most Students on a Brief Questionnaire Demonstrated the Overall Success of Written NDAs.

STEM Writing NDA IV & V: Brain and Behavior Student Responses to an End-of-Term Course Evaluation Survey About the Semester-Long Research Project					
Q1: It was helpful to have multiple, low stakes assignments to shape my writing practice and the final research products (presentation, paper).					
Response Options	Strongly agree	Somewhat agree	Uncertain	Somewhat Disagree	Strongly disagree
Percent Chosen	76%	24%	0%	0%	0%
Q2. Interrogating the literature for a specific topic enhanced my learning of brain and behavior.					
Response Options	Strongly agree	Somewhat agree	Uncertain	Somewhat Disagree	Strongly disagree
Percent Chosen	45.15%	46.15%	7.69%	0%	0%
Q3. My semester-long experience of technical research and writing left me feeling more empowered, knowledgeable, or prepared for any future explorations of careers in STEM.					
Response Options	Strongly agree	Somewhat agree	Uncertain	Somewhat Disagree	Strongly disagree
Percent Chosen	46.15%	38.46%	11.53%	0%	3.84%

Among student respondents, 76 percent strongly agreed, and 24 percent somewhat agreed with the statement: "It was helpful to have multiple, low stakes assignments to shape my writing practice and the final research products (presentation, paper)." Among the 24 percent who "somewhat agreed," the following comments highlight the overall positive experience of working on these assignments, notwithstanding problems encountered by their individual groups:

> I think having the research project being completed in sections over the course of the semester was helpful in making it better quality.
>
> The semester-long research project was fun to have because we got to research a topic of our interest. Since it was a semester-long project, it was not as overwhelming as a normal project

will have. In addition, it was a good idea to work in a group of 5 because we had the chance of getting to know each other.

At first, it was a bit intimidating to hear about this semester-long research project. However, because it was divided into multiple parts, the process didn't feel overwhelming. Overall, I enjoyed the process of studying a specific scientific phenomenon and working with my classmates.

46.15 percent strongly agreed, 46.15 percent somewhat agreed, and 7.69 percent were uncertain about the statement, "Interrogating the literature for a specific topic enhanced my learning of brain and behavior." The following representative quotes were from students who strongly or somewhat agreed with this statement:

I thought it was interesting to [focus] on a topic and a specific subset of that topic to be able to become an expert in that field. I felt as if this did enhance my learning of brain and behavior. . . .

This was the first research paper I worked on in a [STEM] field explicitly, which I did actually enjoy. I was able to elaborate on my knowledge of schizophrenia in a psychological sense and expand on [its neuroscientific bases [sic]]. . . .

I think that it was very interesting to carry out a semester-long research project because I was able to connect each topic that I learned in class to what my research project was based on. This enabled me to gain a greater understanding of both the course material, and my research. Understanding the fundamentals of neuroscience throughout the course helped me to communicate in a scientific way that was focused [on Post Traumatic Stress Disorder]. I think that applying this knowledge is a skill that I will continue to use in my scientific writing.

When asked to evaluate the statement "My semester-long experience of technical research and writing left me feeling more empowered, knowledgeable, or prepared for any future explorations or career in STEM," 46.15 percent strongly agreed, 38.46 percent somewhat agreed, 11.53 percent were uncertain, and 3.84 percent strongly disagreed. Three students who reported uncertainty about this added:

I like the idea of a semester long research project. I liked the ability to choose topics. However, I wish the groups were smaller. At times I felt that people had clashing ideas for the project and what we wanted to focus on. It was also hard to write a paper

> and try to get 5 people in the same place at once due to conflicting schedules.
>
> Carrying out a semester long research project in the background of weekly quizzes, weekly Pearson assignments, recorded lectures, forums, and exams was far from ideal. Although the premise of a research project enhancing our scientific reasoning and writing was well-intended . . .
>
> I felt that the process would have been quite interesting, but I felt that the communication / organization between group members made the process quite stressful and disorganized. . . . I felt that if there were an established platform for communication between group member, group planning and working would have [gone] more smoothly. However, [it] was fulfilling to have a side project / interest during this course.

A student who chose both options "strongly agree" and "strongly disagree" for statement #3 said:

> I really liked having a semester long research project. I think that it allowed me to become closer with people in the class who I wouldn't have otherwise and thy was useful when trying to study or putting faces to names in the discussion forums. I also think these projects were very useful to develop my ability to write scientific papers and prepare my research to be presented. I'm a biology major, so I will definitely have to do more papers like this in the future, so having the opportunity to work together for the whole semester with check-ins along the way made it really easy to get it done. The feedback from my group mates, Professor Seraphin, and the [Trinity College] writing center allowed me to go back and edit my writing in a way that would allow me to [perform] better on the next paper that I write.

The following comments were made by a large number of students who chose "strongly agree" across the board:

> I thought that the semester-long project was extremely valuable in that I learned so much about how to properly research, source, and write literature pertaining to the topic of substance abuse. Being able to engage with the material over the entire semester allowed for my understanding to deepen as [I] continually got to engage with the material in different forms. . . .

I think it was a good way to learn about neuroscience in a new way. Our class was so fast paced and lecture heavy, that having a chance to research something on our own was very helpful. I think it was a good way to meet other peers as well!

I really liked having it be a semester long because it allowed me to learn and take my time without an additional stress to make a final paper in a week. I also was able to connect with classmates and discuss class topics and work. [Although I was extremely nervous,] I felt accomplished that I was able to present my work.

I feel the semester-long research project was quite fun. I found it enjoyable to learn more about a specific aspect of brain and behavior and then share what I learned with the class. I also found it valuable to learn how to work in a group. Developing communication skills and the ability to work as a whole rather than separate parts was a good learning experience.

I liked doing this project even though it was more challenging to write a group paper compared to working independently. It was extremely helpful for the course overall to have a group of peers that I could talk to.

It was very helpful to have had multiple assignments for this semester long research project, as it provided a lot of feedback to help my group work on aspects of the paper (for example) in addition to allowing us to have a lot of time to not only grasp the material deeply but also to enhance our interests in the topics by exploring literature and research conducted about them.

I really enjoyed having a semester long research project, I think one of the main reasons why I enjoyed it so much is because we were able to turn in small portions of the assignment as we went along. [N]ot having to turn in the full project at a specific due date alleviated a lot of stress and allowed for me to plan ahead and produce my best work.

[I] thought it was very helpful to have a structured plan in doing this project especially considering it was done in a group. [Also I] think that the lecture time we spent talking to the library research staff was really helping in finding informational and credible sources. [H]aving Professor Seraphin check and give feedback on our work was really helpful in guiding us in the right direction. [T]he mandatory writing center appointment was also a good way to have our papers checked. [Overall, I] think the whole process was great and very helpful in completing this project.

Students and instructors can be resistant to the adoption of NDAs. Despite potential benefits, student reluctance to engage in group work represents an obstacle for implementing group NDAs. One barrier to working in collaboration is the unequal effort and different costs incurred by group members (Terras et al., 2013). For example, social loafing within a group can discourage sharing among the more conscientious students demonstrating progress on group NDAs. While most groups in my course managed to work together very well by the end of the semester, there were one or two situations, largely exacerbated by student illness, where things remained shaky. In this regard, one student wrote:

> It has been an interesting experience carrying out a semester long research project. I think one thing that stands out is that through this project I was able to work with a diverse group of people and we had to really learn how to work efficiently together. I think this is a great experience as group work does not end in the real world. We had to step up and be leaders and each play a role which was sometimes difficult to navigate.

Written Non-disposable Assignments Represent a Means for Diversifying STEM

We should address inequities in STEM through the adoption of written NDAs because they subvert the structures that reinforce hegemony. We should also embrace them as a means for greater equity because of their ability to enhance learning. Students may struggle to find the purpose or meaning in traditional assignments, which they not only experience as rote and mundane but are tiresome to grade (Jhangiani, 2015). According to Allan and colleagues (2018), well-being and productivity can be enhanced by doing work that benefits people other than yourself. In a study of students, working adults, and public university employees, it was found that people who do work to benefit others experience greater task meaningfulness and increased work meaningfulness over time (Allan et al., 2018).

As knowledge workers, future students will 'think for a living' (Fontana et al., 2015). Thus, a soft skill educators should impart on students is self-regulated learning (SRL), or the ability to assume responsibility for one's professional development by self-regulating one's personal learning needs in an increasingly knowledge-intensive workforce (Fontana et al., 2015). NDAs enhance SRL by simulating the process by which future workers must gain and manifest expertise in a supportive educational environment.

NDAs offer instructors the opportunity to increase students' self-efficacy as they target the development of three general motivational beliefs (Pintrich, 1999 & 2000), including self-efficacy beliefs, task value beliefs, and goal orientations. According to Albert Bandura (2002), self-efficacy not only supports our potential for success and feelings of well-being in a variety of life situations but also impacts the development of media literacy skills. This is particularly important as academic achievement and media literacy are becoming increasingly linked (Terras et al., 2013). Higher-order media literacy skills are needed to push student learning horizons beyond the old limitations of time and space (Terras et al., 2013). To optimize the learning potential of OER, instructors must attend to the psychological dimensions of media literacy skills in their students. Many cognitive (e.g., student's cognitive load, mental representation of internet searches, recall of linear versus non-linear websites, pairing of learning goals with navigational skills), developmental (age), and psycho-social factors (introversion-extroversion, meta-cognition, self-regulation, self-efficacy, self-esteem, and motivation) influence e-Learning by enhancing or impairing the acquisition or maintenance of media literacy skills (Terras et al., 2013). Metacognition is marked by the ability to evaluate, regulate, and monitor what one knows. The effective learner is not only aware of their knowledge but can recognize learning and speak to their learning process as this unfolds (Terras et al., 2013). By scaffolding stages of completion in NDAs, we train student's metacognitive ability while stimulating the three critical phases of self-regulation (Zimmerman, 2002): forethought, performance, and self-reflection.

As they write for different purposes, my students develop media literacy through discussion forums and the creation of audio-visual components for their final research presentations. Using NDAs can help level the playing field for students from underrepresented groups. While those with developmental exposure, through gaming, etc., can easily transfer this experience to the educational task at hand, others having less ease with technology may struggle to meet the competing demands of two separate academic challenges: the learning activity and the technology (Terras et al., 2013). Prior life experience with technology can limit the potential for achievement in using and generating OER because this is associated with different cognitive profiles (Terras et al., 2013), possibly via enhanced visual-spatial skills and lower higher-order processing skills as observed in video game players compared with non-games players (Green & Bavelier, 2006). Eszter Hargittai and Gina Walejko (2008) observed a reduction in typical gender differences, for sharing on social media once internet user skill was controlled. Thus, as psychological enablers, NDAs represent an opportunity for instructors to impact development of a highly demanded vocational skill (i.e., media literacy) typically associated with relatively fixed characteristics such as socioeconomic status (Hargittai & Walejko,

2008), educational opportunity, or age (Terras et al., 2013). By lessening barriers to participation through open pedagogical practices that foster media literacy, the instructor could equalize the playing field for students from underrepresented groups. If other identifying student features (e.g., race, gender, ability, etc.) remain constant, we would thus expect to observe greater richness in the learning objects generated by a more inclusive and now diverse body of STEM practitioners.

Why address inequities in STEM through the adoption of written NDAs? My personal answer to this question is informed by my positionality as a Black female teacher-scholar with personal hindsight and deep aspiration for change. First, while reflecting on how I have felt enabled, writing emerged as a particular source of confidence that kept me grounded in the pursuit of education. My writing journey probably started in high school, where I excelled in Advanced Placement English, but my sense of being a good writer was instilled during a brief college experience at a historically Black institution, Howard University. In keeping with my initial identification as a pre-med major, I chose to enroll in a technical writing course to satisfy the English composition requirement. By mechanisms I cannot now recall, I was therein endowed with the knowledge of how to decipher and produce writing in the manner typical of science communication. I learned not only to decipher but also to adopt jargon as a second language. I came to embrace the rhythmic structure used for reporting empirical research as a stable, orienting device. In a History of the Black Diaspora course intended to fulfill the humanities requirement, I explored my Haitian ancestry through an ethnographic research paper that allowed me to experiment with writing infused with my personal voice. These two intensive writing experiences left me feeling capable and competent in writing for various purposes. Long after I had transferred from Howard to the University of Massachusetts-Boston, where I ultimately earned my bachelor's degree as a commuting student, I observed writing to be the way I could effectively signal my accumulating mastery of scientific concepts and principles—even when momentary changes in my work schedule or family demands periodically prevented top performance on quizzes requiring rote memorization. Eventually, it was my writing—and especially the innovative thinking that it revealed—which stood out, earning me admission to graduate school after a less-than-stellar undergraduate record. While studying human biology at Oxford University, I composed essays in preparation for weekly individual or group tutorials. This experience demonstrated to me that writing is not only a means for communication but also a device for thinking. The confidence I developed in writing helped me to distinguish between writer's block, where emotions interfere with my productivity, and writing difficulty rooted in technical problems around preparation, focus, or confusion about the process. Eventually, my comfort with writing made completing my thesis less daunting.

Second, as the child of immigrants, the plight of poor and marginalized communities within and outside of the academy particularly resonates with me. While advocating for institutional, infrastructural changes to help retain minority students in STEM at the colleges where I have worked, I realized that my underrepresented minority neuroscience students are educated in a STEM context that is predominantly white, cis-gender, affluent, and also views itself as the gatekeeper for future opportunities in research and clinical practice. Although technically a part of the academy, they are tacitly maintained as separate and divided in a way that surely impacts their ability to learn and thrive in the disciplines. Over many years of teaching anthropology, biology, and psychology to class sizes of 1 to 300 students at small liberal arts colleges and large public or private universities, I have also recognized that behavioral sciences education presents special opportunities for an educator to engage students on the bio-cultural bases of human experience and its implication for important social issues, such as racial and economic health disparities. I also noticed the second tension—that between town and gown, or people who pay tuition and Others in their surroundings who are denied access to that commodified knowledge. In addition to advocating for minority students within the college walls, this inspired an interest in open pedagogy, which has the effect of enhancing the equitable dispersal of information—through the pedagogical innovation of NDAs for STEM teaching. There are endless possibilities for fostering gains in social justice (Bali et al., 2020), diversity, equity, and inclusion in STEM by sharpening the tool of writing through NDAs. In other words: Putting STEM in black and white.

References

Allan, B. A., Duffy, R. D. & Collisson, B. (2018). Task significance and performance: Meaningfulness as a mediator. *Journal of Career Assessment, 26*(1), 172–182. https://doi.org/10.1177/1069072716680047.

Alvarez, I. (2013). High aspirations: Transforming dance students from print consumers to digital producers. *Journal of Interactive Media in Education, 1*, 1–16. https://doi.org/10.5334/2013-16.

Baldwin, D. L. (2021). *In the shadow of the ivory tower: How universities are plundering our cities*. Bold Type Books.

Bali, M., Cronin, C. & Jhangiani, R. S. (2020). Framing open educational practices from a social justice perspective. *Journal of Interactive Media in Education, 2020*(1), 1–12. https://doi.org/10.5334/jime.565.

Bandura, A. (2002). Growing primacy of human agency in adaptation and change in the electronic era. *European Psychologist, 7*(1), 2–16. https://doi.org/10.1027/1016-9040.7.1.2.

Basu, A. C. (2021a). Are we ready? The future of inclusive excellence in STEM. *The Thinking Republic*, Fulcrum Issue. March 21. https://www.thethinkingrepublic.com/fulcrum/are-we-ready.

Basu, A. C. (2021b). *Cultivating inclusion in STEM: Imagining the future.* Seminar Presented at the Barnard College Center for Engaged Pedagogy. December 14, New York, NY.

Basu, A. C., Mondoux, M. A., Whitt, J. L., Isaacs, A. K. & Narita, T. (2017) An integrative approach to STEM concepts in an introductory neuroscience course: Gains in interdisciplinary awareness. *Journal of Undergraduate Neuroscience Education. 16*(1): A102–A111. https://www.ncbi.nlm.nih.gov/pmc/articles/PMC5777831/.

Concatelli, A. J. (Winter 2023). Bearing the brunt of bullying: 'Social neuroscience' course delves into the psychological pain. Trinity College Reporter. https://commons.trincoll.edu/reporter/features/bearing-the-brunt-of-bullying/.

Eodice, M., Geller, A. E. & Lerner, N. (2017). *The meaningful writing project: Learning, teaching, and writing in higher education.* Utah State University Press.

Falconer, I. & Littlejohn, A. (2007). Designing for blended learning, sharing and reuse. *Journal of Further & Higher Education, 31*, 41–52. https://doi.org/10.1080/03098770601167914.

Fisk, C. & Hurst, B. (2003). Paraphrasing for comprehension, *The Reading Teacher, 57*(2), 182–185. https://www.jstor.org/stable/20205336.

Fontana, R.P., Milligan, C., Littlejohn, A. & Margaryan, A. (2015), Measuring SRL in the workplace. *International Journal of Training and Development, 19*, 32–52. https://doi.org/10.1111/ijtd.12046.

Green, C. S. & Bavelier, D. (2006). Effect of action video games on the spatial distribution of visuospatial attention. *Journal of experimental psychology: Human perception and performance, 32*(6), 1465–1478. https://doi.org/10.1037/0096-1523.32.6.1465.

Hacker, D. J., Keener, M. C. & Kircher, J. C. (2009). Writing is applied metacognition. In D. J. Hacker, J. Dunlosky & A. C. Graesser (Eds.), *Handbook of metacognition in education* (pp. 154–72). Routledge. https://doi.org/10.4324/9780203876428.

Hargittai, E. & Walejko, E. (2008). The participation divide: Content creation and sharing in the digital age1, *Information, Communication & Society, 11*(2), 239–256, https://doi.org/10.1080/13691180801946150.

Harris, K. R., Santangelo, T. & Graham, S. (2010). Metacognition and strategies instruction in writing. In H. S. Waters & W. Schneider (Eds.), *Metacognition, strategy use, and instruction* (pp. 226–256). Guilford Press.

Hirvela, A. & Du, Q. (2013). Why am I paraphrasing? Undergraduate ESL writers' engagement with source-based academic writing and reading, *Journal of English for Academic Purposes 12*(2), 87–98. https://doi.org/10.1016/j.jeap.2012.11.005.

Hungwe, V. (2019). Using a translanguaging approach in teaching paraphrasing to enhance reading comprehension in first-year students. *Reading & Writing, 10*(1). https://doi.org/10.4102/rw.v10i1.216.

Jhangiani, R. (2015). Pilot testing open pedagogy. https://tinyurl.com/56555fm3

Jhangiani, R. (2017). Ditching the "disposable assignment" in favor of open pedagogy. https://doi.org/10.31219/osf.io/g4kfx.

Klucevsek, K. M. & Brungard, A. B. (2016). Information literacy in science writing: How students find, identify, and use scientific literature. *International Journal of Science Education, 38*(17), 2573–2595. https://doi.org/10.1080/09500693.2016.1253120.

McGilchrist, I. (2019). *The master and his emissary: The divided brain and the making of the western world* (2nd ed.). Yale University Press.

National Center for Science and Engineering Statistics (NCSES). 2020. Survey of Earned Doctorates. https://ncses.nsf.gov/pubs/nsf22300/data-tables.

Pintrich, P. R. (1999). The role of motivation in promoting and sustaining self-regulated learning. *International Journal of Educational Research, 31*(6), 459–470

Pintrich, P. R. (2000). Multiple goals, multiple pathways: The role of goal orientation in learning and achievement. *Journal of Educational Psychology, 92*, 544–555.

Retalis, S. (2003). Commentary by Symeon Retalis on Littlejohn, A. (2003) Reusing online resources, chapter 3: Keeping the learning in learning objects, by Dan Rehak and Robin Mason. *Journal of Interactive Media in Education, 2003*(1), Art. 5. https://doi.org/10.5334/2003-1-reuse-05.

Riegle-Crumb, C., King, B. & Irizarry, Y. (2019). Does STEM stand out? Examining racial/ethnic gaps in persistence across postsecondary fields. *Educational Researcher, 48*(3), 133–144. https://doi.org/10.3102/0013189X19831006.

Seraphin, S. B., Grizzell, J. A., Kerr-German, A., Perkins, M. A., Grzanka, P. R. & Hardin, E. A. (2019). Conceptual framework for non-disposable assignments: Inspiring implementation, innovation, and research. *Journal of Psychology Learning and Teaching, 18*(1), 84–97. https://doi.org/10.1177/1475725718811711.

Seraphin, S. & Stock, S. (2020). Non-disposable assignments for remote neuroscience laboratory teaching using analysis of human data. *The Journal of Undergraduate Neuroscience Education, 19*(1): A105–A112. https://pubmed.ncbi.nlm.nih.gov/33880097/.

Taczak, K. & Robertson, L. (2017). Metacognition and the reflective writing practitioner: An integrated knowledge approach. In P. Portanova, J. M. Rifenburg & D. Roen (Eds.), *Contemporary perspectives on cognition and writing* (pp. 211–29). The WAC Clearinghouse; University Press of Colorado. https://doi.org/10.37514/PER-B.2017.0032.2.11.

Terras, M., Ramsay, J. & Boyle, E. (2013). Learning and open educational resources: A psychological perspective. *E-Learning and Digital Media. 10.* 161. https://doi.org/10.2304/elea.2013.10.2.161.

U.S. Bureau of Labor Statistics. *May 2020 national occupational employment and wage estimates: United States.* https://www.bls.gov/oes/2020/may/oes_nat.htm.

U.S. Bureau of Labor Statistics. *May 2022 national occupational employment and wage estimates: United States.* https://www.bls.gov/oes/current/oes_nat.htm.

Zimmerman, B. J. (2002). Becoming a self-regulated learner. *Theory into Practice, 41*, 65–70. https://doi.org/10.1207/s15430421tip4102_2.

Exploring Ungrading in a Biochemistry Laboratory Course

Jennifer Newell-Caito
UNIVERSITY OF MAINE, ORONO

Imagine we have two first-year college students, A and B. Student A comes into an English course having spent a summer at a writing institute. All summer, they were learning and developing their skills in writing. Student B has spent the summer working to make sure they have money for books and housing in the fall. At the end of the semester, student A produces a paper that exceeds your expectations for a first-year student and would be considered upper-collegiate level. In comparison, student B has developed and worked on their writing skills to produce a quality paper, but still has areas for improvement. Who deserves the A? It can be easy to unintentionally marginalize students with less privilege than their peers, which is why it is important to assess student work with equity and consideration of the whole student. But how do we do that?

This chapter sets out to describe the pedagogical philosophy of "ungrading" proposed by Susan Blum (2020), but that builds on work by Alfie Kohn (1999) and others, which is a teaching style focused on removing grades from classrooms. Specifically, this chapter focuses on ungrading in a writing-focused junior-level undergraduate analytical biochemistry laboratory course at the University of Maine. I will begin this chapter by describing the background and inclusive strategies used in ungrading. Then, I will address how select strategies were employed in my biochemistry lab course and have become my standard approach in the course. I will finish with assessing the use of ungrading in my classroom using open-ended student self-reflections.

When thinking about ungrading, it is equally important to think about why we grade. What does a grade represent? What is it to give a grade or to be graded? The way higher education in the United States perceives grades is that they represent the instructor's evaluation of student work for the duration of the course (International Affairs Office, 2008). Grades are usually represented as letters (A, B, C, D, and F), numbers (0–100), or even a final grade point average (GPA). Ultimately, the intention of giving a grade or "grading" is the act of distilling all student work into a simplified representation (letter or number). It is hard to imagine that one letter or number could possibly encompass all of a student's work or growth during a semester or even, for that matter, on one assignment. James Felton and Peter Koper (2005) argue that grades are "inherently ambiguous evaluations of performance with no absolute connection to educational achievement" (p. 2). Ungrading sets out to look at different ways we can assess student work without using these traditional grading systems.

Why Grades are Not Effective for Assessing Student Learning

When I have asked students to reflect on how they have been traditionally graded in a classroom, many strong feelings arise. Students often reflect that they feel anger, anxiety, fear, and disgust. This is troubling since grades frequently guide educational pathways, as students are often motivated in their coursework by subjects they feel they are "good at." Yet, research has shown that grades are not useful tools for incentivizing students in a classroom. In fact, college students avoid challenging assignments (Milton et al., 1986), are dissuaded from learning (Butler & Nissan, 1986), and have reduced creative thinking on course content (Milton et al., 1986). I often hear from students that grades in high school motivated them to choose their majors in college. These grades are a deciding factor in the career path of students and the potential jobs that they are going to pursue later in life.

Grades continue to influence students in college, as those students who receive higher grades in first-year science, technology, engineering, and mathematics (STEM) courses are more likely to continue in STEM fields (Thompson, 2021). This fact is particularly important for women and ethnic minorities as they have lower persistence rates in STEM majors and often have lower GPAs after the first year (Cimpian et al., 2020; Griffith, 2010). It is not a surprise, then, that these students are underrepresented in STEM, as students will likely stay in a STEM major if their ratio of GPA in STEM courses is higher than non-STEM courses (Griffith, 2010). It is an institutional problem when women and ethnic minorities are dropping out of STEM courses at a faster rate than their white male counterparts (Suran, 2021; Thompson, 2021). STEM, and I argue any field, can only benefit from a diversity of perspectives and backgrounds. Grades can have a negative impact at every level of student learning where they are utilized.

In thinking about how grades are meant to work and how grades work operationally, I argue that there are five ways in which grading falls short in assessing learning in a course. In my experience teaching at the collegiate level, I find that:

1. Grades do not take into consideration the whole student. They don't reflect the knowledge a student brings with them into the classroom and how much they learn over the course of a semester. Consider the student example described in the beginning of the chapter: a holistic approach to education seems more equitable because it accounts for the growth of the

individual rather than relying on skills taught before students enter the classroom. Overall, grades are not always representative of the skills a student has gained over the course of the semester.

2. Grades alone do not provide any meaningful feedback for students. A grade does not tell a student what could be improved upon on an assignment or where they are doing well. In fact, students tell me that the first thing they do when they get an assignment back is to look at the grade and then file the paper in their notebooks. They completely dismiss the comments given or the ways they could improve their learning. The grade appears to supersede feedback and demoralize students. Students often reflect to me that they are an "A" student or a "B" student. They appear to categorize themselves as a grade rather than a learner capable of growth. The question is, if there is feedback on an assignment and all the student does is just look at the grade and not the feedback, then is putting a grade on an assignment even worth doing? For me, it is important to focus on learning as a collaborative dialogue between student and instructor and not on grades. This shift to focusing on feedback as a tool for learning, rather than adding a grade, helps me to shift student mindsets to be more learning-focused.

3. Grades are not necessarily directly linked to our student learning outcomes. As part of our syllabi, we list carefully crafted student learning outcomes and student learning goals. Instructors often use two modes of assessing students on these outcomes: summative (cumulative) and formative (any feedback on improvement) assessments. If an assessment is linked with the learning goal (as we hope it is), there are several questions we can ask. Does giving a grade on that assessment help the student improve and meet the learning goal? When a student receives a grade on the assignment, does that give clarification on a sticking point? Would students be less motivated to improve if you left the grade off and just gave feedback? If I tell you that a student got a "B" on an assignment or learning outcome, does that tell you anything about a skill or knowledge that a student has developed? Most often, the answer to these questions is "No." I would argue that grades do not help guide learning as we may intend; it is the feedback and the growth from that feedback that is connected to our learning outcomes.

4. Grading can demoralize instructors. It severely underappreciates the amount of effort it takes to effectively give students feedback on completed work. Giving a grade requires that the instructor effectively describes expectations for student work, how those expectations align with course objectives, how the instructor will assess effort and learning based on the skill set of the student, and how the instructor will give effective feedback that will lead to student learning. Grading can turn the course culture from one focused on

learning into one focused on competition. Not only is there competition amongst students for the best grades, but there is competition between the teacher and the student for getting a higher grade. The focus of the class becomes on the grade and not on the learning outcomes. Moreover, there is often a deep mistrust between teacher and student. The student may not feel as though the teacher has their best interest at heart. Conversely, the teacher may not feel as though the student is putting their best effort forward and is constantly worried about ways to inhibit cheating.

5. Grading doesn't create a positive culture in our courses as it does not incorporate the whole welfare of the student. It doesn't take into consideration their background and mental health. College student mental health issues have doubled in the past ten years and are especially a problem in ethnic minority students (Colarossi, 2022). Grades do not encourage students to be comfortable in a classroom and are anxiety-inducing. As mentioned above, the culture is not one focused on mental health but on competition. Grades pit students against one another and do not provide students with a safe learning environment where they can take risks, make mistakes, and learn from those mistakes.

Alternative Methods in Ungrading

If the traditional grading scheme has negative impacts on student learning, can alternative methods be used to give positive impacts? Ungrading is the use of alternative methods to remove the focus on grades and switch the focus to learning. Changing the way we educate from traditional methods, as seen with Madison Brown's vignette (this collection), moves education to incorporating many modes of learning and supporting a variety of students in the classroom. If we can support women and ethnic minorities in STEM, we can create a space that supports and rewards creativity and learning rather than focusing on "correct" solutions. In the ungrading approach, students don't have to be perfect to be successful. Students can learn through mistakes and feel pride in their work and in their learning. Changing the approaches for assessment moves the classroom conversation from grades to feedback. Similar to what Janelle Johnson et al. describe (this collection), the ungrading approach seeks to avoid the "weed-out" approach and create a classroom that celebrates diversity and considers a more holistic approach to education. STEM classrooms typically have traditional formats, which often ignore other modes of assessment. The methods listed below, initially described by Jessie Stommel (2020), were those implemented in my classroom and could be used in any STEM undergraduate classroom to make the assessment process more transparent to students. When students are included in the conversation of grading, they are more likely to feel like it is simple and fair. Not all of these methods will fit in every classroom, nor should they, but educators should choose the methods that work best with their teaching style. Additionally, the list below is not exhaustive, and instructors shouldn't limit themselves; they could create assessment strategies that work for them in their classrooms. The best relationships I have cultivated with my students have been when I am authentically myself in the classroom and don't pretend to be someone I am not. For example, I am naturally introverted, so I am not going to be a loud, joking personality type in my classroom. Additionally, I utilize many teaching practices that allow students to engage in self-reflection and anonymous course engagement. For example, using clicker questions and using think-pair-share to answer questions or reflect on learning instead of raising hands gives other introverted people a way to participate in the course other than directly asking questions. Furthermore, creating a classroom community is important to me, so focusing on relationship-rich teaching (Felten & Lambert, 2020) and pedagogy of kindness by Cate Daniel (2019) resonates with me as an instructor.

Minimal Grading

Minimal grading is using scales with fewer gradations. There are several methods that can be used, including strong/satisfactory/weak [three gradations], pass/fail [two gradations], +/- [two gradations], and turned in/not turned in [one gradation]. This accomplishes clarity in the classroom. First, there is wide variability between instructors grading the same work (Meadows & Billington, 2005; Schinske & Tanner, 2014). Simplifying the grading scheme can produce more consistent results between instructors. Second, it can be hard for students to understand how they performed and what they need to improve upon with number grades. Lastly, this approach focuses students on the learning rather than grades, as students will look at the feedback rather than the grade itself. This is especially powerful if an instructor allows students to resubmit work in combination with the use of minimal grading.

Grade-Free Zones

A zone is a defined period of time in a course. There are many different types of grade-free zones that can be implemented in a college course. An instructor can give grades on just a few assignments or not grade for two or three weeks, or it could be more extensive where students would not be graded for a third or half of a semester. It is up to the instructor to decide the length of time that grading will not occur and how it fits into the semester. This approach may seem a little perplexing to conceive, but the time that students are not graded could be spent simply letting them engage in course content before moving on to more traditional assignments. This time is often spent giving feedback and not grades.

Contract Grading

In contract grading (see Mallette, this collection), the course outlines in the beginning exactly what students need to do to earn specific grades. There are concrete criteria given for each grade a student could potentially achieve. Students can work toward whichever grade they would like to achieve based on the work they complete. The advantage for students is in the clarity of the expectations: there will be no additional work added or sudden removal of work from the grading scheme. If this approach is combined with flexibility where students can resubmit the work until they can get a satisfactory grade, it focuses the class on the quality of the work completed. There is less conflict over grades and less competition between students for the best grade: everyone can work toward their own individual goal.

Authentic Assessment

In this strategy, students apply course content to their real-life communities. The definition of community could be broad or narrow, as it could be for the town/city, college, or classroom community. The most important aspect is involving students in designing an assessment that conveys information to a real audience. Not only does this involve students in the decision-making of the course (course buy-in), but these types of assignments are important to students' sense of identity. Every person has multiple identities based on differences that include, but are not limited to, socioeconomic status, age, gender, religion, race, and sexual orientation. Research has shown that creating a classroom where students can celebrate their identity can directly improve student motivation and learning (Lowe, 2019). The expression of identity in a classroom is important for all students, but it is especially important for helping low-income, first-generation, and racial/ethnic minorities (Harackiewicz & Priniski, 2018). Students who express their identity in the classroom have an increase in student persistence through tough course material and continued participation in STEM courses (Murphy & Destin, 2016; Gurin et al., 2002; Dewsbury & Brame, 2019).

Self-Assessment

Self-assessment utilizes the approach of metacognition, or thinking about learning. This is a cross-disciplinary approach that focuses on student awareness of their problem-solving skills, ability to judge how well they understand course material, and understanding their level of learning as the course progresses. As a part of the self-assessment, a growth mindset, or the idea that learning ability is not fixed, can be explored. Exploring growth mindset in the classroom has been found to especially benefit women and underrepresented minorities in math and science (Rattan et al.,

2015; Kricorian et al., 2020). For students coming into a course with insecurities in course content, it is important to instill in our students that they can improve their comprehension of course content with practice and time. Many studies have shown that students have increased learning gains when completing self-assessments (Andrade, 2019). More specifically, Heidi Andrade (2019) describes many benefits, including helping students take responsibility for their learning, development of critical thinking skills, and the ability to set achievable goals for a course. Self-assessment is a powerful tool that puts the ownership of learning back onto the student.

Process Letters

This strategy asks students to reflect on their learning and the work they have completed over the course of the semester. In these reflections, students detail, with examples, what grade they should receive. This typically takes the form of an essay or formal letter. This approach focuses on student reflection on the learning that has occurred over the course of the semester and creates a space for persuasive writing. Usually, there is a meeting with the professor to discuss the process letter and decide together on a final grade in the course. Students can feel empowered in the classroom by being able to take an active role in deciding their own grade.

Background on the Course

The Course

The current form of the Analytical and Preparative Biochemical Laboratory is a course-based undergraduate research experience (CURE). This is an upper-level biochemistry lab for juniors at the University of Maine and is required for all the majors in the Department of Molecular and Biomedical Sciences. The goal for the class is to do original research by answering a research question with no known outcome. To conduct research, we have a two-hour lecture that is discussion-based and a four-hour lab per week with two sections of the course. The purpose of the course is to purify a known enzyme from a new organism. This is novel research for which there are no protocols or data. The class must work together as a group to develop assays (or experiments) for expressing the enzyme, detecting the enzyme, creating the protein purification procedure for the enzyme, and characterizing enzyme function. This course, where faculty and students work together to research and create knowledge, reflects the critical pedagogy described by Ann Fink (this collection).

An assignment is given prior to each lecture period where students research how an assigned assay works (the chemistry behind it) and bring a protocol to the lecture class for the assay from the primary literature. We discuss the background as a class, where everyone contributes to the discussion. Since we are designing our own experiments, this is a calculation-heavy course. As such, every class period, students break up into designated lab groups to work on practice problems related to data analysis they will encounter that week. After we have completed calculations, students are then given a loose experimental protocol to aid in protocol refinement. They work together in groups of two or three to complete the protocol and calculations to prepare them for their laboratory session. Often, students must meet outside of class time to finalize the procedure for the experiment in the week ahead.

During lab, students carry out the experimental protocols they developed and analyze the data. Sometimes, students will not have enough time to analyze data and must do this before the next lecture period. It is important that even if the student does not feel like an expert in the data analysis process, they try it on their own. I emphasize that they will learn a lot through mistakes or incorrect analysis, and they will not be graded on correctness, just completion.

The next week in lecture, we work in a group to go over data analysis from the prior week. This approach gives students the ability to make changes and corrections to their data analysis. We focus on learning through making mistakes. Also, students can analyze data in different ways: there is often not a yes/no (black/white) answer to the analysis, but there is gray area where we discuss different approaches used in the field. In research, we often stay in the gray area until we get more data that makes the path clearer. Working as a group, we try to reach some general consensus on the data analysis, but there is often not one correct way to approach the problem.

The major assessments in the laboratory are notebook checks and a final manuscript. The manuscript (described later) contains publishable quality figures and includes the traditional format of abstract, introduction, methods, results, and discussion with properly formatted references. While challenging, the manuscript represents a deep analysis and understanding of the context of the student's work in the larger scientific community.

Motivation for Ungrading

I have taught this course traditionally graded for three years, but the COVID-19 pandemic was announced, and I needed to pivot my laboratory course to an online experience. This caused me to completely switch my assessment strategies in the course. Coming out of the pandemic, I wanted to keep the changes I made because I saw decreased anxiety and increased performance on assessments. Then I read the ungrading book (Blum, 2020), and I knew that I needed to take the next step toward being a more inclusive course by implementing this teaching philosophy. The catalyst to use ungrading in my classroom has, and always will be, my students. For example, in my spring 2022 analytical biochemistry laboratory, a student said to me: "My entire life I have tried to learn in an environment that seems like it was set up for me to fail." I wish this sentiment was the only time I had heard this type of comment, but I have increasingly seen an uptick of students with anxiety, depression, ADHD, and those with a variety of classroom accommodations. The feedback I get from these students is that they struggle mastering course content and managing workloads. As with most of us in the teaching profession, I want my classroom to be a supportive environment where my students can succeed regardless of their background or current life experiences. Unfortunately, many students feel like they are trying to learn in environments that are not geared toward their success. As Ann Fink describes (this collection), the COVID-19 pandemic upheaved our lives but also allowed us to upheave the way we approached education. I have always been willing to try new approaches in my classroom that can benefit students, but the COVID-19 pandemic certainly motivated me/let me grant myself permission to radically change the way I teach. It led me to think holistically of my students' needs and make sure they were included in classroom and grading decisions.

Implementation of Ungrading

One of the guiding principles of ungrading is to engage students in their learning and make them the conductors of their learning train. In my classroom, I wanted my ungrading journey to focus on flexibility, self-assessment, authentic assessment, and direct student involvement in the grading process. In what follows, I describe how I incorporated each one of those changes in the course.

Flexibility

Students' lives (as our own) can be very complicated with many moving parts. Rigid deadlines and a lack of flexibility in turning in assignments can impact student learning and feelings of success (Yoo, 2015). The goal with this course was to switch from a grade-focused course to a learning-focused course. To assist in this approach, instead of allowing students to turn in assignments once, each assignment and set of data analysis can be turned in as many times as needed to obtain full credit on the assignment. This includes the notebook checks and the final manuscript. A minimal grading system is used where students are given three levels of grades: 50 percent (weak), 75 percent (satisfactory), and 100 percent (strong) on assignments. Feedback is given within our learning management system (LMS) to allow students to make changes to those assignments.

In the lab notebook assignment, students analyze their data to the best of their ability, and then during our lecture time, we discuss the results. This allows all students to weigh in on the interpretation of the data and make corrections as a group. Individualized feedback is provided to students through the LMS. If the feedback is not sufficient or students need more help, then they are able to meet with the instructor to get additional assistance. Toward the end of class, there are three to four weeks where students are not receiving grades but focusing instead on the generation and analysis of data.

The manuscript was a large undertaking since it was modeled on a journal article, including creating publishable quality figures (well-communicated, correctly formatted, with high-resolution). In terms of teaching how to write a manuscript, the assignment was broken down into two parts: the figures and the text. Every week, I would teach about the multi-step process of making publishable quality figures using professional software, and students would practice using the software to create figures with their data. Again, feedback was given through the LMS. Teaching about writing the text of a manuscript was broken down into sections: abstract, introduction, materials and methods, discussion, and overall specific journal formatting. Students were able to work on the first draft of the manuscript at their own pace, making individual appointments if they required immediate feedback. After completion of the first draft, feedback was given, and changes could be made until the due date. This approach allowed students to see that STEM as a whole, but specifically data analysis, making publishable figures, and writing a manuscript, are all iterative processes. The approach was used to demonstrate to students that revision is normal in science: despite our best attempts, perfection is rarely achieved the first time we try something new.

Another way I incorporated flexibility was by moving deadlines for students. Throughout the semester, I got to know my students and understand the complexity of their lives. I moved deadlines around for students who had significant personal struggles since I knew that other faculty at the university would not likely be as flexible. For example, I had a student with a concussion, and once the two-week period passed for healing (doctor's allowance), other professors made them turn in all of the missing work. This required the student to continue to work through the two weeks "off" even though their brain was still healing. As soon as I knew about the concussion, I told the student that I would be flexible with them on deadlines. They were very reluctant to move deadlines because they didn't want it to appear as though they didn't care about the class and wanted to appear "normal" (their words, not mine). I made sure the student knew they were going to be supported and could learn the material at their own pace so that they didn't feel so anxious about coursework. Being flexible on deadlines not only allowed the student to heal properly but also allowed for this student to feel less anxious overall because they knew they could get all the work done. If I had not moved deadlines, the amount of work and strict deadlines that other courses required would mean that this student would have been completing the work but not focusing on learning content in my course. At the end of the semester, this student was incredibly grateful for this approach, but more importantly, very successful in learning course content as gauged by the quality of the final manuscript.

Self-Assessment

In order to pivot the course to student learning, I give students four self-reflection assignments to analyze their growth mindset, metacognition, overall learning, and group dynamics (Appendix). Each self-reflection starts with a metacognition awareness inventory which consists of several metacognition-oriented questions and then gives students the choice on a Likert scale. Additionally, each self-assessment asks for comments on concrete skills students could develop during the course. The reason I focus on skill sets is due to student feedback saying, "I am not going to be a protein biochemist. Why do I need this course?" Every year the National Association of Colleges and Employers surveys employers across the country for qualities they are looking for in college graduates (Koncz & Gray, 2022). I list the top ten skills and ask students to reflect on which skills they want to develop. As the semester progresses, I ask students to reflect on the skills they have developed in the course.

Other than the consistent questions and themes discussed above, the reflections often change in content throughout the semester. In the first reflection (first week of classes), students are asked open-ended questions on what knowledge and strengths they are bringing into the course and some weaknesses they would like help working on during the semester. The second assessment, given in week five, asks students about the hardest concept to master in the course so far and where they have received help on that concept. It also asks them to consider strategies that would improve their learning and one course norm they would change. In the third reflection in week ten, students are asked similar questions to the second assessment but also to comment on their progress so far in the course and think about assessing their grade in the course with evidence. This approach helps prime them for their last assessment in week fifteen, where they are asked about the structure of the course, struggles and successes, and, more importantly, where they write their process letter.

One crucial part of every self-assessment is an open-ended question where students can communicate to me any issue regarding their learning. Having a self-assessment where there is open dialogue between the instructor and students is essential. I have the opportunity and power to change the course based on student opinions, and this approach celebrates the critical pedagogy described by Fink (this collection). This creates an opportunity for discussion directly with me and lets students know that their opinions and perspectives are being valued. The feedback is often that students feel empowered when course norms change, and they feel like they are included in course decisions.

Authentic Assessment

To incorporate authentic assessment into my course, I give students a creative project. The authentic assessment described here is similar to Johnson et al.'s "Call to Action: Cultivating Activism Among Teacher Candidates" project described in this collection, as it was created to include and celebrate student identity and allow for flexibility in assignments. I also want this to be a student-driven assignment, and I focus the creative project around the theme of scientific communication. The communication of science from scientists to non-scientists is essential for both the advancement of science, as well as for human health. As a scientist, it is imperative that students are able to understand and explain primary scientific research. In this assignment, students have the opportunity to create their own project and rubric for peer grading centered around this theme. The overall purpose of the assignment is for students to have a direct contribution to the course in a way that celebrates their individuality and perspectives.

In this course, there are a variety of projects submitted; some people present pieces of art using various mediums including embroidery, digital art, acrylic painting, or charcoal/pencil on paper. Also, in the artistic category, students have created comic strips, children's books, and board games to convey scientific information. Other students have opted for a more traditional science approach with a five-minute lightning talk on a scientific topic of their choice or a poster advertisement. The breadth of the projects has been vast, but the personal connection to the material has been clearly evident through student feedback. One student remarked: "The creative project was so much fun to do!! It was a good break from normal work and made me think and do something I enjoy in my free time."

Process Letters

At the end of the semester, students have an assignment to write a process letter to determine their grade for the course, which is their final self-reflection. Students are given a detailed list of grading criteria at the beginning of the semester so that they know what they have to do to receive an "A," "B," or "C" as a grade. There is no option for a "D" or an "F" as these grades reflect that there is no meaningful learning taking place, and that isn't something that is acceptable in the course. In the self-reflections, there have been students who described grades that were not consistent with the posted criteria for that grade. Moreover, there have been students who did not engage properly in the course. As a result, I hold individual meetings throughout the semester (the more often, the better) to discuss how they are not meeting course expectations. I explain in the grading criteria that my expectations for receiving an "A" are high but that I am on their side and am not trying to trick them into getting anything less than what they feel they deserve. In addition, to receive an "A," students don't have to fulfill all the criteria, but they do need to fulfill most. The process letter reminds them of the grading criteria, but they are able to find and argue for other criteria that allow them to demonstrate their learning. Overall, in their process letter, students are asked to use concrete examples to show their understanding of the biochemistry content, how they engaged with the course, and how those correlated with grading criteria.

Assessing Success of Ungrading

It is incredibly nerve-racking to make large changes in a course, especially without guaranteeing they will result in increased learning gains for students. In summary, the ungrading experiment in my biochemistry laboratory course has been a success. After reflecting on my use of ungrading in this course, the major themes that emerged from the analysis included student trust building, appreciation of flexibility and feedback, increase in confidence, and gratitude for the ungrading approach.

In terms of building relationships with my students, I experienced more meaningful connections than in any prior time I have taught the course. Here is an example from a student:

> Yes, I confidently believe she does care [about my learning]. I think out of all the professors I have had she cares the most, which is so refreshing to have since she is very nurturing. I feel comfortable talking to her about my problems and ask for help, which I rarely do out of discomfort.

Students welcomed me into their lives and trusted me with their insecurities and struggles in STEM. Students were more comfortable focusing on learning course content and also healed some emotional wounds from interactions with previous instructors. One student commented that they "absolutely believe everyone involved with the course cares deeply about my learning of the material and not just assigning me a grade, which I can say is refreshing compared to other classes I have taken." Not only did students interact with me in more positive ways but with one another as well. This was especially evident with group work:

> I have noticed communicating with my lab partners and doing additional research has been excessively helpful to my learning. It helps me feel more comfortable in the classroom. I also really like the environment the TA's and the professor create and the kindness they show. It makes me feel more relaxed, which in turn makes the class more enjoyable for me, so I have noticed that I am doing better.

As the instructor, I have noticed more camaraderie, connection, and eagerness to interact with one another over the course of the semester.

Incorporation of flexibility into the course was a major goal in my ungrading approach. In an analysis of student reflections, students perceived my flexibility as caring more about their learning than their grade:

> I really enjoyed this semester. One of the best parts was how obvious it was that all the instructors [sic] are TRULY passionate about teaching and helping improve our learning experience. I felt completely comfortable asking for help and not knowing the answer 100% of the time. I could tell you all LOVE the topics in BMB 464 and really enjoy teaching and helping us to appreciate analytical biochemistry! I'm grateful for how pleasant the course experience was! Thanks to all for a great semester!

I have found that students consistently encounter obstacles during the semester that are outside their control. Being flexible when other courses were not allowed students to recover from these events. Moreover, students were more focused on learning than their grade in the course. As a student commented:

> The flexibility and level of understanding you have shown has actually allowed me to learn the material and complete the assignments with my best effort, rather than to turn in assignments just to check them off the list and get a grade. So, thank you again for all of the help and for being so understanding throughout the semester because it really had made such a difference in my semester and with all of my classes.

Overall, I believe that being flexible allowed students to capitalize on their strengths and work on their weaknesses.

Confidence was another theme that presented itself during the analysis. During prior iterations of the course, students were very anxious about their grades/performance. My perception of student anxiety over grades was less in this ungraded course. Students loved the design of the course:

> I could not effectively perform work due to the types of tests and assignments provided. I went from being very depressed (. . .) to enjoying and getting to know my professor and class. This class did not focus on tests but learning and developing confidence with the work. I learned that I am, in fact, prepared for a career and will do well in whatever career I chose.

This connection to themselves and the course could be a result of the ungrading approach but could also be, in part, due to the focus on metacognition (understanding themselves as learners) in the self-reflections:

> As a student and learner, I discovered that I am in a position to positively affect others. The conversations I have with my instructors has shown me that learning is an ongoing process, which only verified something that I believed. My instructors were honest with me and helped guide me down a pathway of growth. My peers taught me that I can (. . .) help improve their understanding on the material or I could learn from them.

Either way, the overwhelming feeling was of personal growth and confidence:

> I feel like I am suited for the major and this field. Before taking this class I was really lacking confidence and was second-guessing my decision to go into the field of biomedical sciences, but now I feel a lot more confident. The entire lab was amazing and it is a lab I would retake in a heartbeat if I could. I really think this is the way labs should be run because we actually are learning and I think students would be a lot more successful and want to go to the lab if more courses were taught like this. I wish our department had more courses like this.

Students appreciated the design of the course and appeared to gain confidence as biomedical scientists.

When I read their process letters and self-assessments, I was surprised to see that my students felt the same way that I did about the success of our course. With the ungrading approach taken, students still wanted to learn and seemed to want to learn more enthusiastically. During the course, they were able to focus on learning rather than grades:

> Asides from giving me the freedom to not stress about what my grade will be, it also gave me the option to make mistakes and try new things and learn from them. I was not afraid to get some questions wrong on my assignments or ask for help because I know that they will be learning moments and not a penalty to my grade. I know that I freeze up sometimes because I have the need to do everything perfectly and then, I get anxiety from that and so I don't even make the first step. Ungrading helped me in the sense that it slowly brought down my walls and had me not worry about messing up but instead put myself outside my comfort zone and helped me learn.

Even my strongest students felt like they had changed their approach to learning:

> I feel that my work ethic has actually increased—as a type A person, I honestly hadn't thought that was possible. I feel like my

> approach to work is more balanced at the same time—while I have been putting in more effort, it has also been more efficient. I have really enjoyed being able to not worry about grades and just stick to learning the material, which has been a relaxing change from the norm.

Students were also incredibly observant that the course was focused on personal growth. One student gave advice to students taking this ungraded course:

> If unsure about how to answer a question or analyze a data set: start by doing what you can. This will tell you what you truly do or don't understand. Don't give up! Give your best effort, and don't be afraid to speak up when you don't understand something and just ask for help!

In my observations, students were kinder to themselves by letting themselves make mistakes and then learning from them.

Challenges of Ungrading

I encountered some challenges with the ungrading process that were both expected and unexpected. One expected challenge was that since the course was focused on feedback rather than grades, there was increased feedback on assignments compared to past years. This resulted in more time spent providing written notes to students from both the graduate teaching assistant and me. Another challenge was preparation for the course in the form of making metacognition surveys, creating the grading rubric for the course, and designing the process letter criteria. Regardless of the preparation time I spent on the course, I expect each year will capitalize on preparations made in past years. For example, I was able to copy and paste comments from feedback given on assignments into a Word/Google document. I should be able to use many of those comments going forward. I also plan on re-using in-class problem sets and data analysis templates. These documents should speed up the preparation process and assignment feedback in the future.

Another challenge that I expected was that some students would not show up and/or complete the work. My approach was to meet with these students individually and learn why they hadn't participated to the level of the expectations of the course. Unsurprisingly, discussions with students often uncovered complicated challenges outside my control. I encouraged those students to engage with the course and helped them make a plan for makeup work and course completion. In one case, a student was very far behind, but what had been completed was excellent. We ended up settling on a grade that took into consideration how much work was completed and how much learning had occurred based on the process letter rubric.

When it came to self-assigning grades, I expected all students to give themselves an "A." To my surprise, they didn't. Maybe it was a result of the high expectations and clear goals of the course, but students were honest in their reflections. Some students were harder on themselves than I would be, and others were more generous than I would have been (e.g., A versus B). I scheduled individual meetings with both types of students (higher or lower than expected) to discuss their overall growth, and together, we settled on a grade.

The biggest challenge to ungrading that I have encountered, which also happened to be an unexpected surprise, challenging the mindset of students towards grading. It took a lot of effort to convince students to trust me and to focus on learning instead of grades. For example, one of my students said that he had been graded since middle school and didn't know another way to think about learning. I had to continually repeat that if they focused on learning, the grade would follow. I also had to reiterate that the onus for learning was on them: the effort they put into the course would be reflected in their learning and their final grade. I eventually won most of them over to ungrading, but it surprised me that the ones most resistant to ungrading were my top performers. They were worried about grade inflation and that everyone would get an A. This simply was not the case. In this unique grading process, I learned to trust my students, and I believe I earned their trust as well.

In summary, I came away from my ungraded course with the knowledge that my students really love learning. Also, they wanted me to be a part of that journey. In the end, I had a classroom that was built on trust, appreciation, and student-teacher collaboration. One student remarked:

> You [instructors] are amazing people, and I genuinely don't know if I can encapsulate my gratitude to you in words. I am so thankful that I took this class and even when I was lost or had no idea what was going on, I could count on you [instructors] to always help me though! This class has been very transformative in how I learn and perceive myself and what I am capable of, and it is thanks to the amazing people I had for my classmates, my TAs, and my instructor! My only complaint now that it is the last week of the semester, is that it ended too soon.

I will continue to ungrade in this course and try more ungrading approaches in all my courses.

Institutional Changes in Ungrading

Since teaching this course as ungraded, I have built relationships with my peers in the department and at the institution, surrounding the positive impacts of this work in my classroom. The first way I have built relationships is with another instructor in the Department of Molecular and Biomedical Sciences. We both read the ungrading book (Blum, 2020) at the same time over the summer and were so inspired that we implemented different ungrading approaches in our courses the very next semester. Since then, we have talked about our successes and challenges in ungrading. We have formed a small community where we help one another develop our courses, troubleshoot problems, and strengthen our program. Excitedly, we both have expanded this ungrading approach to other courses that we teach and have been invited to talk about our teaching pedagogy during our departmental meetings. Perhaps as we discuss the success of our classes, we can normalize the perception of ungrading approaches and inclusive teaching.

The second way I have built relationships is with participation in several communities of practice that are offered through the Center for Innovation in Teaching and Learning at the University of Maine. Through interactions with faculty there, I have developed a pedagogical research project on ungrading. Additionally, I am applying this ungrading philosophy in an internally funded institutional grant focused on first-year undergraduate retention.

The last way I have built relationships is by talking with faculty in other departments. One example is that I presented a workshop on ungrading at a Maine Center for Research in STEM Education Conference. This book chapter is a direct result of giving that workshop. Further conversations about ungrading led to another STEM major at the University of Maine considering this course for incorporation into their degree path. Overall, it has not just been the interactions with my students that have been overwhelmingly positive and life-enriching, but also the interactions with my peers. I have simply no regrets about the incorporation of the ungrading philosophy into my life and my courses.

Small Changes, Big Impacts in Ungrading

While the methods employed above were major changes to a course, there are many small steps that anyone can make to move a classroom to one focused on learning rather than grading. One change would be to grade less often using grade-free zones. If there is a way to simplify or remove some grading, this could be an easy way to make a course modification. A second change would be to involve students in the discussion of course expectations and grading. This approach gives a voice and some control over the course to students. It will empower them. A third change would be to have students complete self-reflections. Remember this method increases learning outcomes for all students. This could be as simple as asking students about their learning over the course of the semester or as complex as using validated methods for measuring satisfaction and self-confidence (Bray et al., 2020). A fourth change would be adding flexibility to a course. Some suggestions for being flexible include giving students two options on an assignment, making flexible deadlines on an assignment, or engaging with concepts in multiple ways. Some examples of choices could be allowing students to work alone or in groups, letting students watch videos or read transcripts, and, last, having students complete a writing assignment or a presentation. This choice is engaging and encourages student course buy-in. A fifth change would be to listen to and trust students when they are facing conflicts in their lives. Having an open and safe relationship between the instructor and students will help everyone feel comfortable learning in the course.

My hope is that these changes are seen as manageable and can be included in any STEM course. However, it is important to remember that not all of these above-mentioned changes need to be implemented at one time. Small, meaningful steps to incorporate ungrading can make big impacts in any classroom. Everyone can ungrade in their own way, using their own timeline.

Acknowledgments

The student data used in this book chapter has University of Maine IRB approval (#2023–01–03) under the title "Efficacy of Ungrading in Course-Based Undergraduate Research Experiences on Belonging, Self-Directed Learning, and Growth Mindset." Permission has been granted for quotes gathered outside this IRB. I would like to thank all of the amazing educators who have supported me in my teaching journey and to all my students who taught me to be a better educator.

References

Andrade, H. L. (2019). A critical review of research on student self-assessment. *Frontiers in Education*, *4*(87). https://doi.org/10.3389/feduc.2019.00087.

Blum, S. D. (2020). Why ungrade? Why grade? In S. D. Blum & A. Kohn (Eds.), *Ungrading: Why rating students undermines learning (and what to do instead)* (pp.1–21). West Virginia University Press.

Bray, A., Byrne, P. & O'Kelly, M. (2020) A short instrument for measuring students' confidence with 'key skills' (SICKS): Development, validation, and initial results. *Thinking Skills and Creativity*, *37*(June), 100700. https://doi.org/10.1016/j.tsc.2020.100700.

Butler, R. & Nisan, M. (1986). Effects of no feedback, task-related comments, and grades on intrinsic motivation and performance. *Journal of Educational Psychology*, *78*(3), 210–216. https://doi.org/10.1037/0022-0663.78.3.210.

Cimpian, J. R., Kim, T. H. & McDermott, Z. T. (2020). Understanding persistent gender gaps in STEM. *Science*, *368*(6497), 1317–1319. https://doi.org/10.1126/science.aba7377.

Colarossi, J., (2022, April 21). *Mental health of college students is getting worse.* Boston University: The Brink. https://tinyurl.com/bdkpzpb3.

Daniel, C. (2019). *A Pedagogy of Kindness.* Hybrid Pedagogy. https://tinyurl.com/48hzpnx9.

Dewsbury, B. & Brame, C. J. (2019). Inclusive teaching. *CBE-Life Sciences Education Review, 18*(2), 1–5. https://doi.org/10.1187/cbe.19-01-0021.

Felten, P. & Lambert, L. M. (2020). *Relationship-rich education: How human connections drive success in college.* Johns Hopkins University Press.

Felton, J. & Koper, P. T. (2005). Nominal GPA and real GPA: A simple adjustment that compensates for grade inflation. *Assessment and Evaluation in Higher Education, 30*(6), 561–569. https://doi.org/10.1080/02602930500260571.

Griffith, A. L. (2010). Persistence of women and minorities in STEM field majors: Is it the school that matters? *Economics of Education Review, 29*(6), 911–922. https://doi.org/10.1016/j.econedurev.2010.06.010.

Gurin, P., Dey, E. L., Hurtado, S. & Gurin, G. (2002). Diversity and higher education: Theory and impact on educational outcomes. *Harvard Educational Review, 72*(3), 330–366. https://doi.org/10.17763/haer.72.3.01151786u134n051.

Harackiewicz, J. M. & Priniski S. J (2018). Improving student outcomes in higher education: The science of targeted intervention. *Annual Review of Psychology, 69(1),* 409–435. https://doi.org/10.1146/annurev-psych-122216-011725.

International Affairs Office (2008). *Structure of the U.S. education system: U.S. grading system*s. U.S. Department of Education. https://tinyurl.com/ydm4w37a.

Kohn, A. (1999). From degrading to de-grading. *High School Magazine*, 38–43.

Koncz, A. & Gray, K. (2022). *The attributes employers want to see on college students' resumes.* National Association of Colleges and Employers. https://www.naceweb.org/about-us/press/the-key-attributes-employers-are-looking-for-on-graduates-resumes.

Kricorian, K., Seu, M., Lopez, D., Ureta, E. & Ozlem, E. (2020). Factors influencing participation of underrepresented students in STEM fields: Matched mentors and mindsets. *International Journal of STEM Education, 7*(16). https://doi.org/10.1186/s40594-020-00219-2.

Lowe, A. N. (2020). Identity safety and its importance for academic success. In R. Papa (Ed.), *Handbook on promoting social justice in education.* Springer. https://doi.org/10.1007/978-3-319-74078-2_128-1.

Meadows, M. & L. Billington. (2005). *A review of the literature on marking reliability.* [Unpublished AQA report produced for the National Assessment Agency]. https://filestore.aqa.org.uk/content/research/CERP_RP_MM_01052005.pdf.

Milton, O., Pollio, H. R. & Eison, J. A. (1986). *Making sense of college grades.* San Francisco: Jossey-Bass.

Murphy, M. & Destin, M. (2016). *Promoting inclusion and identity safety to support college success.* The Century Foundation. https://tcf.org/content/report/promoting-inclusion-identity-safety-support-college-success/?agreed=1.

Rattan, A., Savani, K., Chugh, D. & Dweck, C. S. (2015). Leveraging mindsets to promote academic achievement: Policy recommendations. *Perspectives on Psychological Science, 10*(6), 721–726. https://doi.org/10.1177/1745691615599383.

Schinske, J. & Tanner, K. (2014) Teaching more by grading less (or differently). *CBE Life Sciences Education, 13*(2), 159–166. https://doi.org/10.1187/cbe.cbe-14-03-0054.

Stommel, J. (2020). How to ungrade. In S. D. Blum & A. Kohn, (Eds.), *Ungrading: Why rating students undermines learning (and what to do instead)* (pp. 25–41). West Virginia University Press.

Suran, M. (2021). Keeping black students in STEM. *Proceedings of the National Academy of Sciences, 118*(23), e2108401118. https://doi.org/10.1073/pnas.2108401118.

Thompson, M. (2021). *Who's getting pulled in weed-out courses for STEM majors?* Brookings. https://tinyurl.com/523embss.

Yoo, J. H., Schallert, D.L. & Svinicki, M.D. (2015). The meaning of flexibility in teaching: Views from college students and exemplary college instructors. *Journal on Excellence in College Teaching, 26*(3), 191–217.

Appendix: BMB464 Self-Reflection #1

Metacognition is thinking about the way you think and learn. It is a very important strategy for success in college. It is so important that I want this to be a weekly habit for you throughout your college career. Please take your time and answer the questions thoughtfully and truthfully. You are not graded on correctness, just honesty.

1. Please check the box that best describes you.

Metacognition Awareness Inventory*

	I NEVER do this	I do this INFREQUENTLY	I do this INCONSISTENTLY	I do this FREQUENTLY	I ALWAYS do this
I ask myself periodically if I am meeting my goals.					
I consider several alternatives to a problem before I answer.					
I try to use strategies that have worked in the past.					
I pace myself while learning in order to have enough time to learn the material.					
I understand my intellectual strengths and weaknesses.					

*Questions selected from Gregory, S.; Sperling D.R. (1994) Assessing Metacognition Awareness. *Contemporary Educational Psychology. 19*(4), 460–475.

2. Please check the box that best describes your opinion.**

	Strongly Agree	Agree	Neither Agree or Disagree	Disagree	Strongly Disagree
You have a certain amount of science ability and you can't do much to change it.					
Memorizing formulas will make you a good scientist.					
You can greatly change your ability to do science.					
Practice exercises are the best way to learn science.					
Watching a teacher do examples is the best way to learn science.					
Trying a problem I don't know how to solve is the best way to learn science.					
Teaching someone how to solve a problem is a good way to learn science.					
Knowing why an answer is right is just as important as how to find it.					
Being able to build protocols from literature will be important in my future.					
Being an independent researcher will be important in my future.					
Being able to solve complex problems will be important in my future.					

**Questions adapted from a Growth Mindset Survey by Dweck, C.S. (1999)* Self-theories: Their role in motivation, personality, and development. Psychology Press. *and Dweck, C.S. (2006)* Mindset: The new psychology of success. Random House.

3. What do you already know about biochemistry and research that could guide your learning this semester?
4. What was one of the hardest concepts for you to master in a prior biochemistry course?
5. What is research? Describe what it means to you. How is research important in your life?

6. Please identify one or two strengths as a student that you think that you are bringing to this class?
7. Please identify one or two weaknesses as a student that you would like to work on this semester? Please indicate what they are and how you aim to improve.
8. Thinking back on your education so far, how do you learn best?
9. Please check the box that best describes your behavior prior to BMB464.***

	I NEVER do this	I do this INFRE-QUENTLY	I do this INCONSIS-TENTLY	I do this FRE-QUENTLY	I ALWAYS do this
I preview lecture material before coming to class.					
I attend class on time.					
I take notes in class by hand.					
I review my notes after each class.					
I study biochemistry with concentrated time and specific goals.					
I work/ study in groups.					
I understand the lecture and classroom discussion while I am taking notes.					
I try to determine what confuses me.					
I try to work out the example calculations problems without looking at the example problems or my notes from class.					
I review the lecture notes and practice problems before coming to class.					

***Questions adapted from a Study Skills Questionnaire from the University of Houston Clear Lake UHCL Counseling Services (2021) Study Skills Assessment Questionnaire [The University of Houston Clear Lake]. https://www.uhcl.edu/cmhc/resources/documents/handouts/study-skills-assessment-questionnaire.pdf

10. Please indicate what actionable tasks (1–2) you are going to do to be successful in the course this semester.
11. Every year the National Association of Colleges and Employers surveys employers across the country to rate the top skills/qualities that employers seek in new college graduates. Here is the list:
    - Ability to verbally communicate with persons inside and outside the organization.
    - Ability to work in a team structure.
    - Ability to make decisions and solve problems.
    - Ability to plan, organize, and prioritize work.
    - Ability to obtain and process information.
    - Ability to analyze quantitative data.
    - Technical knowledge related to the job.
    - Proficiency with computer software programs.
    - Ability to create and/or edit written reports.
    - Ability to sell or influence others.

    In BMB464 we are going to be working on all of these skills. Please comment on which above skill you are most excited about developing and why.
12. Please list class members you would like to work with in a group (if any).
13. Please list class members you would NOT like to work with in a group (if any).
14. Anything you would like to communicate to your Instructor or TA in regards to your learning? Anything I should know to help you succeed in the course this semester?

A Call to Action for More Inclusive STEM

Janelle M. Johnson, Kimberlee Bourelle, Adrian Clifton,
Mary Coleman, Parker Edingfield, Amanda Myers,
Madeline Onstott, Joseph Schneiderwind, and Katie Weaver
METROPOLITAN STATE UNIVERSITY OF DENVER

Teachers have been described as the keystone species of the STEM ecosystem, yet institutions of higher education often struggle to produce enough teachers to meet the demand for K–12 STEM courses (Bergin, 2018; Beyond100K, 2024; Milgrom-Elcott, 2023; Zhang & Zhu, 2023). The only avenues for secondary teaching in STEM fields at many institutions of higher education, including ours, are in mathematics and science (Johnson et al., 2023) rather than integrated STEM (Berisha & Vula, 2023). This means that students major in their content area and take a handful of education courses. Therefore, the majority of preservice secondary teachers' instructional contact hours are spent with professors who may not consider that some of their students are future teachers. Unfortunately, unlike Madison Brown's vignette (this collection), many STEM course instructors pride themselves on "weeding out" so-called weaker students. As teacher educators, we work to counter the weed-out approach most future STEM teachers have experienced during their educational experiences, especially at the university level (McCoy et al., 2017; Weston et al., 2019; Hatfield et al., 2022). We aim for teacher candidates to develop an awareness of how such tracking and sorting mechanisms can marginalize learners, especially those who represent underserved identities (Hung et al., 2020). Our larger goal is to develop teacher candidates who see themselves as advocates for students and their families, for public schools, and for the teaching profession.

This chapter is co-authored by the instructor and students, exploring the ways these future mathematics and science educators responded to a writing project in their multicultural education course. Their reflections on the writing project illuminate how their work deepened their own funds of knowledge; an unanticipated benefit was broadening their instructor's understanding of inclusive STEM. We share this experience with you in the spirit of welcoming further discussion and recommendations from the larger STEM community.

The Multicultural Education Course

With an overarching goal of closing opportunity gaps for learners (Ladson-Billings, 2013), one of the required courses for all K–12 and secondary teacher education students at Metropolitan State University of Denver (MSU Denver) is Multicultural Education. This course aims to help students develop critical awareness and a multicultural framework for viewing classroom interactions and curricula (Banks & Banks, 2019; Howe & Lisi, 2024). The course addresses *racial and ethnic inequality* and *social stratification* as primary lenses for understanding language, economic class, and other forms of difference in schools. As Ann Fink reminds us later in this section, "Liberatory processes position teachers, students, and their ecologies as companions in scientifically understanding and acting on the world" (this collection). The course syllabus emphasizes the roles teachers as decision-makers play in meeting the educational needs of "learners from diverse backgrounds" (Fink, this collection), though other than a brief introduction to Universal Design for Learning (UDL), disability had not been included in the definition of diversity in the course; all general licensure students are required to take just one course on exceptional learners offered by a different department. Similar to most teacher preparation programs, deep learning about disability seems to be limited to special education majors and segregated from general licensure (Schneiderwind & Johnson, 2021; Shume, 2023). Students in my (Janelle's) classes have often described the exceptional learners course anecdotally as a catalog of disabilities rather than one that helps them develop inclusive pedagogies. Our teacher preparation program has, therefore, reflected a siloed approach to inclusive teaching, counter to our stated social justice-based course objectives (Ogodo, 2024). It was the social justice advocacy by students in the class that raised the awareness of the instructor. This advocacy emerged during one of the key assignments of the multicultural education course, a "Call to Action" (CTA) project that includes both a written paper and a public service announcement. The written portion of the CTA is discussed in the following section.

A Call to Action for Social Justice

The overall goal of the CTA is to help students utilize peer-researched literature as a foundation for improving their self-efficacy for teaching *and* to cultivate their own voices as social justice advocates. While a theoretical framework of educational equity underlies the overall course, I have found that students have varying degrees of awareness of how theory shapes education and their own potential contributions to theory; they overwhelmingly tend to view themselves only as "receivers" of theory (Edelen et al., 2023; Rutten, 2021). The CTA project pushes students to ground a macro-level issue in a specific context at the micro level, planting the seeds of an ethnographic lens meant to cultivate their awareness and empathy with specific communities of students (Moll, 2013; Pérez-Castejón, 2023). Over the course of the CTA project each semester, I have witnessed evidence of a shift in teacher candidates' mindsets from *student* and *subject* toward *teacher* and *agent*. The introduction to the assignment reads:

> This is a chance for you to connect what you are learning in class to an action research project. Research a topic of interest related to this class, the content you will teach, and your field experience. Learn about it from multicultural perspectives and build your own knowledge for your future educational work.

As an instructor, I have found this assignment challenging for students for multiple reasons. One tension I consistently observe is that in the content courses for students' majors, they tend to focus on fulfilling the professors' requirements, completing the homework, and scoring well enough to pass exams. The teacher education courses, on the other hand, push the candidates to think as professional teachers rather than students, which can be a difficult transition (Moran et al., 2023). I feel that this dichotomous positioning between subject and agent helps to explain the challenges for teacher candidates' development of their own theoretical lenses. They have been trained to be compliant as students, and developing their own identity-agency seems out of reach to most of them (Berisha & Vula, 2023; Ruohotie-Lyhty & Moate, 2016). To become social justice advocates, they need to recognize the power of their own voice as well as being part of a larger collective that calls attention to marginalization of many students and communities by the educational system (Cochran-Smith et al., 2009; Grant & Agosto, 2008; Picower, 2012).

To cultivate that identity-agency, the CTA asks students to write in a professional tone and to utilize their *own* voice as they write about an issue they are passionate about. The CTA project is also meant to cultivate the future teachers' capacity for social justice advocacy writ large, so, in the written paper, I push them to use the headings I provide not only for clarity but as a strategic move to align with readers or funders who may be reviewing the work. Over time, I have developed what I have found to be an effective approach to supporting students' confidence to tackle this project by providing feedback on drafts at every step and requiring one-on-one meetings with me for coaching.

The sections of the paper (included in the Appendix) are inequity and rationale, sociocultural and sociohistorical roots of the issue, current context, action plan, and self-reflection. I scaffold the project by first inviting students to select and share a possible topic, which we discuss in small groups. Next, they complete a brief written outline that corresponds to the sections of the paper. I give them feedback on the outline to help them better utilize an ethnographic lens (Moll, 2013). I then assign a draft of one section of the paper at a time, starting with the works cited section. This allows me to help them continually narrow their topic and focus as needed. In the inequity and rationale section, students name the issue they are tackling and describe why it is personally significant to them. I have found that

I have to offer many students encouragement to insert themselves in the writing since much of their previous writing has asked them to be "neutral."

The section on the issue's sociocultural and sociohistorical roots allows students to write about policies and practices that have shaped their issue systematically over time. This aligns with course goals of rethinking uneven student learning outcomes as an intergenerational opportunity gap rather than an achievement gap. The current context section asks students to ground their issue locally, tapping into a range of resources, including their own clinical field sites and local news media. This helps students understand the varying scales of social justice issues, grounding them in the community where they will be working as teachers. The action plan of the CTA asks the student to generate ideas about steps they can take that will be appropriate to their position as novice teachers. Some possible avenues for action include creating a class project on the issue, helping students reach out to policy makers, finding ways to engage reciprocally with students' families, advocating at the district level, or writing an op-ed.

The self-reflection section of the CTA has changed over time. When I started assigning this project around ten years ago, I would simply assign it as the final section of the paper. Over time, I learned that I could better help students document their metacognition by having them write reflection and process notes at the bottom of each draft section of the paper as they composed them one by one. They write about being frustrated during their searches, often not finding exactly what they are looking for. This provides us with the opportunity to discuss what is meant by holes in the research and the need for triangulation. What follows are samples of the teacher candidate co-authors' own CTAs.

Future STEM Teachers' Advocacy for More Inclusive STEM: Reflections on the CTA

"Understanding Gender Bias in Disability Presentation" by Science Teacher Candidate Kimmie Bourelle

After many years of working closely with students with severe special needs, I have noticed that students' intelligence and potential for success in STEM subjects is often overlooked. Students with hidden disabilities, including learning disabilities (LDs), emotional disorders, and mental illness, have an entirely different level of potential dangers and bullying than they must consider. Because they do not physically show their disability, they may get teased for being "weird," "awkward," or even "dumb." Engaging in the work during the CTA helped me reflect on my biases when working specifically with girls with disabilities.

Using a lens of intersectionality in the CTA paper, I examined the experiences of girls and women with "invisible" disabilities such as autism or ADHD because they physically appear "neurotypical." The gender bias women and girls face takes shape through certain social stigmas and assumptions made about them in their education and is undoubtedly amplified in STEM pathways. The implicit bias that favors men in workplaces and within the education system harms *all* females. However, it is crucial to bring the specific issues of women and girls who have a hidden disability into focus if we hope to close opportunity gaps in STEM and beyond.

"The Importance of Math Skills" by Mathematics Teacher Candidate Parker Edingfield

Mathematics is essential for many career pathways. It is, therefore, recognized as a gatekeeper subject for all STEM fields. From arithmetic to linear algebra and beyond, the world functions as a product of math. Teaching students high school math effectively opens the door for any aspirations the student wants to pursue. Basic algebra is a necessary skill for higher education and many trade skills. By not encouraging universal math literacy in the way reading and writing are universally emphasized, students become unable to pursue their passions and career opportunities. Math literacy currently exists within most communities as "optional literacy," while language skills are prioritized as "mandatory literacy." This cultural relationship with mathematics makes it incredibly difficult to reach struggling students and creates generational struggles with mathematics.

Including specific pedagogies in my CTA action plans helped me reflect on how content can be taught in an equitable and fair way for all students. One way to do this is through UDL. UDL operates with the assumption that all students receive instruction differently and deserve a fair chance at participation and assessment. Using UDL as a framework allows flexibility in the thinking, expression, and reflection of learning. This type of flexibility is significant in helping students feel valuable, capable, and confident in the classroom. The CTA started a very new line of thinking for me. It began with reflecting on how my experience as a suburban white student who has gone through an education in mathematics with few obstacles was so different from most students' experience. And the more I investigated, the more I realized that many distinct types of people are marginalized in public schools. How do I, as a teacher, try to put myself in a position to empathize and understand enough to help them with whatever they need to help transition academically, emotionally, socially, or whatever? And so that's disability, equity, and culturally responsive work.

"Autism Spectrum Disorder in the Educational System" by Science Teacher Candidate Mary Coleman

There is a need for a greater understanding of students on the autism spectrum in all fields, but especially in STEM fields. Since the introduction of the Individuals with Disabilities Education Act, many changes and reforms have taken place to better include and serve individuals with autism in the public school system. This same system created to protect and serve these individuals, however, has continued to be overwhelmed with deficits. My rationale for advocating for a greater understanding of autism spectrum disorder in the education system is that this spectrum is a central challenge in our education system. The funding, necessary steps, and participation of schools are vital to the care and success of individuals on the spectrum. Not only is it essential to recognize that everyone is on a case-by-case basis for programs, but these individuals may also have abilities not generally seen in neurotypical children. In educational history, it has been unintentionally ignored that individuals with disabilities also have power. My STEM teacher preparation coursework needs to address this.

We must focus not just on disability but on learners' *abilities* from a strengths-based perspective. For example, many students may struggle with social interactions but excel in math and science. As a teacher, I may have to use different approaches than I do with neurotypical students, but it is essential to recognize the students' cognitive strengths. My key takeaway from the CTA was developing the perspective that I could combine two things: my passion for becoming a science teacher and my appreciation for individuals who are neurodivergent. This process has helped me create classroom engagement strategies since the accommodations I may make for neurodiverse students are often adequate for a range of learners.

"Challenges of Pandemic Mask Use for Deaf and Hard of Hearing Communities" by Science Teacher Candidate Katie Weaver

Many STEM teachers think they only need to know their content area well to be effective teachers, but they must consider the needs of all their students. Deaf and hard-of-hearing students have historically been underrepresented in STEM fields, and the pandemic surfaced as another contributor to that inequity. My CTA tackles the additional challenges the deaf and hard of hearing experienced with mask use during the COVID-19 pandemic due to the importance of lip-reading skills, facial expression, and body language for comprehensive communication in addition to or instead of using Sign Language. A common misconception that wearing a mask would not affect deaf and hard-of-hearing students because these individuals don't use their mouths to speak is a dangerous and harmful idea that further marginalizes these students and teachers. When signing (ASL), facial cues and expressions are an extremely important component of communication and tone. Overall, deaf and hard-of-hearing individuals face the challenge of putting themselves and others at risk by not wearing masks or facing the stark fact that they cannot communicate effectively. This enormous inequity that deaf and hard-of-hearing students and teachers face increases an already present opportunity gap.

I am incredibly grateful for this process and to have the opportunity to uncover this inequity. Throughout this process, I was expecting to find more information and more research on the effects that this inequity has had on deaf and hard-of-hearing students. However, it is a relatively new inequity, as COVID has only recently created the masked world we currently live in at the time of this writing. Ultimately, there is no official "research" on the effects of this inequity and if and how deaf and hard-of-hearing students are affected in terms of academic achievement, social disparities, or language loss.

"Expectancy Effect" by Mathematics Teacher Candidate Joseph Schneiderwind

I live with an extremely difficult and progressive physical disability that was not noticeably prohibitive until after achieving a graduate education; I was ABD (all but dissertation) in an acoustical physics program until the challenges of my disability did not allow me to continue that program. But this makes me think back and realize how little I knew about, or was exposed to, anybody with a disability during my own years as a student.

Social upbringing, the media, and self-reinforcement contribute to the problem of students who are statistically not "supposed" to do well tending not to. This leads to lower test performance, less interest in pursuing studies in science and mathematics, and reduced effort to pursue counter-stereotypic skills, amongst other things. The studies that I have read mentioning the expectancy effect are in relation to racial and ethnic minorities or women in STEM fields. However, such an effect can easily include students with disabilities. However, when viewing this through the expectancy effect lens, none of the research I encountered specifically addressed students with disabilities in STEM. Statistics that I found have primarily been census data and not related to a specific study. Further, many authors write in pedagogical terms about how a classroom or subject should be approached with respect to students with disabilities. However, they do not write about the effects of implementing that approach. The studies that have been done are largely funded by agencies looking at the accessibility of students with disabilities in postsecondary education. The notable lack of research seemed to be indicative of the little importance society places on this issue.

"Collective Action for Educational Inequities" by Science Teacher Candidate Maddie Onstott

More equitable access to quality STEM education for students with disabilities in rural schools matters to me because I believe all students have the right to an education/school system that will support them as they are and not forget about them because of their ability. As a future science educator, I want to help provide my students with learning opportunities to participate in hands-on activities that fire up their critical thinking skills so that they clearly understand a concept, rather than having them look up definitions of science terms online and write them down. I would like to have access to technology and materials that will help my students learn more engagingly, but I may not be able to do so if the school I teach at lacks adequate access. The importance of access to technology and materials that will help students with disabilities learn in more engaging ways is crucial.

After researching this topic and interviewing key leaders in the state, I feel like I have learned so much. I could fill in gaps in the information I was finding in my research about funding for inclusive STEM education in rural schools. It also seemed like even though the resources are available, not many know about them or are not interested in teaching in rural schools. A large part of this is due to the lack of support networks available to educators in these rural areas. If there were more PD opportunities for rural teachers and strong, localized support networks in these communities, in addition to funding opportunities for schools and teachers, more educators would student-teach in rural areas and continue to teach there. I also learned about a toolkit that provides support and structure for schools to build teams that can better serve students and their communities. This was a great way to give a voice to the local community and students because the people living and working in an area truly know what is going on there.

"Societal Importance of Gaps in the Research" by Science Teacher Candidate Adrian Clifton

I have learned many different things in the process of this CTA project and the work I have done over the last year. As a student with disabilities, I always knew that going through the school system was tough, but I did not realize how widespread this issue was and how little help these kids get until I saw this while working at a public school. These kids were not prioritized, and neither were the workers who were supposed to be helping them. I have not yet seen enough articles and studies directed toward how teachers can equip themselves to reach and teach students with disabilities. Since the amount of research on a topic can often reflect its societal importance, it is apparent that this topic has not been prioritized. Teachers who want to educate themselves will have to do their own digging and will likely have to spend good money to access the content they want (and most teachers are overworked and underpaid as is).

Learning to help kids with disabilities is difficult because there is no one-size-fits-all way to teach kids with disabilities, as each disability is different. The existing resources are buried in journals that most teachers probably have never heard of, much less have access to, since so many of these journals are expensive to access. If I want to advocate for students with disabilities and argue that teachers should be trained to help them, I, too, need to be trained to help students with disabilities. Serving students with disabilities was only addressed during one course in my teacher preparation program. The fact that I am on the autism spectrum and have OCD gives me firsthand knowledge about how to help similar students, but there are many more disabilities out there that I am untrained about. Ultimately, this issue is a systemic failure of the entire structure of society and the schooling system, and there is thus no quick and easy fix to the problem.

An Instructor Learning from Their Students and the Process

From an instructor's perspective, utilizing this CTA as one of my course's key assignments has been an incredible source of learning on multiple levels. It has certainly been an effective approach to supporting future teachers' social justice identity-agency (Jacobs & Perez, 2023), but it has also had a profound effect on my own thinking. I describe my entire career trajectory as having a focus on STEM equity, but that focus had not included disability until I learned from my own students and their CTA projects. As a teacher and an academic, I shied away from any topic I did not have recognized expertise in; I had minimal background in so-called special education. But I faced the harsh realization that I was underserving students with disabilities in my own classes by ignoring the topic. And I underprepared all my students as future educators by not including learning about disability as an integral part of the equity lens. Luis Moll (2013) writes about Vygotsky's call to understand that all students are part of a continuous spectrum. As these future educators tackled topics related to disability in STEM in their CTA research, I learned with and from them. It has been a humbling and consciousness-raising experience. It has helped me be transparent with my own students about the value of ongoing learning as a teacher and recognizing our students' funds of knowledge (Gonzalez et al., 2006). I genuinely learn from my students' CTAs every semester. Reflections of this nature on centering the needs of our own students are also reflected in the chapter by Jennifer Newell-Caito (this collection).

I would like to encourage instructors in other higher education contexts to experiment with creating a CTA in their own courses. The scaffolded structure certainly makes it more approachable for instructors to adapt this writing project. Start with an open discussion of inequities in your field of study. What are students passionate about? What experiences with this issue do they bring with them into your classroom community? You can capture students' initial ideas about possible topics with an online poll, for example. Help them build out their ideas a bit more, writing bullet points or short paragraphs for each of the sections—history of the issue, where they see it locally, some ideas they have to tackle it, and any resources they find that may support their learning. Make sure to give yourself time to give them feedback each step of the way and coaching as needed. We stretch out this project over more than half of the semester; students see the coherence of the work since the topics they choose overlap with the content of the course. While this chapter focused on the written portion of the CTA, students also create a public service announcement (PSA) once they have finalized their written papers. The PSA is a wonderful opportunity for the students to publicly present their work and get feedback from their peers. We've had success with the PSAs both in person in a "gallery walk" style and shared remotely through an online platform such as Padlet. Students learn from each other and have yet another opportunity to cultivate their identity-agency.

References

Banks, J. A. & Banks, C. A. M. (Eds.). (2019). *Multicultural education: Issues and perspectives*. John Wiley & Sons.

Bergin, K. (July 18, 2018). Remarks during 2018 AAAS/Noyce Summit, *Towards a 2026 STEM education: Implications of convergent science for K–12 STEM teacher preparation in the face of changing student demographics*. Washington, D.C. https://tinyurl.com/26mejwnd.

Berisha, F. & Vula, E. (2023). Introduction of integrated STEM education to pre-service teachers through collaborative action research practices. *International Journal of Science and Mathematics Education*, 1–24.

Beyond100K. (2024). 2024 trends report: Trends and predictions that are defining STEM in 2024. https://tinyurl.com/y4kprxp8.

Cochran-Smith, M., Shakman, K., Jong, C., Terrell, D. G., Barnatt, J. & McQuillan, P. (2009). Good and just teaching: The case for social justice in teacher education. *American Journal of Education, 115*(3), 347–377.

Edelen, D., Cox Jr., R., Bush, S. B. & Cook, K. (2023). Centering students in transdisciplinary STEAM using positioning theory. *Electronic Journal for Research in Science & Mathematics Education, 26*(4), 111–129.

González, N., Moll, L. C. & Amanti, C. (Eds.). (2006). *Funds of knowledge: Theorizing practices in households, communities, and classrooms*. Routledge.

Grant, C. A. & Agosto, V. (2008). Teacher capacity and social justice in teacher education. In *Handbook of research on teacher education* (pp. 175–200). Routledge.

Hatfield, N., Brown, N. & Topaz, C. M. (2022). Do introductory courses disproportionately drive minoritized students out of STEM pathways?. *PNAS nexus, 1*(4), 167.

Howe, W. A. & Lisi, P. L. (2024). *Becoming a multicultural educator: Developing awareness, gaining skills, and taking action* (4th ed.). Sage Publications.

Hung, M., Smith, W. A., Voss, M. W., Franklin, J. D., Gu, Y. & Bounsanga, J. (2020). Exploring student achievement gaps in school districts across the United States. *Education and Urban Society, 52*(2), 175–193. https://doi.org/10.1177/0013124519833442.

Jacobs, J. & Perez, J. I. (2023). A qualitative metasynthesis of teacher educator self-studies on social justice: Articulating a social justice pedagogy. *Teaching and Teacher Education, 123*, 103994. https://doi.org/10.1016/j.tate.2022.103994.

Johnson, J. M., Konuk, N. & Koester, M. (2023). Using lesson study to build interdisciplinary STEM collaborations at an urban commuter university. In S. Dotger, G. Matney, J. Heckathorn, K. Chandler-Olcott & M. Fox (Eds.), *Lesson study with mathematics and science preservice teachers* (pp. 137–147). Routledge. https://doi.org/10.4324/9781003326434.

Ladson-Billings, G. (2013). Lack of achievement or loss of opportunity. In P. L. Carter & K. G. Welner (Eds.), *Closing the opportunity gap: What America must do to give every child an even chance* (pp. 11–22). Oxford University Press.

McCoy, D. L., Luedke, C. L. & Winkle-Wagner, R. (2017). Encouraged or weeded out: Perspectives of students of color in the STEM disciplines on faculty interactions. *Journal of College Student Development, 58*(5), 657–673.

Milgrom-Elcott, T. (2023, September 27). *Ending the STEM teacher shortage*. STEMM Opportunity Alliance. https://tinyurl.com/5832fxps.

Moll, L. C. (2013). *L.S. Vygotsky and education.* Routledge.

Moran, R. M., Robertson, L., Tai, C., Ward, N. A. & Price, J. (2023). Developing pre-service teachers' adaptive expertise through STEM-CT integration in professional development and residency placements. *Frontiers in Education, 8.* https://doi.org/10.3389/feduc.2023.1267459.

Ogodo, J. A. (2024). Culturally responsive pedagogical knowledge: An integrative teacher knowledge base for diversified STEM classrooms. *Education Sciences, 14*(2), 124.

Pérez-Castejón, D. (2023). Practices and intellectual requirements for attaining inclusive education and social justice in initial teacher education: Ethnography. *Ethnography and Education, 18*(1), 112–126.

Picower, B. (2012). Teacher activism: Enacting a vision for social justice. *Equity & Excellence in Education, 45*(4), 561–574.

Ruohotie-Lyhty, M. & Moate, J. (2016). Who and how? Preservice teachers as active agents developing professional identities. *Teaching and Teacher Education, 55,* 318–327. https://doi.org/10.1016/j.tate.2016.01.022.

Rutten, L. (2021). Toward a theory of action for practitioner inquiry as professional development in preservice teacher education. *Teaching and Teacher Education, 97,* 103–194. https://doi.org/10.1016/j.tate.2020.103194.

Schneiderwind, J. & Johnson, J. M. (2021). Broadening the equity lens for STEM teacher education: The invisibility of disability. *AAAS Bulletin.* https://tinyurl.com/h625zhff.

Shume, T. J. (2023). Conceptualising disability: a critical discourse analysis of a teacher education textbook. *International Journal of Inclusive Education, 27*(3), 257–272. https://doi.org/10.1080/13603116.2020.1839796.

Weston, T. J., Seymour, E., Koch, A. K. & Drake, B. M. (2019). Weed-out classes and their consequences. In E. Seymour & A.B. Hunter (Eds.), *Talking about leaving revisited: Persistence, relocation, and loss in undergraduate STEM education* (pp. 197–243). Springer.

Zhang, Y. & Zhu, J. (2023). STEM pre-service teacher education: A review of research trends in the past ten years. *Eurasia Journal of Mathematics, Science and Technology Education, 19*(7), em2292. https://doi.org/10.29333/ejmste/13300.

Appendix: Colorado Call to Action Project: EDS 3150: Multicultural Education, by Dr. Janelle M. Johnson

This is a chance for you to connect what you are learning in class to an action research project. Research a topic of interest related to this class, the content you are going to teach, and your field experience. Learn about it from multicultural perspectives and build your own knowledge for your future educational work. Some ideas include: teen homelessness, the foster care system, unequal school funding, transgender student rights, inclusion of students with disabilities, educational access for English Language Learners, school choice, school discipline, equity of school lunch programs, funding for the arts, immigration, girls in STEM fields, poverty in schools, desegregation policies, bullying, overrepresentation in special education; the need for art and music as core classes, bias in the curriculum, etc.

Steps of the assignment:

1. Clearly identify an **educational inequity**. It must be something specific we can see evidence of in schools or classrooms, including challenges and problems of students and their families.
2. Research the issue using multicultural lenses (race, ethnicity, language, culture, disability, sexual orientation, etc.). Think of how the needs of different populations of students may be **underserved in a specific context**. What have been the institutional and educational blind spots?
3. Write a 10–12 pages total (double-spaced) paper on the topic with 5–10 references and reflection. **You must include page numbers and the provided headings.**
4. Create a public service announcement (PSA) that raises awareness about the educational inequity you researched.
5. Share your PSA with the class and the world.

Part I: Specifics for the PAPER (50 points)

Your document should be no more than 12 pages total **(number your pages)**, including a works cited page, with appropriate referencing for your content area. **You must use the following as your section headings, and you can add additional headings or subheadings if you choose:**

Educational Inequity and Rationale (minimum one page)

[Examine the effects of **bias, prejudice, and/or discrimination** and how they have affected ac**cess and opportunity to academic success** of one or more of the groups of color in United States society. (TQS 2.B, 2.C, 2.D, 3.A, 3.D, 4.C)]

What is your rationale for advocating this particular issue over others? **WHO** is suffering because of this inequity? Be specific. Here, you want to make as strong a case as possible for WHY this is a central challenge. Include your own voice. Why does this matter to YOU? Relate the issue to educational issues facing Colorado and your current or past field placement school. To the extent possible, lay out your "thinking/decision-making" process. Cite any references. (10/50 points)

Sociocultural/Sociohistorical Roots of the Inequity (2–3 pages)

[Examine structures of power, control, and governance in schools in relation to race, ability, age, ethnicity, gender identity/expression, religion, sexual orientation, and socioeconomic status. (TQS 1.C, 2.B, 2.C, 4.C, 4.D)]

In other words, what do we need to understand, contextually, in order to make sense of this project? What is the macro context of this issue? These roots can be *economic, political, social,* and/or *demographic*. What policies shaped this issue? How has it changed over time? **Use references.** (10/50 points)

Current Context (2–3 pages)

[Analyze the impact that race, ability, age, ethnicity, gender identity/expression, religion, sexual orientation, and socioeconomic status have upon learning, and explain the roles of teachers, administrators, parents, and the community in the pursuit of multicultural goals in education. (TQS 2.A, 2.B, 2.C, 2.D, 3.A, 4.A, 4.C, 4.D)]

What is the current state of events (*economic, political, social, demographic, environmental. . . .*) that needs to be understood to address this challenge? Use any statistics you can find, especially if you are discussing a challenge in a specific school or district. News stories are an excellent source for this section. Feel free to describe conditions you have observed firsthand. Cite any references, including conversations. (10/50 points)

Action Plan (2–3 pages)

[Develop strategies/methods that lead to the formation of and continuation of multicultural education in schools. (TQS 2.A, 2.B, 3.A, 3.D, 4.A, 4.B, 4.C, 4.D) Identify methods of reducing prejudice and racism in the classroom. (TQS 2.A, 2.B, 2.C, 2.D, 3.A, 3.D, 3.E)]

What will be the most useful actions YOU can take as an educational advocate? You do not need to "solve" the problem but need to find a way to address it. This can include raising awareness about the issue. Try to ALIGN the actions with the problem—if it's funding, how could you advocate for funding to try to write grants? If it's an issue faced by families, can you organize some kind of parent group that brings these issues to the table? Defend your choice based on the specifics of your field, the nature of the educational equity issue, and based on your own self-reflection about what you will actually USE. Include the Teacher Quality Standards this issue addresses. (10/50 points)

Resource: https://tinyurl.com/mvcfe8yj

Self-Reflection (1 page)

[Explain how personal views and experiences may influence attitudes and behaviors as an educator. (TQS 4.A, 4.B, 4.C)]

Your own voice should come through most clearly in this section. Apply critical thinking dispositions to your own thinking, especially concerning issues of race, ethnicity, language, culture, sexual orientation, and/or disability. Describe what challenges you had tackling this topic and what kind of information you did or did not find. What have you learned? (5/50 points)

Works Cited (1 page)

Whole paper must be spell checked with formatting that reflects professionalism. Must include page numbers, headings, works cited page (5/50 points) [Use technology to access, organize, interpret and present information. (TQS 3.C, 3.F)]

Some Resources:

- Colorado school data http://highered.colorado.gov/Data/DistrictHSSummary.aspx
- Teaching Tolerance.org
- Colorado Public Radio
- Chalk Beat
- Denver Post
- The equality of opportunity project http://www.equality-of-opportunity.org
- Annie E. Casey—Kids count data http://datacenter.kidscount.org/data#CO/2/0
- Multicultural education: Issues and perspectives http://www.slideshare.net/mariaferc/0470483288multicul
- NEA Grant writing http://www.nea.org/home/10476.htm

Rubric for Colorado Call to Action Part I

	Exceeds Expectations	Meets Expectations	Partially Meets Expectations	Does Not Meet Expectations
Inequity and rationale: 1 page (TQS 2.B, 2.C, 2.D, 3.A, 3.D, 4.C,)	Clear explanation of one major educational equity challenge. Clear explanation of which aspects are most important to this challenge. Clearly argues why this is an important inequity. 9–10	Adequate explanation of one major educational equity challenge. Indicates which aspects are most important to this challenge. Argues why this is an important inequity. 7–8	States one major educational equity challenge. Confusing or not compelling reason why this is an important inequity. 5–6	Does not clearly state one major educational inequity. Lack of rationale provided. 4–0
Roots of the inequity: 2–3 pages (TQS 1.C, 2.B, 2.C, 4.C, 4.D)	Clear explanation of roots (economic, political, social, demographic, environmental) of this inequity. 9–10	Adequate explanation of roots (economic, political, social, demographic, environmental) of this inequity. 7–8	Minimal explanation of roots (economic, political, social, demographic, environmental) of this inequity. 5–6	Does not address roots (economic, political, social, demographic, environmental) of this inequity. 4–0

	Exceeds Expectations	**Meets Expectations**	**Partially Meets Expectations**	**Does Not Meet Expectations**
Current Context: 2–3 pages (TQS 2.A, 2.B, 2.C, 2.D, 3.A, 4.A, 4.C, 4.D	Clear explanation of current conditions (economic, political, social, demographic, environmental) that affect this inequity. 9–10	Adequate explanation of current conditions (economic, political, social, demographic, environmental) that affect this inequity. 7–8	Addresses only one current condition (economic, political, social, demographic, environmental) that affect this inequity. 5–6	Does not address current conditions (economic, political, social, demographic, environmental) that affect this inequity. 4–0
Action Plan: 2–3 pages (TQS 2.A, 2.B, 2.C, 2.D, 3.A, 3.D, 3.E, 4.A, 4.B, 4.C, 4.D)	Well-aligned action plan to address the situation. 9–10	Action plan fairly well aligned. 7–8	Minimally aligned action plan. 5–6	Action plan not aligned to inequity. 4–0
Self-reflection: 1 page (TQS 4.A, 4.B, 4.C)	Clear explanation of choice of topic and process. 5	Adequate explanation of choice of topic and process. 4	Lack of explanation of choice of topic and process. 3	Limited to no explanation of topic and process. 2–0
Headings, Numbering, Spelling, Citations (TQS 3.C, 3.F)	Headings correct, pages numbered, no misspelled words, proper citations. 5	Most headings correct, 1–2 misspelled words or improper citations. 4	Some headings correct, 3–5 misspelled words minor citation errors. 3	No headings and/or page numbers, Many misspelled words, major citation errors. 2–0

Part II: Specifics for the PSA (30 points)
Public Service Announcement and Q & A session

1. Choose a format for your PSA: television (30–60 second animated PowerPoint with recorded audio); radio (30–60 seconds); two-sided brochure; billboard or bus stop ad.
2. This PSA should NOT be an oral recitation of your written report. It should be designed to raise awareness about your educational inequity issue and **encourage some specific action**. You are encouraged to use a variety of media formats and to include photos and "voices" from the field.
3. Include charts or data in the PSA.
4. Ask friends and colleagues—to watch/listen and provide feedback on how professional it is.
5. Prepare for the Q & A session.

Rubric for the Colorado Call to Action Part II: PSA

	Exceeds Expectations	**Meets Expectations**	**Partially Meets Expectations**	**Does Not Meet Expectations**
Clarity of Educational Inequity	Clear explanation of one major educational inequity. Clear explanation of which aspects of diversity are most important to this challenge. Excellent use of charts and data 9–10	Adequate explanation of one major educational inequity. Indicates which aspects of diversity are most important to this challenge. Good use of charts and data. 7–8	States one major educational inequity. Confusing or not compelling reason why this is an important equity challenge. Charts and data included but do not support call. 5–6	Does not clearly state one major educational inequity. Minimal or no charts and data. 4–0
Call to Action	Very clear call to action to address the situation. 9–10	Somewhat clear call to action. 7–8	Unclear call to action. 5–6	Minimal or no call to action. 4–0
Professional	Extremely professional quality in selected medium. No misspelled words, no obvious grammatical errors. 9–10	Professional quality in selected medium. 1–2 misspelled words or obvious grammatical errors. 7–8	Somewhat professional quality in selected medium. 3–5 misspelled words or obvious grammatical errors. 5–6	Unprofessional quality in selected medium. Many misspelled words and obvious grammatical errors. 4–0

Teaching Neuroethics in a Time of Crisis: Lessons in Liberatory Pedagogy

Ann E. Fink

Lehigh University

Introduction

Crises lay bare the core values around which institutions are organized. A crisis represents a state of overwhelm, a system pushed to the limit of its capability to adapt. Crises may occur within the lives of individuals or at broader societal and ecological levels. They may take the shape of an acute illness, floods on a warming planet, or a pandemic. Disasters may originate within the natural world, yet a state of crisis is crafted and perpetuated within human activities. Crisis invariably creates opportunities for power to assert itself.

Nevertheless, crisis may also present opportunities for liberatory and humanist modes of change, and education is central to such capabilities. It is a testament to this potential of education that recent crises have elicited such reactionary attacks on educational institutions and educators, including crusades against the teaching of critical race theory and LGBTQIA+ educators and students. An intentional, values-informed approach to STEM fields and to education is needed to counter reactionary narratives and responses to crisis.

STEM teachers and researchers may feel a sense of remove from political problems (science has long been touted as objective, detached, and apolitical); nevertheless, science is a fundamentally social activity, anchored in time and place and requiring substantial collaboration. I echo Helen Longino's (1990) observation that "it is the social character of scientific knowledge that both protects it from and renders it vulnerable to social and political interests and values" (p. 12). At present, problems such as climate change, disparate illness outcomes, and revelations about scientific racism, sexism, and ableism challenge STEM educators to more explicitly consider values in their work.

Students in STEM classes will become future scientists and educators or will otherwise make use of scientific knowledge in their lives. These realities further invite teachers to act responsibly in their roles and to welcome values, ethics, and critical methodologies into the STEM classroom. I respond to these challenges as a neuroscientist, ethicist, and educator with a deep commitment to liberatory pedagogies. Neuroethics, as a discipline, centers the critical analysis of the scientific study of brain and mind and informs the totality of my teaching in neuroscience.

In this chapter, I describe how teaching neuroethics through a liberatory lens has facilitated students' ability to think critically, name and examine values, and take nuanced and historicized perspectives on neuroscience-related topics, even as the COVID-19 pandemic and other recent crises unfolded. I expand on how my own theoretical foundation in liberatory pedagogies and understanding of critical consciousness shaped the design of course curricula and classroom spaces during the pandemic and illustrate how such critical methodologies may alter engagement with power in the sciences. Finally, I provide practical insights and applications from the neuroethics classroom. Throughout, the neuroethics view, informed by critical methodologies, allows for novel conceptualizations of accountability and justice (Hue, 2020).

Author Positionality

I am a multiracial, Asian American, light-skinned, bisexual, queer person. I live with chronic illness. I benefit from a high-quality education spanning community college and a Ph.D. I have been engaged in learning, memory, and mental health-related research and teaching for almost 20 years and have also become a practicing clinician. Together, my social identities, interdisciplinary knowledge, and values have shaped my presence in science as well as my teaching philosophy and methods. Being raised across cultures, in an ambiguous skin, made me more comfortable with ambiguity and complexity, and being without generational wealth helps me to recognize financial limitations on time, attention, and options. I confront oppressive legacies and the workings of power in the sciences, and I have learned to be curious about how dimensions of identity shape the science classroom and students' experiences therein. My dimensions of positionality confer special awareness for marginalized perspectives that are similar to my own while obscuring other marginalized experiences and knowledge forms. The concept of cultural humility (Abe, 2020) reminds me that each student is the expert in their own social identity and experience. With awareness of my own positionality, I can attend to their perspectives and craft spaces more conducive to their learning.

Critical Pedagogy and Liberatory Process

In *Pedagogy of the Oppressed* (1970/1993), Paulo Freire inscribed key theoretical and practical foundations of critical pedagogy upon which others continue to expand. This approach toward pedagogy may encompass goals of inclusion but transcends them. Critical pedagogies seek transformed relationality between teachers and students, with a flattening of the hierarchies that characterize traditional classrooms.

Central to critical pedagogy is *conscientização*, a practice of critical thinking, collective awareness, theory building, and action. This stands in contrast to the "banking model" of education (Freire, 1970/1993, pp. 53–54, 90), where a supposedly omniscient and omnipotent teacher deposits information into the minds of passive students. Critical pedagogy rejects the assumption that learners are so fundamentally blank, proposing instead that students and teachers both shape the learning process. Through a critical dialogue that encompasses their socio-historical context, they may come to understand the world together.

An attention to power dynamics that subverts the authoritarian structure of traditional classrooms is one key principle of critical pedagogy. Another is concurrent engagement with theory and praxis; theory is meaningful inasmuch as it connects to the material conditions of the world. As students and teachers name the world together, they also act together. This process requires dialogue and mutuality, a reflexive collaboration in the classroom and beyond.

In accordance with these principles, critical pedagogy requires learning in communion with one another and with surrounding ecosystems (Freire, 1970/1993, pp. 62–63). Freire argues that critical pedagogy is a relational practice requiring *intersubjectivity*, a recognition of the value, personhood, perspective, and agency of the other (e.g., teacher and student). This also engenders spontaneity, liveliness, and creativity, in contrast to the deadness of authoritarian, dominating forms of education (Freire, 1970/1993, p. 58). The liberatory classroom is a living organism.

Education for critical consciousness is playful at its best but has serious stakes. Teachers and students embark together on a project of intellectual and material liberation. In this process, one does not liberate the other (an impossible endeavor; Freire, 1970/1993, pp. 75–76). Rather, the liberatory process entails "doing with" rather than "doing to." This critical, relational process also differentiates between the revolutionary process of *conscientização* and either bureaucratic or rigid sectarian thinking. The liberatory process is itself fraught with problems of power, including dangerous hero fantasies about idealized leaders. Akin to this is the danger of investment in a tokenized few representatives of oppressed identities who may enforce the norms of unchanged oppressive structures. A liberatory goal, rather, is that all participants become "subjects who meet to name the world in order to transform it" (Freire, 1970/1993, p. 148).

The scholar and educator bell hooks (Gloria Jean Watkins) was one of the most prolific U.S. practitioners of critical pedagogy. In her own indispensable work, *Teaching to Transgress* (1994), hooks explores both her connection to and ambivalence about Freire; she is known for expanding Freire's analysis of power and oppression to include the "interlocking forms of domination" (hooks, 1989a, p. 25) known as gender, race, and class. hooks' theorizing dovetails with her contemporary Kimberlé Crenshaw (1989, 1994), who introduced the framework of *intersectionality* within legal analysis. hooks recurrently explored love as a defining feature of the feminist, liberatory classroom (hooks, 1989b; 1994, pp. 198–199) while simultaneously embracing productive conflict and confrontation, which she saw as central to the project of critical consciousness (1989b; 1994).

Other authors, such as Maurianne Adams and Barbara Love (2010), examine how liberatory educational practice aligns with contemporary "diversity" rhetoric within universities. These authors emphasize that simply adding diversity in admissions is not enough; the quality of the educational environment must be cooperative, consistent, and supportive. Regardless of institutional rhetoric, these authors also note that many faculty continue to teach from familiar, authoritarian traditions predicated on the "banking model." The authors employ a social justice argument for adopting liberatory educational practices, and they place the leveling of power at the center of meaningful equity and inclusion.

Adams and Love (2010) propose a four-quadrant model that may be used to analyze and assess liberatory, social justice-focused teaching and learning. These four quadrants include:

> (1) what our students, as active participants, bring to the classroom, (2) what we as instructors bring to the classroom, (3) the curriculum, materials, and resources we convey to students as essential course content, and (4) the pedagogical processes we design and facilitate and through which the course content is delivered. (p. 7)

This framework may be used to identify the positionality and strengths of students and instructors and to help classroom participants to reflect on social identities and exercise perspective-taking skills. Such models may be helpful in linking theory to pedagogical practice. Nevertheless, Adams and Love (2010) also argue that social justice goals require transformation at institutional and societal levels, in addition to self-reflexive practice and changes within the classroom. They note particular difficulties with introducing social justice perspectives in STEM fields while recognizing potential rewards in doing so: greater insight, connectedness, and well-rounded views on important problems.

As a group, these theorists converge on the importance of flattening power differentials, developing more authentic interpersonal connections, and joining theory with praxis. In recent U.S. culture, however, the language of justice has come to be framed largely in terms of diversity, equity, and inclusion (DEI), a set of activities often overlapping with the pre-existing academic language of "multiculturalism." It is important, however, to be cognizant of the limitations in DEI frameworks and the dangers of co-opted social justice language, what Vijay Prashad (2010) terms the "bureaucratic approach to the problem of diversity" (p. 121).

Contemporary efforts toward diversity, equity, and justice within U.S. education developed as a result of civil rights demands and have consistently received pushback since its implementation in the 1960s, its goals unsurprisingly at odds with monied white power structures (hooks, 1994, pp. 29–31; Prashad, 2010). These efforts, however, not only faced outright suppression but additional co-optation as, in the words of hooks (1994), "the stuff of a colonizing fantasy" (p. 31). Prashad (2010) explores how multiculturalism was diluted into colorblind ideologies and model minority myths, adopted as institutional commodity and used to forestall transformation within institutional hierarchies (pp. 123–124).

Feminist scholar Sarah Ahmed (2012) describes "diversity" as a potential exercise that "evokes the pleasures of consumption" (p. 69), as "a form of public relations" (p. 143), and inclusion as a potential "technology of governance . . . those who in being included are also willing to consent to the terms of inclusion" (p. 163). In the absence of *conscientização* as process, DEI language and technologies may be manipulated to sustain oppressive institutions and practices. Those seeking to engage in liberatory practice would be wise to remain aware of such dynamics.

Critical Consciousness in STEM Education

Contemporary liberatory pedagogies were born within the brutal, mid–20th century constructions of economic and political crisis so well documented in Naomi Klein's (2007) *The Shock Doctrine*. Such tactics of crisis, shock, and political and economic restructuring were repeated worldwide, informing events from the overthrow of Latin American democracies to responses to natural disasters such as 2005's Hurricane Katrina in New Orleans (Klein, 2007). The common threads in these states of crisis, per Klein, include both explicit actions and selective inaction and institutional neglect that disproportionately target marginalized populations and consolidate power in fewer hands.

Freire was one of many Latin American radical educators and artists facing oppression by military regimes employing US-backed tactics of economic "shock." A military coup toppled Brazil's government in 1964; the junta quickly escalated to overt torture and terror. Freire fled Brazil, living in exile within the US. This "shock therapy" allowed the architects of the coup to dismantle the Brazilian state and open the nation for economic exploitation; the regime recognized liberatory educators as enemies. Liberatory pedagogies must indeed be understood as a confrontation against cultivated states of crisis and their accompanying ideologies. When hooks explicitly orients herself against the racism, sexism, and "alienation" of the university (1989a; 1989b; 1994, pp. 5–8), for instance, she assumes a deliberate position of resistance. In this tradition, I argue that critical pedagogy is urgently needed in the sciences to resist narratives of alienation and domination within crisis.

In a heated state of crisis, it may be easy to ignore the complex social and historical context of key problems precisely at a time when they are most important.

The history of science is rife with oppressive constructs, yet I have found that many STEM majors learn little, if any, of this history. Students may understand eugenics, for instance, as the opinions of individual bad actors rather than frameworks that justify oppressive social structures. Students with marginalized identities may feel estranged from scientific subfields without knowing why; those born with power and privilege may struggle with defensiveness when engaging this history. Liberatory frameworks can build the intellectual and relational capacities for a critical, historicized engagement with the sciences.

Many contemporary oppressive practices have their historical roots in biological theorizing about social categories such as race and gender. These narratives may elevate certain populations as ideal and construe others as less human or inferior. They often do so by referencing a point of physical difference (e.g., skin color, reproductive physiology) and making spurious claims about evolutionary processes (see Gould, 1996; Schiebinger, 1993; Saini, 2019). These arguments may involve both inept and selective interpretations of evolutionary theory (Fuentes, 2012, pp. 3–4) and outright fabrications.

An example: the 18th-century European anatomist Blumenbach took a liking to the skull of a young woman from what is now Georgia (Schiebinger, 1993, pp. 150–153). Blumenbach's praise for the beauty and symmetry of this skull, taken from the Caucasus mountains, provided fuel for the construction of an idealized race of people. "Caucasian" became a categorical term for white people of European descent and a reference point for a destructive race science that continues to this day.

Another example: the woman known as Saartje Baartman (her true name is unknown) was bribed into traveling from what is now South Africa to Europe with so-called European men of science. She was abused: subjected to bizarre, sexualized curiosities about her body and exhibited in Europe's human "zoos" (Schiebinger, 1993, pp. 168–172; Saini, 2019, pp. 39–41). The writings of these men provided more fuel for Europeans' generalizations about, and denigration of, the bodies and minds of Black Africans and contributed to the scientific construction of racialized "others."

The practice of measuring skulls (Gould, 1996, pp. 105–141) lingered for many years as a malignant subculture of race science aimed at reifying supposed categorical differences in intelligence and mental fitness. More recent efforts in the same lineage invoke essentialist claims about average brain differences and disingenuous uses of genomics to justify racial and gender hierarchies (Fine, 2012; Saini, 2019, pp. 103–124). Such scientific narratives are not morally neutral or apolitical. They provide intellectual cover to white supremacy and related cis-heterosexisms, ableisms, and other oppressive philosophies. It is no coincidence that these dehumanizing narratives flourished during the aggressive expansion of European colonies (Saini, 2019, pp. 21–24), where scientific construction of racial/gender hierarchies was used to justify a "civilizing" mission and, thus, Europe's claims to the lands and resources of others.

Science education is inextricably entangled with this history of crisis and conquest. One need only consider the periodic resurgence of racialized, gendered, and classed hierarchies of intelligence (Gould, 1996, pp. 26–50; Saini, 2019, pp. 90–92, 95–102), often used to argue that educational resources are wasted on "lesser" classes of people (Panofsky, 2015; Comfort, 2018). Yet the construction and usage of STEM knowledge does not need to be fundamentally oppressive; this usage is a choice. What would it look like to make different choices concerning the structure of scientific disciplines, power relations within them, the questions asked, people retained, and employment of scientific narratives? What would it look like to abandon questions about who belongs at the top of the heap? To transcend these patterns of power and domination requires a different approach. This is what critical pedagogy offers.

A liberatory STEM education requires a foundation in history and critical analysis. Liberatory process also emphasizes the interconnectedness of people with each other, their environment, surrounding natural phenomena, and scientific processes. These processes of sociohistorical analysis and relationality go hand in hand, countering the artificial distance of students and instructors from their social positioning, countering states of alienation and dissociation. Liberatory processes position students, teachers, and their various ecologies as companions in scientifically understanding and acting on the world. This is, in short, the binding of theory and praxis in *conscientização*.

Recent years have seen efforts to address discrimination in STEM fields, with arguments tending toward the multicultural. Diverse views, it is argued, allow science to be increasingly relevant, accurate, and useful. These are fine arguments, but these fields still struggle to move beyond DEI platitudes and travel the road toward liberatory practices. This may reflect frank antipathy toward such practices, yet it may also reflect a worry that they are too difficult to implement. To such educators, I would offer hope. Critical pedagogy is an ongoing process and not an endpoint; change may feel challenging, but it is possible.

Critical pedagogy, as a framework, inspired my earliest teaching days and my research into learning and memory. These methods have allowed deep insight into scientific concepts and a self-motivated interest in learning. They have encouraged sincere connection and creativity in classes. The more immediate changes required in such a liberatory practice include greater self-reflection (for all parties), explicit naming of power, and changes in how power is given and taken in science classrooms. A shift toward liberatory pedagogy at some point requires an overall reassessment of the classroom environment, which is key to engagement with the projects of learning and critical consciousness. Students must be able to take risks and experience both enjoyment and productive discomfort while unlearning punitive expectations of education. Students may also have personal historical connections to oppressive practices within the sciences; trauma-informed practices (Brunzell et al., 2019, expanded on later in this piece) are imperative.

Liberatory practice is often limited by educational institutions that are stubbornly resistant to change. To work in such places requires a constant, critical awareness of how we, students and teachers, comply with hierarchical power relations and exclusionary practices. With this limitation in mind, STEM fields desperately need practitioners from liberatory traditions who are willing to view and construct power differently. Consolidated power, as illustrated by Freire, becomes ever more rigid and short-sighted. This is untenable when addressing problems as expansive as climate change, emerging pathogens, or health care justice. If ever there was a time for a living, dynamic, and liberated science, it is now.

Course Design
Practical Insights from the Neuroethics Classroom

Studying ethics can be a key component of the critical consciousness that students may carry into their social and work lives. Neuroethics casts a critical eye on the brain sciences, requiring not only basic proficiency in neuroscience but the ability to interpret scientific findings and narratives within their socio-historical context. Students in my classroom learn to take multiple perspectives and to question how neuroscientific knowledge and technologies may be used to help or harm. They may ask questions about power and participation in the sciences. They may learn to avoid the traps of biological *essentialism* in their own work. They learn the basics of logic and argumentation, become more skilled in articulating key values, and apply these skills to real-world dilemmas. Such skills are invaluable in understanding and responding to crisis and may confer resistance to misinformation.

From 2019 to 2021, I taught *Neuroethics* as an advanced undergraduate offering. Since the summer of 2020, I have taught a graduate *Behavioral Neuroscience* course, also neuroethics-based, for counseling and school psychology graduate students. In these classes, we have covered topics ranging from cognitive enhancement (Maslen et al., 2014) and implantable neurotechnologies (Mayberg et al., 2005; Kubu & Ford, 2017) to definitions of brain death (Bernat, 2014; Fins, 2016) and environmental neuroethics (Cabrera et al., 2016; Tesluk et al., 2017). We have discussed the neuroethics case against solitary confinement (Lobel & Akil, 2018) and the neuroethical implications of U.S. policies that have separated children from their parents at the US-Mexico border (Teicher, 2018).

In these classes, we discuss how neuroscientific terms and technologies can be "hyped" and sensationalized (Caulfield et al., 2010) and how these concerns inform our obligation to responsible science communication. We observe tentative cross-cultural neuroethics collaborations beginning across the globe (Rommelfanger et al., 2018). We also discuss pitfalls of biological gender essentialism (e.g., Fine, 2012) and consider how oppressive theorizing can be replaced with a more complex, liberatory view on neuroscience, gender, sex, and sexuality (Gupta, 2012; Cipolla & Gupta, 2018). Through these conversations, we engage with the history of biology, and in so doing, we enter into a critical analysis of how neuroscience is constructed and used. For future coverage of these historical topics, I highly recommend Angela Saini's (2019) *Superior: The Return of Race Science*. Saini's work not only accessibly covers the historical context of European sciences in greater depth but establishes continuity with ongoing political battles within the biological sciences as they traverse the early 21st century.

A number of practices contribute to an environment conducive to liberatory pedagogy. Course assignments are one key area for intervention. I create a flexible yet structured assignment schedule, with student input, at the start of a course. Students endorse the utility of structured due dates; flexibility then allows for fuller participation of disabled students and those juggling family duties or other important responsibilities. What this means is that I clearly define course assignments and provide tentative "due dates" at the start of the semester. Students and I then have a conversation about how to take extra time when they need it. I offer a standard grace period (from two days to one week) with no explanation needed for most assignments and additional extensions when feasible. While larger classes require different management, approaches that are both structured and flexible can apply to courses of any size; for instance, greater flexibility and control can be achieved simply by providing students with some choice about which assignments to complete (e.g., dropping assignments). Often, students report lowered stress, work is improved, and instead of negotiating due dates, we spend more time discussing ideas.

These strategies begin to move a classroom toward *universal accessibility*, which seeks to remove constructed barriers to learning. Issues of access and disability justice are beautifully addressed by Johnson et al. in this collection. "Universal Design" stands in contrast to the accommodations approach to disability, which usually entails onerous and intrusive documentation. The burdens of accommodation and the urgency of "academic ableism" are addressed at length by Jay Dolmage (2017) who also makes crucial links between ableism in higher education, its history within colonial violence and eugenics (pp. 11–20), and the potential co-optation of Universal Design, like multiculturalism, within the neoliberal university.

The fact is, minor modifications can help a wide range of students to participate more meaningfully in learning. For instance, I design universally extended testing periods for quizzes and exams; time trials are rarely useful or appropriate. A class will usually require at least twice as long as an instructor does to take the instructor's test, and I aim to provide students with time and a half beyond that (e.g., if I finish my test in 25 minutes, most students finish in 50, and time and a half can be given in a 75-min. period). Online, the time window for an exam can be extended even longer (hours or days) with questions of greater complexity that assess understanding. For instance, in such an assessment, I may ask students for an experimental design that would answer a research question or apply ethical concepts such as autonomy and justice to a hypothetical clinical case (e.g., brain injury or dementia).

I most often do not use exams in my neuroethics courses; if I do, I will not use surveillance software for remote testing. This is particularly salient for neuroethicists who are concerned with the unregulated infiltration of digital platforms into people's lives. Such software raises ethical concerns about privacy, consent, and the corrosive impact of widespread surveillance. These may include "lockdown browsers" or more extreme software that captures test taker movements, including eye tracking technologies to detect "cheating." Among other technological and bureaucratic nightmares, such technology has demonstrated racist bias and is untenable for students with certain disabilities (Barrett, 2023).

Similarly, software that detects plagiarism (overlap with available sources) must be used with care. Instructors who use such software must be competent in its use as an instructional tool and aware of its limitations. While instructors express understandable frustration and concern about plagiarism, the problem itself is a complex product of stress within educational systems and online norms and attitudes about writing. Usual responses are punitive in accordance with institutional norms. A liberatory lens, however, may reframe the problem of plagiarism as one of power, motivation, and trust. This shift in perspective also reframes the uses of writing assignments.

From a critical perspective, the teacher must wonder why a student's instinct is to adopt the words of others. I have heard students worry that their own words don't sound polished enough. They may be anxious or numb. Some are even convinced that they don't have ideas worth expressing. Plagiarism signals the student's alienation and despair, marking a systemic failure; widespread cheating and plagiarism are products of rigid, commodified, and impersonal banking models of education. Students in such a system are incentivized to avoid punishment while simultaneously seeking the highest possible grade with the least possible investment. This is capitalistic efficiency; within current societal value systems, the emptiness of plagiarism makes sense.

In my neuroethics classroom, I trust that students can learn while experiencing a range of excitement, discomfort, and ease in the classroom. Not driven by threats, students can enjoy creating and talking about their creations. Assignments based in creativity and used as a basis for discussion can be rich ground for connection, critical analysis, and growth. I have often seen this dynamic at work when making comics to explore thorny neuroethical issues (Fink, 2020b); students may draw one-panel or one-page case narratives and use them for in-class discussions of ethical dilemmas. Students may also complete similar written assignments, which are framed as a semester-long project in learning to express their stance on an issue. Early in this process, students who struggle with expressing themselves benefit from sincere encouragement and validation. Shorter, low-stakes exercises provide them first with the opportunity to create and for the creation to be received with joy. If they move beyond early discomfort, they may gain intrinsic motivation to create, enter into more authentic communication with peers, become better able to receive and give feedback and shape their own intellectual growth. Examples of assignment prompts can be found in Appendix A.

Assignments bring the specter of grading, a practice that is difficult for most instructors to avoid completely. It can be useful, again, to reconsider punitive strategies that place undue focus on a grade rather than the learning. Educators might instead make a habit of asking themselves: 1) What is the important learning that needs to happen? 2) How can students be given adequate and equitable opportunities to demonstrate that learning? This reassessment of values is fundamental to the practice of *ungrading*, the use of which in STEM classrooms is discussed by Newell-Caito in this volume.

Upon re-examination, certain assignments, grading practices, and micromanaging rubrics may appear newly onerous and unnecessary. Many of my assignments, particularly early ones, are graded on full completion and originality, prioritizing learning process over product. For instance, students may be asked to answer a few key questions and offer reflections on class readings. They receive feedback on their answers, emphasizing process, but their grades simply reflect whether they answered each question. I also engage students in discussion about what they would like to express through their work, making sure that they know their ideas and interests are valued. This approach builds competency and confidence; students organically learn to tackle more challenging reading and analysis, and they learn about their own interests.

This classroom approach also represents a thoughtful balancing of emotional arousal based on longstanding insights from stress neurobiology. Stress exists on a continuum (e.g., Herman, 2013); moderate, temporary stress can be beneficial, enhancing learning and engagement. Extreme, unremitting stress, however, is destructive to attention, emotion regulation, and learning. A well-functioning classroom may aim for a window, the peak of this curve. Students should be engaged, alert, and even productively uncomfortable at times, but they should not be stressed beyond capacity or, importantly, outside of their reasonable control. Using this window effectively is an important teaching skill; a neuroethics view might argue for the importance of trauma-informed classrooms based on an awareness of disproportionate exposure to stress in distinct populations of students (Brunzell et al., 2019).

As within health care (Sweeney et al., 2018), trauma-informed classrooms emphasize choice, collaboration, safety, and trust. One component of such a classroom might be content notifications: for instance, noting potentially activating content that depicts racism, sexism or sexual assault, or other forms of violence.

There is no way to avoid (or identify) all individual trauma triggers, but avoidance is not the point. Instead, a thoughtful content notification can normalize the fact that students may have strong emotional reactions to material and open such topics for discussion. They may then prepare and make choices about *how* to engage. The class may also collectively discuss coping strategies. This brings up other key ingredients that can be easy to implement, including an upfront discussion about the classroom environment and the co-creation of a classroom agreement that helps to shape an atmosphere of exploration.

A brief, optional, pre-course survey (see Appendix B) has proven useful for me, and anecdotally, for others, in managing access and participation in virtual and in-person classes. My teaching and learning survey inquires into students' accessibility (technological and disability) needs, concerns about the ongoing impact of the COVID-19 pandemic or other current issues, how they might best engage with the class, and their interests. Using this tool, we can address questions and concerns together in advance. This is also one way in which I convey my respect for and interest in students' experiences. Even partial co-construction of course policies is central to transforming more authoritarian course structures into more equal, discursive ones.

In-class activities breathe life into a course. I rely heavily on student-led discussion and creative methods such as comics, and drawing is central to my teaching practices (Fink, 2020a; 2020b). Creative tools can allow deep engagement with emotionally difficult topics and may allow students who are "stuck" to find their voice. Drawing may also elicit understanding that is not apparent in verbal communication. "Drawing-to-learn" (Quillin & Thomas, 2015) is effectively used by others in biology. In an excellent recent example, Edlund and Balgopal (2021) demonstrated how drawing could be used to communicate cross-cultural and spiritual meanings of neuroscience.

Artistic and narrative methods of learning may also encourage perspective taking, cultural humility, and new ways of considering social responsibility and justice. Creative approaches often provide new avenues toward critical analysis of course material. As an example, I recall using comics to explore students' imaginings of gender and biology. It was only when one student drew their depictions of gender that they noticed the many stereotypical physical features that they unconsciously assigned to their stick figures. They expressed astonishment and a realization that the image revealed a mental representation of gender that their words may have overlooked. Their insight then sparked a transformative class discussion on biological essentialism in neuroscience.

When they are making art, students are laser focused on the material at hand and more open to meaningful, spontaneous, and joyful connection between classroom participants. This is the living thing that Freire wrote of as critical consciousness (1970/1993), and this, in part, is also what bell hooks spoke of as love.

Liberatory pedagogy requires practical actions: a welcoming and vital atmosphere, policies and assignments that allow students to best demonstrate their learning, feedback, and evaluation that emphasize student strengths over punishment, and non-coercive opportunities for interpersonal connection and critical evaluation of course materials. Teaching neuroethics provides unique opportunities for such pedagogy, with key moments of insight about the biopsychosocial process of learning itself and opportunities to discuss real-world issues. When students bring discussion items and artistic creations into class as equal participants, their contributions lead to deeper and more satisfying conversations. These practices have proven their utility throughout the U.S. crises of the 2010s and the ongoing COVID-19 pandemic.

Developing Critical Consciousness During the COVID-19 Crisis

This chapter took shape during the second and third years of the COVID-19 pandemic and was molded by crises specific to this period. As the SARS-CoV-2 virus spread across the world, the WHO in 2020 urged countries to "take urgent and aggressive action" (World Health Organization, 2020). Employers and educational institutions made dramatic shifts to remote activities. Reasonable accommodations previously considered impossible or unfair (Burgstahler, 2021; Pak, 2020) were immediately implemented. The pandemic starkly illuminated the ableism of U.S. institutions and the failure of eviscerated American public health and healthcare systems. Long-standing impacts of structural racism and ableism resulted in disproportionate illness and death in marginalized communities (Acosta et al., 2021; Chowkwanyun & Reed, 2020; Quan et al., 2021).

The summer of 2020 also saw a revitalized movement to repudiate white supremacy and police brutality against Black Americans and to promote the flourishing of historically oppressed populations. Nevertheless, 2021 began with a white supremacist attack on the U.S. capitol, and targeted attacks against Black and Asian Americans continued. Populations with the least wealth and power continue to be most negatively impacted by the COVID-19 pandemic and its ensuing crises.

An orientation toward critical consciousness proved invaluable while teaching during this time, where sequences of crisis and "shock" and the politicized and social nature of science were so apparent. My neuroethics-based classes offered opportunities to contextualize these crises, discuss historical precedents, and build a sense of intellectual community in the face of potentially overwhelming problems. Students also arrived at specific insights through a neuroethics framework. For instance, some students explored the bioethics of inequities in vaccine access. Others found parallels between COVID-19 and the stigma involved in "disease" labeling of mental illnesses (Corrigan et al., 2014) or substance use (Hammer et al., 2013).

Discussing long-lasting and neurological impacts of COVID-19 also facilitated key conversations around disability rights and healthcare accessibility.

Virtual teaching and learning became the norm during the pandemic, bringing both new accessibility successes and pitfalls (Burgstahler, 2021) and highlighting existing barriers to participation in the sciences. In an isolating time, many students and teachers appreciated the safety and flexibility of virtual connections, while some encountered new hurdles in access. During the COVID-19 emergency, the federal government also took the unusual step of making emergency funds available, including resources for digital infrastructure and access. While not perfectly allocated, this aid made a tangible difference for many students. Pandemic-associated services and policies such as expanded internet access, the temporarily expanded U.S. Child Tax Credit, and federal aid for education at all levels provided a glimpse of what is possible; advocating for their continuation and expansion is an unglamorous but needed part of a justice orientation.

Teaching neuroethics in this year also cemented key justice considerations in the classroom. A liberatory approach allowed us to disengage from the frantic pace of the news cycle and to engage in slow, thoughtful analysis. Students reflected honestly on their own presence within STEM fields. They were able to observe how recent crises could be co-opted by those seeking to consolidate their power and how communication about science could be used to political ends. Overall, students expressed appreciation of discussions and assignments that allowed them to exercise their analytical muscle and connect with each other. They also endorsed benefits from drawing and other creative modalities, citing stress relief, opportunities to be more present in classes, and avenues for self-expression.

Summary and Conclusions

To close, I reiterate key features of critical pedagogy within a liberatory STEM classroom: transformation of power from a hierarchical structure to more horizontal forms; cultivation of student and teacher strengths in place of punitive strategies, critical attention to social and historical context, joining of theory and praxis, and attention to interpersonal connection in building knowledge. The instructor brings important expertise to the table, yet they may also plan to leave the classroom transformed. As recounted in this chapter, these critical methodologies were also born of crisis and present a hope for equitable, transformative, liberatory action. While this is indeed a significant undertaking, the process can begin with concrete, actionable steps. Prashad (2010) lists key ideas for the practice of activism on campuses (pp. 125–127). Similarly, I summarize key practical components toward liberatory STEM classrooms:

- Eliciting student input into course policies and structure. Examples: Pre-course survey, first-day discussions and agreements, opportunities to revisit policies.
- Creative means of learning. Examples: Drawing-to-learn (Quillin & Thomas, 2015; Edlund & Balgopal, 2021) and comics (Fink, 2019; 2020a; 2020b).
- Providing social/historical context when reading and interpreting STEM texts.
- Building awareness of power dynamics within STEM fields and classrooms; naming oppressive structures and working to change them.
- Engaging student agency through student-led discussions. Example: Students bring in a course-related item (ad, news article, etc.) and may lead a class discussion.
- Valuing quality of interpersonal relationships within STEM classrooms, labs, and in application of STEM knowledge.
- Moving toward universal accessibility. Examples: Accessible testing formats appropriate to the course (accounting for topic and size), assignments that focus on understanding, flexible due dates, multiple modes for demonstrating learning.
- Instructor feedback and grading on early assignments that encourage consistent, original engagement and avoid punitive strategies (see Newell-Caito, this collection).
- Meeting material needs of students and their communities. Example: Advocating for higher education funding through federal, state, and campus mechanisms.

These practices may be risky. They are difficult to standardize and align with traditional (banking) rubrics of academic or career achievement. Instructors using these methods also encounter risk and discomfort in sharing control of the classroom. Nevertheless, all classroom participants may benefit from pedagogical methods that enhance the agency of students and engage their intrinsic creativity, interest, and ability to build relationships.

Teaching neuroethics during the COVID-19 pandemic, above all, highlighted the importance of compassion, particularly in an atmosphere that pushes productivity amidst widespread death and suffering. Students and instructor alike worked to name what was happening and to articulate pressing moral problems and mental distress arising from the crisis. Because students with marginalized identities (lower-income students, disabled students, students of color, and LGBTQIA+ students) are more heavily impacted, academic spaces that can adequately serve these students gained even greater importance (e.g., Gilbert et al., 2021). Students in the neuroethics classroom engaged compassionately in a way that is too often inaccessible in the sciences.

Humility is warranted when making claims to liberatory practice within Western educational institutions. Practitioners must decide how and when they will resist oppressive practices around them, knowing that this also, inevitably, involves risk. Additionally, academia abounds with buzzwords that deflect from needed radical restructurings; this requires that teachers and students take stock of efforts that operate on tokenism or serve a public relations purpose. Practitioners must confront their limits and the ongoing tension between their liberatory aspirations and institutional inertia; this, too, is praxis.

As educators, it is crucial to hold the hope that any class can erupt in moments of transformation and connection, even within imperfect classrooms and history-bound institutions. And it is important to think beyond the institution. Liberatory theory and praxis in STEM fields cannot be confined to a single classroom or the goals of academic career advancement. Instead, the success of liberatory pedagogy can be observed by the extent to which students and teachers can make sense of the wider world and act on it and with it. Through such joint action, they might come to enact humanizing narratives and technologies within and beyond states of crisis.

References

Abe, J. (2020). Beyond cultural competence, toward social transformation: Liberation psychologies and the practice of cultural humility. *Journal of Social Work Education*, 56(4), 696–707. https://doi.org/10.1080/10437797.2019.1661911.

Acosta, A. M., Garg, S., Pham, H. Whitaker, M., Anglin, O., O'Halloran, A., Milucky, J., Patel, K., Taylor, C., Wortham, J., Chai, S. J., Kirley, P. D., Alden, N. B., Kawasaki, B., Meek, J., Yousey-Hindes, K., Anderson, E. J., Openo, K. B., . . . & Havers, F. P. (2021). Racial and ethnic disparities in rates of COVID-19–associated hospitalization, intensive care unit admission, and in-hospital death in the United States from March 2020 to February 2021. *JAMA network open*, 4(10), e2130479–e2130479. https://doi.org/10.1001/jamanetworkopen.2021.30479 .

Adams, M. & Love, B. J. (2010). A social justice education faculty development framework for a post-Grutter era. In Skubikowski, K., Wright, C. & Graf, R. (Eds.), *Social justice education: Inviting faculty to transform their institutions*. (pp. 3–25). Stylus Publishing.

Ahmed, S. (2012). *On being included: Racism and diversity in institutional life*. Duke University Press.

Barrett, L. (2023). Rejecting test surveillance in higher education. *2022 Michigan State Law Review*, 675. http://dx.doi.org/10.2139/ssrn.3871423.

Bernat, J. L. (2014). Whither brain death? *The American Journal of Bioethics*, 14(8), 3–8. https://doi.org/10.1080/15265161.2014.925153.

Brunzell, T., Stokes, H. & Waters, L. (2019). Shifting teacher practice in trauma-affected classrooms: Practice pedagogy strategies within a trauma-informed positive education model. *School Mental Health*, 11(3), 600–614. https://doi.org/10.1007/s12310-018-09308-8.

Burgstahler, S. (2021). What higher education learned about the accessibility of online opportunities during a pandemic. *Journal of Higher Education Theory & Practice, 21*(7). https://doi.org/10.33423/jhetp.v21i7.4493.

Cabrera, L. Y., Tesluk, J., Chakraborti, M., Matthews, R. & Illes, J. (2016). Brain matters: From environmental ethics to environmental neuroethics. *Environmental Health, 15*(1), 20. https://doi.org/10.1186/s12940-016-0114-3.

Caulfield, T., Rachul, C. & Zarzeczny, A. (2010) "Neurohype" and the name game: Who's to blame? *AJOB Neuroscience 1*(2), 13–15. https://doi.org/10.1080/21507741003699355.

Chowkwanyun, M. & Reed, A. L. (2020). Racial health disparities and Covid-19—Caution and context. *New England Journal of Medicine.* https://doi.org/10.1056/NEJMp2012910.

Cipolla, C. & Gupta, K. (2018). Neurogenderings and neuroethics. In Johnson, L. S. M. & Rommelfanger, K. S. (Eds.), *The Routledge handbook of neuroethics* (pp. 381–393). Routledge.

Comfort, N. (2018). Genetic determinism redux. *Nature, 561,* 461–463. https://doi.org/10.1038/525184a.

Corrigan, P. W., Druss, B. G. & Perlick, D. A. (2014). The impact of mental illness stigma on seeking and participating in mental health care. *Psychological Science in the Public Interest, 15*(2), 37–70. https://doi.org/10.1177/1529100614531398.

Crenshaw, K. (1989) Demarginalizing the intersection of race and sex: A Black feminist critique of antidiscrimination doctrine, feminist theory and antiracist politics. *University of Chicago Legal Forum, 1*(8), 138–167. https://tinyurl.com/a3wubcrp.

Crenshaw, K. (1991). Mapping the margins: Intersectionality, identity politics, and violence against women of color. *Stanford Law Review, 43*(6), 1241–1299. https://doi.org/10.2307/1229039.

Dolmage, J. T. (2017). *Academic ableism: Disability and higher education.* University of Michigan Press.

Edlund, A. F. & Balgopal, M. M. (2021). Drawing-to-learn: Active and culturally relevant pedagogy for biology. *Frontiers in Communication,* 203. https://doi.org/10.3389/fcomm.2021.739813.

Fine, C. (2012). Explaining, or sustaining, the status quo? The potentially self-fulfilling effects of 'hardwired' accounts of sex differences. *Neuroethics, 5,* 285–294. https://doi.org/10.1007/s12152-011-9118-4.

Fink, A. E. (2019). Fanon's police inspector. *AJOB Neuroscience, 10*(3), 137–144. https://doi.org/10.1080/21507740.2019.1632970.

Fink, A. E. (2020a). *Graphic neuroethics: A comics-making curriculum (Part I of II).* The Neuroethics Blog. [archived]. https://tinyurl.com/yc3aw7c6.

Fink, A. E. (2020b). *Graphic neuroethics: A comics-making curriculum (Part II of II).* The Neuroethics Blog. [archived]. https://tinyurl.com/52pt7js3.

Fins, J. J. (2016). Giving voice to consciousness: Neuroethics, human rights, and the indispensability of neuroscience. *Cambridge quarterly of healthcare ethics, 25*(4), 583–599. https://doi.org/10.1017/S0963180116000323.

Freire, P. (1993). *Pedagogy of the oppressed* (Ramos, M.B., Trans.). Continuum. (Original work published 1970).

Fuentes, A. (2012). *Race, monogamy, and other lies they told you: Busting myths about human nature*. University of California Press.

Gilbert, C., Siepser, C., Fink, A. E. & Johnson, N. L. (2021). Why LGBTQ+ campus resource centers are essential. *Psychology of Sexual Orientation and Gender Diversity*, 8(2), 245–249. https://doi.org/10.1037/sgd0000451.

Gould, S. J. (1996). Measuring heads. In *The mismeasure of man* (2nd ed). (pp. 135–141). W.W. Norton and Company, Inc.

Gupta, K. (2012). Protecting sexual diversity: Rethinking the use of neurotechnological interventions to alter sexuality. *AJOB Neuroscience*, 3(3), 24–28. https://doi.org/10.1080/21507740.2012.694391.

Hammer, R., Dingel, M., Ostergren, J., Partridge, B., McCormick, J. & Koenig, B. A. (2013). Addiction: Current criticism of the brain disease paradigm. *AJOB neuroscience*, 4(3), 27–32. https://doi.org/10.1080/21507740.2013.796328.

Herman, J. P. (2013). Neural control of chronic stress adaptation. *Frontiers in Behavioral Neuroscience*, 7, 61. https://doi.org/10.3389/fnbeh.2013.00061.

hooks, b. (1989a). Feminism: a transformational politic. In *Talking back: thinking feminist, thinking black*. (pp. 19–27). South End Press.

hooks, b. (1989b). Toward a revolutionary feminist pedagogy. In *Talking back: thinking feminist, thinking black*. (pp. 49–54). South End Press.

hooks, b. (1994). *Teaching to transgress: Education as the practice of freedom*. Routledge.

Hue, G. E. (2020). Justice, justification, and neuroethics as a tool. *AJOB neuroscience*, 11(4), 221–223. https://doi.org/10.1080/21507740.2020.1838165.

Klein, N. (2007). *The shock doctrine: The rise of disaster capitalism*. Macmillan.

Kubu, C. S. & Ford, P. J. (2017). Clinical ethics in the context of deep brain stimulation for movement disorders. *Archives of Clinical Neuropsychology*, 32(7), 829–839. https://doi.org/10.1093/arclin/acx088.

Lobel, J. & Akil, H. (2018). Law & neuroscience: The case of solitary confinement. *Daedalus*, 147(4), 61–75. https://doi.org/10.1162/daed_a_00520.

Longino, H. E. (1990). *Science as social knowledge: Values and objectivity in scientific inquiry*. Princeton University Press.

Maslen, H., Faulmüller, N. & Savulescu, J. (2014). Pharmacological cognitive enhancement—how neuroscientific research could advance ethical debate. *Frontiers in Systems Neuroscience*, 8, 107. https://doi.org/10.3389/fnsys.2014.00107.

Mayberg, H. S., Lozano, A. M., Voon, V., McNeely, H. E., Seminowicz, D., Hamani, C., ... & Kennedy, S. H. (2005). Deep brain stimulation for treatment-resistant depression. *Neuron*, 45(5), 651–660. https://doi.org/10.1016/j.neuron.2005.02.014.

Pak, C. (2020, April 30). *Disability, visibility and the COVID-19 crisis*. Medical Humanities. https://tinyurl.com/2wjzmxr7.

Panofsky, A. (2015). What does behavioral genetics offer for improving education? *Hastings Center Report*, 45(S1), S43–S49. https://doi.org/10.1002/hast.498.

Prashad, V. (2010) On commitment: Considerations on political activism on a shocked planet. In Skubikowski, K., Wright, C. & Graf, R. (Eds.), *Social justice education: Inviting faculty to transform their institutions*. (pp. 117–127). Stylus Publishing.

Quan, D., Luna Wong, L., Shallal, A., Madan, R., Hamdan, A., Ahdi, H., Saneshvar, A., Mahajan, M., Nasereldin, M. Van Harn, M., Opara, I. N. & Zervos, M. (2021).

Impact of race and socioeconomic status on outcomes in patients hospitalized with COVID-19. *Journal of General Internal Medicine, 36*(5), 1302–1309. https://doi.org/10.1007/s11606-020-06527-1.

Quillin, K. & Thomas, S. (2015). Drawing-to-learn: A framework for using drawings to promote model-based reasoning in biology. *CBE—Life Sciences Education, 14*(1), es2. https://doi.org/10.1187/cbe.14-08-0128.

Rommelfanger, K. S., Jeong, S-J., Ema, A. Fukushi, T., Kasai, K., Ramos, K. M., Salles, A. & Singh, I. (2018). Neuroethics questions to guide ethical research in the international brain initiatives. *Neuron, 100*(1), 19–36. https://doi.org/10.1016/j.neuron.2018.09.021.

Saini, A. (2019). *Superior: The return of race science*. Beacon Press.

Schiebinger, L. L. (1993). *Nature's body: Gender in the making of modern science*. Beacon Press.

Sweeney, A., Filson, B., Kennedy, A., Collinson, L. & Gillard, S. (2018). A paradigm shift: Relationships in trauma-informed mental health services. *BJPsych Advances, 24*(5), 319–333. https://doi.org/10.1192/bja.2018.29.

Teicher, M. H. (2018). Childhood trauma and the enduring consequences of forcibly separating children from parents at the United States border. *BMC Medicine, 16*, 1–3. https://doi.org/10.1186/s12916-018-1147-y.

Tesluk J., Illes J. & Matthews, R. (2017). First nations and environmental neuroethics: Perspectives on brain health from a world of change. In Illes, J. (Ed.), *Neuroethics* (pp. 455–476). Oxford University Press. https://doi.org/10.1093/oso/9780198786832.003.0023.

World Health Organization. (2020). *Timeline of WHO's response to COVID-19*. https://www.who.int/news-room/detail/29-06-2020-covidtimeline.

Appendix A: Examples of Assignment Prompts for Undergraduate Neuroethics Course

"Neurobiological definitions of mental health and illness" (an earlier writing assignment)

This week's readings ask us to consider neurobiological definitions of mental health and illness. You will identify some important ethical questions relating to the personal, social, and clinical implications of such biological definitions. Address the following:

1. **Start with an introductory paragraph.**
2. **Consider the three papers and briefly describe:**
    - What is the primary problem that each set of author(s) raises (i.e., why have they written the article)?
    - What is / are the main argument(s) or prescription(s) offered in each paper?

3. **Provide your analysis and reflect on the papers:**
    - What seems to be the scholarly background of the author(s) of each paper? Reflect on how the authors' field of expertise shapes their questions and conclusions.
    - After reading these papers, what do you think is (at least one) potential harm and (at least one) potential benefit of using neuroscience-based information to define mental health and illness? *Be specific.*
    - What recommendations do you have for the responsible use of such information? Name the ethical principles that lead you to argue for these recommendations.

"Neuroethics of education and child development" (an earlier graphic / comics assignment)

The readings for this week explore how neuroscience and technology can specifically impact the lives of children, adolescents, and young adults. You may also see recurring themes from earlier course material. This time, you will draw your responses to the readings. In this assignment, we have the chance to think about these topics creatively.

1. **The authors discuss "raising children" versus "designing children":**
    - Draw what you think "raising children" looks like.
    - Draw what you think "designing children" looks like.
2. **Illustrate an ethical concern that you have regarding neuroscience-based educational interventions.**

"Neuroethics of gender, sex, sexuality, and love" (late in semester assignment):

YOUR CHOICE! You can submit a written reflection or a graphic reflection. Aim for 2–4 pages regardless of format.

1. **Describe the primary ethical concern raised by Fine (2012) regarding the interaction of neuroscience research with "gender" and "sex."**
    - Here, address: What is "gender"? How does "gender" relate to "sex"?
    - **What is "love"?**
    - Can love ever be seen as an illness? Illustrate or explain your answer.

2. **What is "sexuality"?**
    - How might "sexuality" interact with definitions of "love" and/or "gender"?

Appendix B: Pre-Course Online Teaching and Learning Survey

Welcome to [Course Name]! I am looking forward to getting to know each of you in our (virtual) classroom. Before we start this course, I hope that you will take a few minutes to complete this survey. Your answers will be kept private—I do not share them with anybody else—and this questionnaire can help me to learn more about your interests as well as your learning and accessibility needs.

Item 1: What is your full name? You can also let me know here if your chosen name is different from your roster name.

Item 2: If you'd like to share your gender pronouns with me, please do:

Item 3: Do you have reliable internet and computer access?

Item 4: Will you be able to participate reliably in class sessions on [days] at the scheduled times? If not, please tell me more.

Item 5: Do you have access to [required software, e.g., . . . suite]? [*If applicable, provide information about how to obtain software through institution or course site*].

Item 6: Do you have any accessibility concerns regarding course readings or other materials? [*Provide other relevant info here: e.g., "All readings will be provided as PDFs . . . "*].

Item 7: Do you have any specific concerns or needs regarding online learning [*if virtual*]?

Item 8: Are there any issues that you would like to share with me (e.g., COVID-19 or other illness, work, family responsibilities) that may impact your participation in the class?

Item 9: What will help you to engage successfully with this course?

Item 10: How can this _____ course contribute to your growth as a _____ student? Any topics of special interest?

Item 11: Is there anything else that you would like me to know?

Conclusion: Lessons from the Front Lines

LaKeisha McClary
THE GEORGE WASHINGTON UNIVERSITY

Heather M. Falconer
UNIVERSITY OF MAINE

This book has focused heavily on the theme of increasing feelings of belonging in STEM, particularly for individuals from historically marginalized populations in these disciplines, but not exclusively. The approaches discussed throughout the text are to the benefit of all STEM students, not just those who have historically encountered additional challenges to access. For us, belonging is about more than just being welcomed into a space. It's about having a secure attachment to the community. It is about feeling as though you can speak, push back, and offer new approaches without suffering consequences. We help facilitate a sense of belonging for students by constructing spaces that illustrate the variety of viewpoints and backgrounds individuals within the community hold. We make visible in our assignments and curriculum the diversity that exists so that we don't create an impression that only one group of individuals participates in the procedural and knowledge-making tasks of our field. We clear a path to belonging when we create spaces that allow all of our students to thrive, regardless of how they arrived; we meet students where they are, not where we think they should be, and we help them to grow. Belonging happens when we listen to the many voices that have contributed to our discipline's ways of knowing over time and share those voices with our students, when we create space for multiple epistemologies. We foster belonging when we are accountable—accountable to our students, to our colleagues, and to our disciplines. This includes holding *others* accountable for harm that is done. And that is challenging but necessary work. By engaging with this text, you have already shown a commitment to (and have begun) doing that work, so we are working from the assumption that you see its value. This concluding chapter offers some reflection and resources for continuing that journey.

    The organization in this chapter was heavily inspired by Rebecca Walton, Kristen R. Moore, and Natasha N. Jones' (2020) *Technical Communication After the Social Justice Turn*—a book we highly recommend educators from all disciplines read, despite the title's disciplinary reference. We begin with some reflection: Amplifying lessons each of us has learned or been inspired by while working on the collection. These are followed by a series of questions that we commonly encounter with individuals starting this work of building awareness, as well as our answers, taking into account our position as members of different disciplines. These answers are followed by short lists of resources that readers can use to further explore the specific topic areas. Not all resources listed explicitly address writing instruction from an equity perspective, but we do believe they illustrate practices that will lead toward that goal. We recognize that these resources are not exhaustive; rather, our goal is to provide entry points.

Reflections from a Chemistry Perspective—LaKeisha McClary

When I was first asked to be a co-editor for this collection, I was hesitant. Yes, I have taught for over ten years a writing in the disciplines (WID) course, but I was not sure what I could contribute. I had no formal training in writing pedagogy, and I learned mostly by teaching a lab-based writing course, CHEM 2123W, semester after semester. I also was not aware of much scholarship and research that existed at the intersection of equity and disciplinary writing. In fact, thanks to the Writing Program at GW providing me with a paid membership to the Association for Writing Across the Curriculum in 2020, I had only recently learned that there were entire areas of research on writing. Of course there are! But I never knew it.

Like most STEM Ph.D.s, I received undergraduate and graduate training in science departments that focused their curricula on lab-based scientific research. It was not until I transferred Ph.D. programs to pursue chemistry education research (which I also was unaware of until I attended an American Chemical Society Conference while pursuing a Ph.D. in organic chemistry) that I learned about the rich legacy of social justice efforts in K–12 spaces through my graduate education courses at The University of Arizona. But my day-to-day in a chemistry department never explicitly considered how social justice frameworks could improve student outcomes in STEM. Even though I was interested in doing research at this intersection of social justice and chemistry, I made a choice to follow a road that would be more likely to lead to a position within a chemistry department. It was already a risk at that time to pursue an academic career in chemistry education research. [What then did it look like for a Black woman with an afro puff to be talking about promoting social justice in chemistry education? Confident though I may be, I self-censored to be employable.]

Now, however, the stakes are even higher in U.S. higher education following the COVID-19 pandemic and forecasts for lower college enrollments taking place among the conversations surrounding college affordability. I choose to no longer be complicit but to become part of the solution to the challenges we face in producing a diverse pool of STEM professionals and STEM educators. I am convinced that many of the challenges we face in STEM can be addressed through effective writing pedagogies that are inclusive and incorporated consistently in higher education so that students are repeatedly provided opportunities to practice different genres of science writing within their interdisciplinary programs of study. Even as co-editor of this collection, I still have barely scratched the surface. I am grateful for each of the authors of this collection for their commitment to shifting the paradigm of what and how we can educate STEM students for a more socially just future. Their work is an accessible entry point to STEM faculty like me, who are deeply committed to equity in STEM but with limited knowledge of how to do it in our disciplinary spaces. I hope that, like me, you have added to your vocabulary and have a framework within which to reflect critically on teaching and assessment practices in your writing and non-writing courses. And most importantly, I hope that, like me, you will continue the conversation with colleagues on your campus and within your disciplines.

I end with a personal call to action for different stakeholders employed in colleges and universities:

> **Writing program directors:** Visibility is crucial. My daily life is in my department, and I forget that we have writing professionals and workshops available to assist me in sharing resources about writing pedagogy and writing studies broadly and within STEM/science. Consider pooling resources to make your workshops available to faculty at other institutions. We talk about inclusive STEM at the classroom or program level, but let's also extend this to a cross-institution level.
>
> **College deans and university provosts:** Increase funding to support the efforts that Writing Centers are engaging in. Provide fellowships for faculty to have course releases to spend time developing or refining a curriculum that supports writing-to-learn (WTL), writing across the curriculum (WAC), and WID within STEM disciplines. Hire faculty with expertise in STEM writing studies. Furthermore, fund graduate level courses targeting Ph.D. STEM students; we need courses that teach professional writing skills and writing studies research. Those graduates who will remain in academia need this valuable and relevant educational experience to better prepare undergraduates whom they will teach and graduate student researchers whom they will mentor. Every Ph.D. graduating with a STEM degree who pursues an academic career should be equipped to effectively teach writing courses with equity built in from the beginning.

And truthfully, regardless of their career path, every STEM student should be required to take at least one science writing or writing-intensive science course every semester as part of their program of study. In a time when university budgets are strained, preparing faculty and students who can meet these challenges is a wise investment.

Faculty: I have yet to attend a faculty meeting on my campus where STEM research faculty spend as much time arguing for resources to support writing in their courses as they do for resources to support their research. What good are discoveries in science if we cannot prepare students to create and consume science communication that can reach a wide range of audiences? Empirical research is clear: Writing is an effective way to help students learn conceptually within their traditional STEM courses and to learn professional writing skills in research-based and WID courses. Students are underprepared to engage in science writing in my third-year WID course (CHEM 2123W) because they do not consistently engage in science writing in their pre-requisite STEM courses, including ones offered by my own department. Fortunately, I do see much improvement when some of these same students enroll in the writing course (CHEM 4195W) that accompanies our undergraduate research course, in part because students are able to practice honing their skills in writing-intensive laboratory courses for which 2123W is a pre-requisite.

Reflection from a Writing Perspective— Heather M. Falconer

In 2021, I had the opportunity to conduct a workshop on incorporating social justice into STEM courses through the Boston Rhetoric and Writers Network (BRAWN). One of the first things we did in that workshop was reflect on what it means to belong; to really think about what that looks like. But to get there, we had to reflect on times we *did not* feel that sense of belonging—mostly because those experiences are often easier to conjure in our minds. Some of the responses in the anonymous Google Jamboard included:

> "Subtle social cues—the unspoken—made me feel out of place. My jokes don't land, and I don't get *their* jokes. We don't care about the same things."

"People were talking at/through/around me. People were not interested in what I had to contribute. I didn't feel comfortable sharing my opinions, thoughts, or feelings."

"There was little interaction or acknowledgement of my presence. It was clear that I had to adapt to the people there; there was a palpable sense of exclusion."

"I did not 'get' what others were talking about, in terms of language, sometimes—but just as often in terms of topics, or activities that they apparently had shared."

As I have been working on this collection, reading the stories and activities that our chapter authors have contributed, my mind has often wandered back to that workshop and the experiences people identified as making them feel unwelcome or not belonging in a space. I can imagine that any one of us reading those comments can immediately recollect a time when we felt something similar. It's easy enough to say, "I don't ever want to make someone else feel that way!"; it is less easy to say, "This is how I make sure students in my class *don't* feel that way." That last part is what this collection has done such a nice job of addressing. The authors have offered us specific, actionable things we can do in the classroom to recognize a diversity of viewpoints, to make sure our students are reflected in the space, and to help our students feel their perspectives and experiences are valid and that they are not out of their depth.

As I have read, though, I have found myself both challenged and inspired. I've wondered about which assignments currently in rotation could realistically be swapped out and still meet my learning objectives. I've found myself stopping to ask whether my pedagogical choices in the last few years have swung too far into the traditional realm after experiencing some pushback in student evaluations of teaching about being *too* social justice-focused. (Blomstedt's chapter, in particular, has caused me to bring back discussions of linguistic bias to my STEM writing classrooms.) I've thought back on my challenges and outright failures of "ungrading" in the classroom and wondered whether I had just done it wrong. I've wondered, as someone who *does* this stuff all the time, whether I have the time and energy to try something new.

Why have I lifted the curtain to show what's happening behind the scenes in *my* mind? Because it's important to acknowledge that this work never gets "easy." Not in the way we might hope, anyway. These collection authors have offered us a way in—a way *through*. We can't go over or under and still land in the same place; we have to reckon with the brambles and mud and mosquitos first.

I have been inspired by the work these authors are doing at their respective institutions, and it gives me such hope for the future. Like stones in a cairn, each one contributing to the spire, we can work collectively to shift the way STEM educators and practitioners (both emerging and seasoned) think about who can do this work and what kind of work they can do. Though each chapter is presented within the confines of highly specific courses, the practices transcend such spaces and are applicable broadly. For example, Barlow and Quave discuss explicitly teaching ontology and epistemology and how that impacts the methodologies we use. Weaving in these different perspectives helps students see that there are multiple ways of creating knowledge—a theme also taken up by Bitler and Oraby and extended to considering multiple accounts of history and considerations of interdisciplinarity. These reflective, contemplative approaches are not relegated to a STEM classroom; they can easily be replicated in any disciplinary space, including rhetoric and composition.

Similarly, thinking administratively, many of the contributions have given me an opportunity to consider ways of practically integrating these ideas, concepts, and activities into our existing structures. Burry et al. remind us how it is possible to build in considerations of equity and inclusion *programmatically* by incorporating explicit questions about power dynamics, erasure, and the reification of inequity within organizations and systems. Callow and Shelton beautifully illustrate the balance between addressing the *content* students need to learn with presenting capacious ways to critically examine that content. At the same time, they remind us that, in addition to designing great courses, it is just as important to work with institutional partners to ensure the overall success and adoption of such courses (an issue addressed in many chapters in the collection, including Bitler and Oraby, Barlow and Quave, Riedner et al., and Mallette).

Having partners and an open dialogue are important, as well, for finding a common language across disciplines. In reading Seraphin's chapter on non-disposable assignments, I couldn't help but think about how similar these are to the meaningful writing activities discussed by Eodice, Geller, and Lerner (2016). While not exactly the same, the fact that both emerged in very different disciplinary spaces, with different names, but never overlapped in scholarship has made me wonder what other kinds of activities might be showing up under different guises throughout our institutions. At a recent discussion about undergraduate research experiences, I was struck by how many different names are used throughout my institution to, essentially, label activities that get students involved in the process of learning (such as the course-based undergraduate research experience that Newell-Caito discusses). Why are we so siloed in this work, and how can we break down those silos so that we all can benefit from shared knowledge? In short, working with these authors has taught me much while raising even more questions. I have learned something from each of them that I am empowered to bring forward into my own teaching and research.

How do I begin to understand inequity in STEM disciplines?

Heather: If we are being honest, inequity in STEM spaces is directly connected to inequity in education broadly. This isn't *just* a STEM issue. What helped me early on in this journey was learning about the educational infrastructures in the US—how they have been shaped historically, the ways in which assessment measures have been implemented and institutionalized, access to education based on gender and race, etc. Understanding, even only superficially, the ways biases have influenced the way we teach, what we teach, and so on helps peel back the curtain and shift responsibility. If students are not performing at a level we expect as educators, then it's on us to figure out why rather than assume a deficit in the student. Blomstedt's chapter in this collection does a wonderful job of highlighting the ways in which language, for example, can impact not only how faculty perceive students but also how students perceive themselves as both writers and scientists.

Reading books like Stephen J. Gould's *The Mismeasure of Man* and Rebecca Skloot's *The Immortal Life of Henrietta Lacks* was eye-opening for me because they unpack, historically, the way race has played a prominent role in scientific knowledge-making (whether that is about who was allowed to do scientific work or the physical exploitation of historically minoritized groups in the name of scientific knowledge-making). The key thing to remember, though, is that these historical accounts are illustrating how ideas become part of the institution and that just because they're historical accounts does not mean that the perpetuation of biases are history. The bias has been built in from the start, so our job as educators is to try to understand which parts need an overhaul and to question our own assumptions as we go. Scholars like Chanda Prescod-Weinstein's Decolonizing Science Reading List (https://tinyurl.com/yjyfwc9u) and Priya Shukla's Diversity, Equity, and Inclusion in Science: A Reading List (https://tinyurl.com/2sprd5rw) are living curations that provide a way into this knowledge. Approaching inequity in STEM from this angle means that we can step away from casting blame on 'a few bad actors' in the past and take an active role and responsibility in remediating that harm ourselves. From a writing studies perspective, that means that I need to actively think about linguistic bias in disciplinary writing spaces and how that is enforced in STEM journals and granting agencies (publishing and funding are currency, after all).

LaKeisha: Understanding the roots and manifestations of inequities in STEM disciplines is one way to begin to chart a path forward. The same approaches that we use when entering a new research area are helpful here: scholarship and good old-fashioned open-minded conversations with knowledgeable people. Heather highlights some great scholarly resources to begin a journey. Sharing the journey with students and colleagues can be equally impactful in moving toward a more inclusive science education. Are you able to start a faculty learning community or a journal/book club around a theme of learning about inequities in STEM disciplines? What opportunities exist on your campus or nearby campuses to learn more from students, colleagues, and outside experts about inequities, their root causes, their manifestations and harm in STEM, and the ways that others are addressing those harms? Are there organizations that you can join that offer such opportunities? Until solutions are as pervasive as the harms, those of us who want to understand inequity in STEM will have to be proactive and seek resources or even create them ourselves within our spaces.

In my own spaces on my campus, I strive to listen to as many voices as possible, particularly student voices, so that I can make informed pedagogical decisions. Being able to hear from students enrolled in courses featured in our collection is something I appreciate and am very grateful for because it really helped me to consider how students in my WID laboratory course might respond to the assignment or assessment practice. As instructors, we are the experts, but students are the experts of their lived experiences. How do we make science education work for more of them? How do we understand how inequities in their prior education—including in other courses taken on our campuses!—influence their experiences in our courses? What can we learn from students about their other courses to make ours more inclusive and just? Such reflexive questions and an inquiry-driven approach are great guides for the journey toward understanding and empathy. Lastly, I will add that seeking to understand does not necessarily mean that you have to solve a great societal problem that has centuries-old roots. But I would argue we can start chipping away with each course we teach. I recommend *Humble Inquiry: The Gentle Art of Asking Instead of Telling* by Ed Schein for a framework to approach creating a dialogue with students, colleagues, and administrators around how to make STEM disciplines more inclusive.

Resources for Continuing the Journey

Bian, L., Leslie, S. & Cimpian, A. (2017). Gender stereotypes about intellectual ability emerge early and influence children's interests. *Science (American Association for the Advancement of Science), 355*(6323), 389–391. https://doi.org/10.1126/SCIENCE.AAH6524.

Falconer, H.M. (2022). *Masking inequality with good intentions: Systemic bias, counterspaces, and discourse acquisition in STEM education.* The WAC Clearinghouse; University Press of Colorado. https://doi.org/10.37514/PRA-B.2022.1602.

Gould, S. J. (1981). *The mismeasure of man.* Norton.

McGee, E. O., Robinson, W. H., Baber, L. D., Chapman, R., Cox, M. F., Madden, K., Pereira, P., Rezvi, S., Trinder, V. F. & Martin, D. B. (2019). *Diversifying STEM: Multidisciplinary perspectives on race and gender.* Rutgers University Press. https://doi.org/10.36019/9781978805712.

McGee, E. O. & Martin, D. B. (2011). "You would not believe what I have to go through to prove my intellectual value!" Stereotype management among academically successful black mathematics and engineering students. *American Educational Research Journal, 48*(6), 1347–1389. https://doi.org/10.3102/0002831211423972.

Schein, E. H. (2013). *Humble inquiry: The gentle art of asking instead of telling.* Berrett-Koehler Publishers.

Skloot, R. (2010). *The immortal life of Henrietta Lacks*. Crown.

Torres, L.E. (2012). Lost in the numbers: Gender equity discourses and women of color in science, technology, engineering and mathematics (STEM). *The International Journal of Science in Society, 3*(4), 33–45. https://doi.org/10.18848/1836-6236%2FCGP%2FV03I04%2F51352 .

How do I find contributions to my field from historically minoritized scholars? Isn't doing this an example of bias?

Heather: These are such important questions for many reasons, most notably because they highlight citation bias, but also because they raise ethical questions about whether we should include someone just because they are from a historically minoritized background in our discipline. Citation bias is a well-documented phenomenon (see resources below, as well as Barlow and Quave, this volume) that involves the conscious or unconscious citation of scholars whose worldview supports our own. This might include the outcomes of a study, but also include citing scholars we know and have faith in their authority, journals we publish in (or wish to publish in), and even citing ourselves. The challenge with citation bias is that it tends to create an insular bubble where the same people are repeatedly cited on certain topics, even when others have successfully published scholarship that agrees with, challenges, or complicates those findings. It isn't a phenomenon exclusive to STEM by any means, but it certainly contributes to the silencing of particular voices in those spaces.

The latter question has some unarticulated assumptions associated with it, though, that we have to confront individually. Do we believe that the systems of publication and recognition in our field are fair and equal? Do we believe that an individual's identity and lived experience create unique lenses that might impact how they view the world and, as a result, the contributions to scholarship they might make? I firmly believe that, no, we should not include an individual scholar's work based solely on their identity markers. They should be included based on the value of the contribution. For me, though, the value of the contribution sometimes lies *specifically* in the different-from-me viewpoint and interpretation that is being offered.

So, how do we find those scholars? Due diligence. There is no simple way to go about this because of the history of citation bias. It takes time and conscious effort, as well as use of your favorite search engine and library database. Begin with the subject area or lesson at hand. Are you teaching about DNA or cell organelles? Consider including Rosalind Franklin or Barbara McClintock in the discussion (with mention of how they are often left out of such discussions). I personally love to start with resources like The Visionlearning Project https://www.visionlearning.com/en/ , which provides open-access educational materials, including learning modules on scientific communication, profiles of underrepresented individuals in

STEM, and the process of science. There is also the science podcast This World of Humans (TWOH) https://www.visionlearning.com/en/twoh , which is dedicated to recent advances in biology and social science and emphasizes scholarship from scientists from communities minoritized in STEM. (TWOH is a collaboration with Visionlearning and includes teaching resources to aid science instructors in using this podcast and its featured science in their classrooms.) Teaching the history of a thing is a lot like doing a review of the literature. How do we know what we know? Only, here we are also highlighting *who* contributed to that knowledge and as many empirically valid perspectives as possible. (Such validity includes qualitative methodologies, like ethnography, as well as Indigenous ways of knowing).

When building your lesson plans and curriculum, include current scholars (including those from your own institution) and look them up. These days, it isn't too hard to find biographical information about scholars from institutional and professional networking websites. If you use social media, consider exploring hashtags like #CiteBlackWomen (which also has a podcast), #BlackInStem, #WomenInStem, or #DisbilityInStem. Consider also looking at national organizations like the National Society of Black Engineers, Society for Advancement of Chicanos/Hispanics and Native Americans in Science, Society of Asian Scientists and Engineers, National Association of Mathematicians, and National Action Council for Minorities in Engineering to see which members are being highlighted in the publications and doing work in your field. Within the field of writing studies, Cana Uluak Itchuaqiyaq maintains the Multiply Marginalized and Underrepresented (MMU) Scholars List (https://www.itchuaqiyaq.com/mmu-scholar-list), as well as the MMU Bibliography (https://tinyurl.com/3z8eh5ek), which provides names and scholarship of self-identified MMU scholars in technical communication and related fields. (As a side note: When you *do* find and include these scholars in your teaching and research, make sure you cite them in relevant publications!)

LaKeisha: Heather unpacks these questions beautifully. I would add that using professional gatherings to seek out and to engage historically minoritized scholars about their work is particularly important. Gatherings can include conferences, symposia, local meetings of professional organizations; build a network of diverse scholars, researchers, and practitioners. Is there an opportunity to collaborate on research projects? Might students or other colleagues at your institution benefit from their scholarship in their learning, teaching, and research? When inviting speakers for departmental seminars or colloquia, consider scholars from minority-serving institutions, Tribal Colleges, and primarily undergraduate institutions. Also, consider scholars and teacher-scholars who are actively working to create inclusive spaces; many will showcase these on their faculty websites. Graduate students from historically minoritized communities are often acutely aware of faculty researchers with whom they share identities, so invite them to share or contribute to a list of speakers that they would like to see at department seminars.

Resources for Continuing the Journey

Cite Black Women Collective. https://www.citeblackwomencollective.org/.

Leng, G. & Leng, R. I. (2020). *Unintended consequences: The perils of publication and citation bias.* The MIT Press Reader. https://tinyurl.com/mwmvy7yc.

Krupnik, I. & Jolly, D. (Eds.). (2002). *The earth is faster now: Indigenous observations of arctic environmental change. frontiers in polar social science.* Arctic Research Consortium of the United States.

Reid, G., Jones, C. E. & Poe, M. (June 7, 2022). Citational racism: How leading medical journals reproduce segregation in American medical knowledge. *Bill of Health: Examining the Intersection of Health, Law, Biotechnology, and Bioethics.* Harvard Law. https://tinyurl.com/4env3k4a.

Itchuaqiyaq, C. U., Jones, N. N. & Franchini, J. (2023, August 12). *Multiply marginalized and underrepresented scholars list.* https://www.itchuaqiyaq.com/mmu-scholar-list.

Itchuaqiyaq, C. U., Jones, N. N. & Franchini, J. (2023, August 12). Multiply marginalized and underrepresented scholars bibliography. https://tinyurl.com/3z8eh5ek.

The Visionlearning Project. https://www.visionlearning.com.

This World of Humans. https://www.visionlearning.com/en/twoh.

Urlings, M. J. E., Duyx, B., Swaen, G. M. H., Bouter, L. M. & Zeegers, M. P. (2021). Citation bias and other determinants of citation in biomedical research: Findings from six citation networks. *Journal of Clinical Epidemiology, 132,* 71–78.

How do I talk about inequality without "politicizing" my class? Do I need to explicitly talk about these things in my courses? I'm not an expert and feel out of my depth.

LaKeisha: Truthfully, until I began working with Heather on this collection, such a question never entered my mind. I did not give a second look to the workshops on campus for faculty to learn how to have "difficult" conversations in courses. The vast majority of STEM disciplinary courses in the US are still lecture-based, and those that are more collaborative in nature focus on the content rather than inequalities. But the very fact that most STEM courses are lecture-based masks inequalities and can make our students feel invisible. Giving students opportunities to explore and develop their science identities through non-technical science writing exposes this myth that science is objective (see Barlow and Quave, as well as Callow and Shelton, in this collection).

Since we started gathering and editing chapters in 2021, I have since redesigned one WID undergraduate research course that I oversee to include a lesson that highlights inequalities in STEM Writing. The lesson incorporates inclusive citation practices using the ACS Inclusivity Style Guide from the American Chemical Society (ACS) as a starting point. We currently use the Inclusivity Style Guide to discuss creating visuals of experimental results that are accessible and, while not addressed directly in the style guide, choosing citations that represent a broader range of ideas than what we typically see. For example, I ask students to search for articles published by authors from institutions outside the US that support their research findings or ones that are relevant to their research but published outside of typical chemistry research journals. Students do not have to include these articles as citations in the required research paper, but most will do so because they recognize that science needs to include more diverse voices, and they appreciate being able to contribute to their field in such a way. Giving students a choice lessens the chance that they might view a particular pedagogical practice as "politicizing" the course. Regardless, as instructors, we should be prepared to justify our practices as a way to be transparent with students.

For courses that are lecture-based, it may not be necessary to have conversations about inequality and inequities in STEM. Including examples or even creating problems on assessments that are culturally relevant can be a way to address inequities (see Callis' chapter, for example). When we used an atoms-first textbook in general chemistry, the chapter that included acid-base chemistry was just before the Thanksgiving holiday. I would share a story about being a little girl and watching my grandmother use a fork tine dipped in baking soda to neutralize some of the oxalic acid in her collard greens. Of course, neither of us knew at the time what "chemistry" was involved. All my grandmother knew was it made her greens less bitter and more tender, and I enjoyed watching the fizzing. My choice to tell this particular story was deliberate. It reinforced my belief that science is something that everyone does, whether or not they know it or understand it, and it was a way for me to invite all of my students to see themselves as capable of "doing science" if only by virtue of making a meal for themselves or their family.

Heather: This is a real tightrope that we have to walk—one that can blow up in our faces if we push too hard, too quickly. At least, that's been my experience. We *can* talk about how historical bias has impacted current inequities without casting blame. As Barlow and Quave noted, it "is not that science is fundamentally, irredeemably flawed, but rather that the sciences are brought to life by humans working within social and individual contexts" (this collection). So often, these things become political when individuals feel that they, personally, are being accused of wrongdoing (e.g., blaming white men for all the ills). This work isn't about specific individuals but the collective working within specific contexts. The Hidden Brain podcast on implicit bias (see Revealing Your Unconscious in the resources below) does a fantastic job of showing how implicit bias works and how inequity isn't about individual people doing bad things but the ways in which beliefs held by the majority of individuals in a space are institutionalized, and as a result cause harm. It's a law of averages; we are trying to shift where the average lies. That podcast is one place to start if you're struggling with understanding inequity on a structural level.

After much trial and error, my approach is to weave in discussions of the worldviews that have historically been held by those in positions of power and how those assumptions have impacted the policies that are made. Fink's chapter is an excellent example of how to do this, as is Callow and Shelton's focus on question-asking. Sometimes, the simplest thing to do is to include a few readings that complement or extend the discussions in the classroom. When I teach science writing, for example, I often include a module on the impact language has on the communication of scientific knowledge. We look at the role of linguistic markers (hedges, boosters, etc.) on knowledge reception through the lens of Andrew Wakefield et al.'s now-retracted *Lancet* article that anti-vaccine groups continue to use as evidence that the MMR vaccine causes autism. We also read Ann Morning's article alongside Cherice Escobar Jones and Genesis Barco Medina's (both listed below) to discuss the ways in which the use of language can lead to conflation of genetic information with race, even when such a connection doesn't exist. In classes focused on writing and engineering, I include articles on indigenous ethics (see Itchuaqiyaq below) and use the Tarot Cards of Tech (https://tarotcardsoftech.artefactgroup.com/) to ask students to challenge assumptions in their group projects. Weave these considerations in as an expected part of the curriculum instead of as an add-on. Normalize the inclusion of diverse viewpoints.

My advice to anyone interested in taking a social justice approach in their classes is simply to start where you are. Yes, we need to lean into our discomfort and challenge ourselves, but *you* need to be comfortable enough with the content in the classroom in order to be an effective guide for your students. Consider small edits to start. If your students are writing memos of project proposals, for example, ask them to incorporate a consideration of ethics (see Burry et al. in this collection). If they are doing user testing, ask them explicitly to consider the perspective of individuals with disabilities in the use of that tool or product. To me, this is about opening perspectives, not forcing students to take on my personal ideology.

Resources for Continuing the Journey

ACS inclusivity style guide. (2021). https://tinyurl.com/4kkeuvyr.
Artifact Group. *The tarot cards of tech*. https://tarotcardsoftech.artefactgroup.com/.
Brandt, C. B. (2008). Discursive geographies in science: Space, identity, and scientific discourse among indigenous women in higher education. *Cultural Studies of Science Education, 3*, 703–730. https://doi.org/10.1007/s11422-007-9075-8.
Brandt, A. M. (1978). Racism and research: The case of the Tuskegee Syphilis study. *The Hastings Center Report, 8*(6), 21–29.
Gawthorp, E. (2023. October 19). *COVID-19 deaths analyzed by race and ethnicity*. https://www.apmresearchlab.org/covid/deaths-by-race.
Cudd, A. E. (2001). Objectivity and ethno-feminist critiques of science. In K. M. Ashman (Ed.), *After the science wars: Science and the study of science*. Routledge. https://doi.org/10.4324/9780203977743.

Gebru, T. (2020, October). *Hot topics in computing: Who is harmed and who benefits?* MIT CSAIL. https://tinyurl.com/yupn24sf.

Hidden Brain Podcast. Revealing your unconscious, Parts 1 and 2. https://hiddenbrain.org/.

Itchuaqiyaq, C. U. (2021). Iñupiat iḷitqusiat: An indigenist ethics approach for working with marginalized knowledges in technical communication. In R. Walton & G. Agboka (Eds.), *Equipping technical communicators for social justice work: Theories, methodologies, and pedagogies*, pp. 33–48. Utah State University Press.

Jones, C. E. & Medina, G. B. (2021). Teaching racial literacy through language, health, and the body: Introducing bio-racial rhetorics in the writing classroom. *College English, 84*(1), 58–77. https://doi.org/10.58680/ce202131452.

Miller, D. I., Nolla, K. M., Eagly, A. H. and Uttal, D. H. (2018), The development of children's gender-science stereotypes: A meta-analysis of 5 decades of U.S. draw-a-scientist studies. *Child Development, 89*, 1943–1955. https://doi.org/10.1111/cdev.13039.

Morning, A. (2008). Reconstructing race in science and society: Biology textbooks, 1952–2002. *American Journal of Sociology, 114*(S1), S106–S137. https://doi.org/10.1086/592206.

I have too much disciplinary content to cover already. How do I make room for this?

LaKeisha: Researchers at the University of Virginia published an article in *Nature* showing that people overwhelmingly will add components to an object, idea, or situation rather than subtract components, even if the subtractive change leads to a better-desired outcome. Gabrielle S. Adams, Benjamin A. Converse, Andrew H. Hales, and Leidy E. Klotz (2021) wrote, "If people default to adequate additive transformations—without considering comparable (and sometimes superior) subtractive alternatives—they may be missing opportunities to make their lives more fulfilling, their institutions more effective, and their planet more livable." (p. 261). I consider Adams et al. and our collection an invitation to reimagine our courses and consider what is not really needed. Chances are there is something.

It is important to recognize what constraints may be placed on your course and the source(s) of the constraints. Professional societies are taking the lead on creating and disseminating discipline-specific tools and strategies to incorporate inclusive teaching and assessment practices as well as providing a space for conference symposia and proceedings. Even if every topic we teach must remain in the curriculum, we can teach those topics differently. Anjali Joshi (2023) recently published an Edutopia article online entitled "5 ways to make your science classroom more culturally responsive." I cannot emphasize enough using existing resources on your campus and through professional networks and organizations to tap into the expertise of folx whose research, teaching, and job roles center on inclusive teaching practices. This is a journey best traveled together!

Heather: Time is a major commodity. We have 6 or 9 or 16 weeks in a quarter, trimester, or semester, and within that time, we have to make sure we tick all of the boxes required by our program and accrediting bodies. It's a lot. Recently, when I was talking with a colleague about the inaccessibility of their presentation materials, they remarked: "I don't have time to go back and do all of that!" The two of us were just talking about slides, but even still, that felt like a huge ask. The thing about inclusion, though, is it isn't a *product*. We don't get to the end of a syllabus or presentation and say, "Fine! Done! Now I'm good." Inclusion is a process. There is always going to be something to alter, and with every move we make to create space for one group, we inevitably may be cutting others out. The goal, then, is to make changes in ways that don't mess the whole thing up. Thinking back to my colleague's slide materials: If they had chosen a slide design that had high text to background contrast, minimal color, and no unnecessary frilly designs *before* they put any content into the file, it never would have been an issue. So, our first thought with this work is: "How can I build it in from the start so that it doesn't feel like a retrofit?"

That said, sometimes, the way to bring inclusion into our class is not in the content. We can't do everything, everywhere, all the time. Sometimes, the best thing to do is to consider *how* we are teaching, not *what* we are teaching. Riedner, Francis, and Paretti in this collection offer us one way to think about this through the lens of engineering judgment. By creating spaces for students to be recognized by their peers and faculty for their capacity to participate *as* engineers, we can create a space for the growth of disciplinary identity and self-efficacy. Similarly, Newell-Caito and Fink also offer ways of approaching the classroom epistemologically to be more inclusive—whether that is in our assessment practices or our pedagogical approaches. Considerations of Universal Design and teaching in multiple modalities is another way of making the classroom more inclusive (see both Callis and Mallette, this volume), particularly for students with learning disabilities or cultural views that prioritize communal work. You might also review the ten rules offered by Suchinta Arif et al. (2021), referenced below, that offer considerations of support for historically marginalized students in science. The faculty development centers of most institutions of higher learning will also be able to assist (e.g., Columbia University has a very useful inclusive teaching guide that is available on their website).

Start with where you are and do what you can manage.

When you can do more, do more.

Resources for Continuing the Journey

Adams, G. S., Converse, B. A., Hales, A. H & Klotz, L. E. (2021). People systematically overlook subtractive changes. *Nature*, 592, 258–261. https://doi.org/10.1038/s41586-021-03380-y.

American Association for the Advancement of Science (2022). *STEMM professional societies' self-assessment for diversity, equity & inclusion: Guidance and criteria*. https://tinyurl.com/yw6zzppu.

Arif, S., Massey, M. D. B., Klinard, N., Charbonneau, J., Jabre, L., Martins, A. B., Gaitor, D., Kirton, R., Albury, C. & Nanglu, K. (2021). Ten simple rules for supporting historically underrepresented students in science. *PLoS computational biology, 17*(9), e1009313. https://doi.org/10.1371/journal.pcbi.1009313.

CAST. (2018). *Universal design for learning guidelines, version 2.2.* http://udlguidelines.cast.org.

Guide for inclusive teaching at Columbia. Columbia University. https://tinyurl.com/2ywctw7c.

Inclusive teaching at the University of Michigan: Resources for STEM Courses. (n.d.). Retrieved November 2, 2023, from https://tinyurl.com/ykpdshc4.

Sathy, V. & Hogan, K. A. (2022). *Inclusive teaching: Strategies for promoting equity in the college classroom.* West Virginia University Press.

If I modify how I teach and grade in my class, aren't I compromising the discipline? How will students learn what they need to be successful?

Heather: As a WAC specialist, these are the questions that I hear the most when talking with disciplinary faculty (even within English Literature). Usually, it's about writing instruction, but the same concern arises when it comes to questions of social justice. The thing I most *want* to say when these questions come up is: "Our disciplines are already compromised by bias! This is *part* of helping students be successful in the 21st century!" But that rarely gets a positive response. So, instead, I usually present it in a more accessible way, through questions. What does it mean to be successful in your field? Are students most successful when they can effectively memorize a list of facts to be selected on a multiple-choice exam? Or are they better off when they can think critically about their subject within different contexts and apply theories in interesting ways? What is it that you want students to be able to do when they leave your class, and how is your curriculum and assessment plan designed to privilege that learning?

When it comes to writing, many instructors wish to emphasize the grading of grammar and mechanics when they do include writing. Much as Blomstedt's work (this collection) shows, we are not compromising student success by leaving room for errors that can be caught in proofreading. If the work a student submits is so fraught with errors that it impedes meaning, that is an issue larger than what any one writing assignment or class is going to be able to address. That is when we include other supports and ask ourselves what role the writing is playing in the course (and how we are instructing the students within it). Including diverse viewpoints or a fuller picture of how we know what we know and how we communicate what we know isn't compromising; it's enhancing. We need students to see the messy parts, not just the cleaned-up, final versions. As instructors, we are always already making choices about what to include and what to skip in our classes–let's make it a priority to be more conscious about those decisions.

LaKeisha: To me, there is irony in thinking that traditional grading does not compromise STEM disciplines. Grades in STEM courses serve as gatekeepers (Gasiewski et al., 2012) to careers in science, engineering, technology, math, and even health fields like public health, medicine, and pharmacy. The rise of the modern grading scheme in the US is rooted in bias (Feldman, 2019). Students struggle to use grades as feedback to adjust study habits (Chamberlin et al., 2023, p. 116). Yet, there are no real alternatives to grades and one-assessment-for-all in large-enrollment courses. So, what are we to do?

Heather's questions to reframe the "student success" concern are very effective in brainstorming ways to create classroom materials that can foster a classroom culture where all students see an opportunity and a pathway toward *mutually defined success*. All of our authors provide a window into productive ways to approach these questions. Mallette (this collection) mentions specifications grading (Nilson, 2014) as a mode of ungrading, and I am eager to incorporate ungrading principles into lectures and writing courses where I have a bit more autonomy to tinker and lower enrollments to co-create with students.

Specifications grading has been used successfully in biology (Katzman et al., 2021), biochemistry (Donato & Marsh, 2023), chemistry (McKnelly et al., 2023; Noell et al., 2023), and physics and engineering physics (Evensen, 2022). Notably, McKnelly et al. (2023) show how this form of alternative grading was implemented in an organic chemistry laboratory course with 1,000 students across several sections. To attract creative minds from diverse backgrounds, we have to adapt how we teach STEM courses, especially introductory courses. In addition to ungrading practices, incorporating writing activities will not only help students make sense of their learning but will provide rich data to help instructors and departments design more inclusive courses that increase engagement and foster meaningful learning for those students who buy in to these alternative forms and methods of assessment.

Resources for Continuing the Journey

Adler-Kassner, L. & Wardle, E. (2022). *Writing expertise: A research-based approach to writing and learning across disciplines*. The WAC Clearinghouse; University Press of Colorado. https://doi.org/10.37514/PRA-B.2022.1701.

Condon, F. & Young, V. A. (Eds.). (2016). *Performing antiracist pedagogy in rhetoric, writing, and communication*. The WAC Clearinghouse; University Press of Colorado. https://doi.org/10.37514/ATD-B.2016.0933.

Chamberlin, K., Yasué, M. & Chiang, I. A. (2023). The impact of grades on student motivation. *Active Learning in Higher Education, 24*(2), 109–124. https://doi.org/10.1177/1469787418819728.

Donato, J. J. & Marsh, T. C. (2023). Specifications grading is an effective approach to teaching biochemistry. *Journal of Microbiology and Biology Education, 24*(2), e00236–22 https://doi.org/10.1128/jmbe.00236-22.

Evensen, H. (2022, August). Specifications grading in general physics and engineering physics courses. In *2022 ASEE Annual Conference & Exposition*. https://peer.asee.org/40676.pdf.

Feldman, J. (2018). *Grading for equity: What it is, why it matters, and how it can transform schools and classrooms*. Corwin Press.

Gasiewski, J. A., Eagan, M. K., Garcia, G. A., Hurtado, S. & Chang, M. J. (2012). From gatekeeping to engagement: A multicontextual, mixed method study of student academic engagement in introductory STEM courses. *Research in Higher Education, 53*, 229–261. https://doi.org/10.1007/s11162-011-9247-y.

Inoue, A. B. (2022). *Labor-based grading contracts: Building equity and inclusion in the compassionate writing classroom* (2nd ed.). The WAC Clearinghouse; University Press of Colorado. https://doi.org/10.37514/PER-B.2022.1824.

Katzman, S. D., Hurst-Kennedy, J., Barrera, A., Talley, J., Javazon, E., Diaz, M. & Anzovino, M. E. (2021). The effect of specifications grading on students' learning and attitudes in an undergraduate-level cell biology course. *Journal of Microbiology & Biology Education, 22*(3), e00200–21. https://doi.org/10.1128/jmbe.00200-21.

McKnelly, K. J., Howitz, W. J., Thane, T. A. & Link, R. D. (2023). Specifications grading at scale: Improved letter grades and grading-related interactions in a course with over 1,000 students. *Journal of Chemical Education, 100*, 3179–3193. https://doi.org/10.1021/acs.jchemed.2c00740.

Nilsen, L. B. (2014). *Specifications grading: Restoring rigor, motivating students, and saving faculty time*. Stylus Publishing.

Noell, S. L., Rios Buza, M. Roth, E. B., Young, J. L. & Drummond, M. J. (2023). A bridge to specifications grading in second semester general chemistry. *Journal of Chemical Education, 100*, 2159–2165. https://doi.org/10.1021/acs.jchemed.2c00731.

Poe, M., Inoue, A. B. & Elliot, N. (2018). *Writing assessment, social justice, and the advancement of opportunity*. The WAC Clearinghouse; University Press of Colorado. https://doi.org/10.37514/PER-B.2018.0155.

Sathy, V. & Hogan, K. A. (2022). *Inclusive teaching: Strategies for promoting equity in the college classroom*. West Virginia University Press.

I am a lecturer/pre-tenure and don't have the authority to change up the curriculum. What can I do to be more inclusive?

LaKeisha: I am in a non-tenure track position at my campus. Early in my career, I was concerned that my contract would not be renewed because course evaluations not tailored to my course were the primary form of my quality of teaching (which is absurd yet common). In my limited experience at my institution, faculty teaching smaller courses—and for STEM courses, these will usually be major courses—rarely have pushback from students when the curriculum changes to be more student-centered and inclusive. It is those of us who teach first-year courses with larger enrollments that often face the greatest resistance from students and parents and, therefore, administrators.

My strategy, learned from experience in chemistry education research, has been to have clear goals around any curricular changes *and* ways to measure the effects of the changes. The measures are, of course, for desired changes, but I would also include items and open-ended responses to capture unexpected and undesired changes to have a fuller picture. Not only would these data help me as the instructor, but I could report them to the students and to administrators. I have used both mid-semester evaluations and ones submitted at the end of the course (but before the university-wide ones were emailed to students). Because the questions align with my curricular goals, I included responses in my annual evaluations.

I enjoy crafting evaluations, so I do not mind spending the time doing them. I often reuse them with modifications to better capture students' experiences in my courses. But if creating an evaluation of your course is not what sparks joy, are there folx on your campus who are available to help faculty not only with instructional design but also ways to measure the impact of the curricular changes? On my campus, we even have an education developer who will come to our classes and facilitate a conversation with students (faculty are not present) around course goals. Depending on the scope of the change and the time you are able to devote, partnering or collaborating with education researchers or discipline-based education researchers who are interested in measuring impacts on students' experiences in STEM is a possibility.

Ultimately, I want to encourage you to be the change you want to see on your campus. Each campus culture is different, each STEM discipline culture is different, and each department culture is different. What works for others, even if it is published research or shared at a conference, may not work in your course or on your campus without modifications. And that's okay. When you go looking, you will find so many colleagues willing to help you create a more inclusive space through your courses.

Heather: This is a real issue and one to consider in light of your institution. The best scenarios are those where the institution has identified a commitment to DEIJ somewhere—in a strategic plan, in institutional priorities, in department curriculum changes, etc. If we can align what we are doing in the classroom with what the institution says it is prioritizing, then there is some leverage for change. But we live in a world where that is not always the case, and so my advice would be to tread cautiously. Use your course outcomes and goals as the lodestar. Sometimes, the simplest thing to do is to add on, rather than replace, content. As noted at the start of Section 2, maybe it's not about altering content but about pedagogy. Do you have room to modify the assessment of student work? Or to explicitly teach an element of disciplinary writing? Or to offer additional resources to assist students in succeeding? Where is the wiggle room?

I agree wholeheartedly with LaKeisha's note above about seeking out like-minded people and strategizing together. I have been very mindful, lately, of just how lonely this work can be. You need to know *why* you are doing what you are doing, how that benefits the students in your class and the discipline, and who is going to have your back. Finding that network of support is important, even if it comes from outside your institution because it reminds you that you're not alone and can lead to some interesting innovation. It also helps to have someone to talk to when you encounter the occasional student pushback.

Resources for Continuing the Journey

Kadmos, H. & Taylor, J. (2023). No time to read? How precarity is shaping learning and teaching in the humanities. *Arts and Humanities in Higher Education, 23*(1). https://doi.org/10.1177/14740222231190338.

Opdycke, K. (n.d.). *A precarious professorate works against an antiracist curriculum*. Boston University, Center for Interdisciplinary Teaching & Learning. Retrieved November 10, 2023, from https://tinyurl.com/2s3f3udn.

Schell, E. (2017). Foreword: The new faculty majority for writing programs: organizing for change. In S. Kahn, W. Lalicker, and A. Biniek-Lynch (Eds.), *Contingency, exploitation, solidarity: Labor and action in English composition* (pp. ix-xx), The WAC Clearinghouse; University of Colorado Press. https://doi.org/10.37514/PER-B.2017.0858.1.2.

I'm not a writing instructor. How do I assess writing fairly if I don't know how to teach and assess writing in general?

Heather: After questions of compromising the discipline, this is the second-most frequent concern that I hear when working with faculty outside of writing studies. If we agree (as we usually do) that being able to write well is a key element of success in your disciplinary career, then we have to ask: Who is actually responsible for teaching students how to write well as members of their discipline? Sure, many institutions have writing-specific courses for STEM majors taught by specialists like me. We see examples of these in Burry et al. and Mallette (both in this collection). But not all institutions have the kind of funding or the commitment to writing that offering such courses requires. Additionally, just like a first-year composition course cannot prepare students for all the writing they will do in higher education, one writing in the disciplines course is not going to make them expert writers in their major. They need multiple points of contact from multiple experts throughout the curricular experience. I may be able to teach engineering students about rhetorical situations and language expectations, some genre considerations, and give them practice at composing a variety of document types, but that doesn't mean that they won't *also* need feedback from their engineering professor when writing a project update or memo in another class. There will be subtle differences across subdisciplines and spaces that students need guidance on, and they need concepts and rhetorical moves reinforced from multiple directions.

Sometimes, I am challenged with: Who has the authority to teach disciplinary writing? Really, for me, we all do—faculty who teach disciplinary courses, as well as those who teach courses in the major. We are all experts in different ways. I am an expert because I study the rhetorical moves, language, and genres of different disciplines; STEM practitioners are experts because they actively employ these things in their everyday work. We come from different perspectives, but all have something to teach our students. *How* we do that, though, will be different. Consider using resources like those listed below and spend time thinking about what you want students to show you in the writing you assign. Don't assess grammar and mechanics if you aren't teaching these things (though you should definitely point out issues if such errors get in the way of meaning). If you teach the structure of a lab report, then it's appropriate to assess that structure in student work. If you use a haiku to assess students' understanding of structural relationships in biology, then assess how the knowledge is conveyed, not how good the haiku is. You don't have to be an expert on writing to know if a student is meeting the goals of your writing assignment, unless your goals are something you aren't teaching.

LaKeisha: I am not a writing instructor. I have embarked on eleven years of intense on-the-job training. I am fortunate that we have amazing faculty in our University Writing Program. Even if my students never write a science manuscript for a peer-reviewed journal or a research paper as part of their undergraduate research experience, there are so many transferrable skills to learn from disciplinary writing. Perhaps the skill that I lean into the most with my students is that writing affords them a level of creativity they will not experience in a lecture or traditional laboratory course.

As Heather mentioned, it is not appropriate to assess grammar unless it is formally re/taught in the course. Since I am not an expert, I do not even attempt this as part of the formal curriculum. We use a science writing textbook (LeBrun, 2011) as a required textbook that includes chapters on grammar and writing mechanics, and each week during the experimental phase of the course, I only teach them about key points of a title, abstract, introduction, materials and methods, results and visuals, and discussion and conclusion. So, those aspects of students' science manuscripts comprise the bulk (25/40) of points on the rubric.

Because my course does not formally re/teach students writing mechanics, other aspects of a science manuscript are re/taught informally through their consultations with me during the writing phase that takes place in the second half of the semester. How to best incorporate evidence from other studies, how to properly format references, and how to know when to include a citation are some topics that I touch upon during instructional time but really drill into during consultations. So, through these consultations—which are as labor intensive as they sound but oh so rewarding for students and me—I am then able to provide personalized instruction for the remaining points of the rubric.

Avoiding potential bias in assessing writing quality is desired and critical to inclusive teaching. So, providing feedback using an established rubric or set of criteria will help students see a path to being successful in the course. Even as we work to decolonize many disciplinary practices that cause harm and erase ways of knowing, having a framework for using these practices already embedded in the course curriculum and made transparent to students is essential.

Resources for Continuing the Journey

Berdanier, C., McCall, M. & Fillenwarth, G. (2021). Characterizing disciplinarity and conventions in engineering resume profiles. *IEEE Transactions on Professional Communication, 64*(4), 390–406. https://doi.org/10.1109/TPC.2021.3110397.

Carter, M. (2016). Value arguments in science research articles: Making the case for the importance of research. *Written Communication, 33*(3), 302–327. https://doi.org/10.1177/0741088316653394.

Fillenwarth, G. M., McCall, M. & Berdanier, C. (2018). Quantification of engineering disciplinary discourse in résumés: A novel genre analysis with teaching implications. *IEEE Transactions on Professional Communication, 61*(1), 48–64. https://doi.org/10.1109/TPC.2017.2747338.

Gopen, G. D. & Swan, J. A. (1990, November-December) The science of scientific writing. *American Scientist, 78*(6), 550–558. https://www.jstor.org/stable/29774235.

Hyland, K. (2011). Disciplines and discourses: Social interaction in the construction of knowledge. In D. Starke-Meyerring, A. Paré, N. Artemeva, M Horne & L. Yousoubova (Eds.), (2011). *Writing in knowledge societies,* pp. 193–214. WAC Clearinghouse; Parlor Press. https://doi.org/10.37514/PER-B.2011.2379.

Lebrun, J. L. (2011). *Scientific writing 2.0.* World Scientific Publishing Company.

Moore, R. (2000). Writing about biology: How rhetorical choices can influence the impact of a scientific paper. *Bioscene, 26*(1), 23–25.

Puruganan, M. & Hewitt, J. (2004). *How to read a scientific article.* Cain Project for Engineering and Professional Communication. https://tinyurl.com/4afvdrxm.

How much of an effect will this really have on students? Do these things stick, or do they fade away?

LaKeisha: Absolutely, yes! We wanted to include student vignettes in our collection to capture some of the lingering impacts that inclusively designed courses had on students. Course evaluations given by departments or institutions often are not meant to capture these sentiments, and the end of the semester may be too soon to see what remains after time has passed. Riya and Madison each describe how pivotal enrollment in such courses influenced choices to persist in a STEM major and to incorporate writing as part of their educational journeys and career paths.

In my own WID laboratory course, students are able to see their transformation as science writers within the course itself. Peer feedback is required for all WID courses at GW, and I facilitate sessions during two lab periods. During the first session, I overhear the students commenting on how they see through peers' papers where they can improve their own writing. Then, during the second session, following an opportunity to revise and resubmit, students compliment their peers on improvements within their papers. I only provide general feedback to the class, so students really take ownership of this process. By the time I meet with them after these feedback sessions, I can then focus on the chemistry, syntax, and minor formatting. And even then, students will spontaneously comment on how they noticed their writing improved from their first consultation to their second. So, when I have chemistry majors from my laboratory WID course in the WID course linked to their undergraduate research experiences one to two years later, they have a solid foundation from which to write a 15–20 page research paper on their own semesters-long research projects. I should also note that these students will have taken one or two writing-intensive laboratory courses in the intervening years. Though the courses are not WID courses (there is no writing instruction, no opportunities for revisions, and no peer feedback), they do serve as opportunities for students to continue to practice their science writing.

Though limited, empirical research also supports these narratives from students featured in our collection and my own WID courses. Gere, Knutson, and McCarty (2018) use three case studies of three STEM students' progressions as writers to describe the ways in which they incorporated aspects of their varied writing courses to create their own concepts of disciplinary writing. More research is needed to support what and how much sticks from disciplinary writing, particularly the influence that inclusive practices have on retention of disciplinary content, disciplinary writing skills, and students themselves in STEM. A great reason to collaborate with others who are committed to inclusive writing practices in STEM courses!

Heather: Examining the ways in which disciplinary spaces have been constructed to welcome some people while keeping others out can be a daunting task, but it's one that is critical if we wish to make educational and workforce spaces more inclusive. So much of what we are doing is about sowing seeds and creating new perspectives to view the world. We don't know how that will show up in the long term, really. It's anecdotal. But that doesn't mean we shouldn't try. Johnson et al. (this collection) show how this work will impact the way new teachers enter the classroom, which is likely to stick and have an impact. Each positive interaction has the potential to counteract the negative ones; our goal is to have more positive than negative in the end.

Relatedly, it did not go unnoticed when we were working on this collection that the vast majority of authors are coming from communities historically marginalized in STEM. In fact, there were at least four additional chapters that were originally planned for inclusion that chose to bow out explicitly *because* that extra labor was cutting into their time and ability to work on scholarship, and those authors did not want to hold up production of the overall book. Authors who identify as female, as disabled, as Black or Latinx, and first-generation college students, who are in contingent roles at their institutions or in non-tenure track positions. The reality, as we pointed out earlier in this chapter, is that this work—this caring, emotionally-laden labor—is typically carried by the individuals with the least amount of power. Maybe it is because of our life experiences on the margins that make us more likely to help the generations that are coming up, but it also takes away our time and attention for the practices that academia privileges (research and publication). It's a double-edged sword. So, yes, this stuff *does* stick, and it is why, more than ever, those who are in positions of power need to step up and carry their share of the load so that we can break these cycles and move toward a more equitable society.

Resources for Continuing the Journey

Bowen, C. L., Johnson, A. W. & Powell, K. G. (2020, October). *Critical analyses of outcomes of marginalized undergraduate engineering students*. 2020 IEEE Frontiers in Education Conference. https://doi.org/10.1109/FIE44824.2020.9273827.

Gere, A. R., Knutson, A. V. & McCarty, R. (2018) Rewriting disciplines: STEM students' longitudinal approaches to writing in (and across) the disciplines. *Across the Disciplines, 15*(3), 63–75. https://doi.org/10.37514/ATD-J.2018.15.3.12.

Gnagey, J. & Lavertu, S. (2016). The impact of inclusive STEM high schools on student achievement. *AERA Open, 2*(2), 233285841665087. https://doi.org/10.1177/2332858416650870.

Katz, J. (2016). Effects of the three-block model of UDL on inclusive teaching. *Journal of Research in Special Educational Needs, 16*(S1), 898–899. https://doi.org/10.1111/1471-3802.4_12347.

Means, B., Wang, H., Wei, X., Young, V. & Iwatani, E. (2021). Impacts of attending an inclusive STEM high school: Meta-analytic estimates from five studies. *International Journal of STEM Education, 8*(1), 1–19. https://doi.org/10.1186/s40594-020-00260-1.

Moten, Q. (2020). *The effects of inclusive teaching practices on the retention of Black Asian minority ethnic students: An organizational case study* (Order No. 28968813) [Doctoral Dissertation, Concordia University] ProQuest Dissertations and Theses Global.

Contributors

Dhatri Badri is a M.Sc. student at Boston University, where she studies Bioinformatics. Her research interests lie in molecular ecology and characterizing microbial communities in soil.

Jameta Nicole Barlow is Assistant Professor of Writing, Health Policy & Management and Women's, Gender and Sexuality Studies at the George Washington University. She teaches courses on writing science and women's health for under/graduate students. Dr. Barlow has 25 years of experience in transdisciplinary federal government, nonprofit, and academic collaborations in diverse settings throughout the world. Her edited collection, *Writing Blackgirls' and women's health science*, was published in 2023, and her research on decolonizing methodologies and Black girls' and women's health appears in *Women's Health Issues, Meridians: feminism, race and transnationalism* and the *American Journal of Health Promotion*. A public scholar, Dr. Barlow's women's health writings have been quoted in *The New York Times, The Washington Post, Essence, Shape, NPR,* and *Healthline*.

Alicia Bitler is a chemistry teacher at Magruder High School in Montgomery County, Maryland. Throughout her career, she held various roles as a STEM teacher educator, professor, and trainer.

Elizabeth Blomstedt is Assistant Professor in the Writing Program at the University of Southern California, where she teaches first-year writing courses themed on sustainability and upper-division natural science writing courses. In her previous role as Assistant Director of the Warren Writing Program at UC San Diego, she developed and taught the first upper-division writing course on that campus, Technical Writing for Scientists and Engineers, and worked with graduate student writing instructors from a variety of disciplines. Her research interests include writing assessment, interdisciplinary writing, and teaching multilingual writers.

Madison Brown, member of the Muscogee (Creek) Nation, was born and raised in Tulsa, OK. She received her B.S. in Physics and Mathematics from Baker University in 2018, where she minored in creative writing. She earned her M.A. in English from the University of Maine, where she completed a creative thesis inspired by her maternal grandfather. Her research and teaching interests include creative writing and contemporary Indigenous and North American literature with an emphasis in magical realism, postmodernism, and digital humanities. Madison most enjoys interdisciplinary studies where her STEM education informs her writing.

Justiss Wilder Burry is Assistant Professor of Professional Writing at Tarleton State University, where he teaches rhetoric and composition courses with a focus on professional and technical writing. His research interests include technical and professional communication pedagogy, programmatic evaluation and improvement, and the rhetoric of health and medicine, particularly community-oriented methodological approaches and writing strategies.

Laura Kyser Callis is Associate Professor in the Department of Natural Sciences & Mathematics at Curry College in Massachusetts. She is the co-principal investigator of the National Science Foundation funded project DISCUS-IS (Discourse to Improve Students' Conceptual Understanding of Statistics in Inclusive Settings), which investigates the statistical conceptions of students with learning and attention differences and the instructional practices that support these students. She teaches statistics, mathematics for teachers, modern algebra, and history of mathematical inquiry.

Megan Callow is Associate Teaching Professor in the English Department's Program for Writing Across Campus at the University of Washington, Seattle. She also serves as campus-wide Director of Writing. Her scholarly interests include STEM writing and writing pedagogy, faculty development, and WID/WAC Studies.

Adrian Clifton has a background in science, having loved the subject from a young age and gotten a B.S. in Geology from University of Nevada Reno. He went on to get an M.A. in Philosophy from Biola University. He recently completed licensure in secondary science education from MSU Denver. He is very passionate about helping kids with disabilities, especially being on the autism spectrum himself. His hope is that all kids have the opportunity to succeed despite any challenges they face.

Mary Coleman is a secondary science teacher at Arvada High School and graduate student at University of Colorado, Denver. Coleman is continuing her mastery of science pedagogy and hopes to support a diverse array of students and show them that science is for everyone.

Kimberlee D'Aquila (Bourelle) is a high school science teacher in Highlands Ranch, Colorado. Her background in working within a severe needs high school classroom has helped to focus on different aspects of teaching for her own school and classroom and originally sparked her passion for inclusion. She currently teaches Zoology, Biology, and Applied Biology with a focus on making science more relevant and applicable to students.

Madeline Dougherty is originally from rural Oregon and graduated from the United States Naval Academy in 2007 with a Bachelor of Science in Honors Oceanography. Upon graduation, she commissioned into the Marine Corps and attended naval flight school in Pensacola, Florida. After serving for ten years as a helicopter pilot (including a tour as a flight instructor), she left active duty. Since then, she has held a number of positions including airline pilot, aviation curriculum writer, and helicopter air ambulance pilot. She is currently a graduate student in the Master of Science in Teaching program through the Research in STEM Education (RiSE Center) at the University of Maine, Orono. She continues to serve as a flight instructor in the Navy Reserves.

Parker Edingfield graduated with a Bachelor of Science in Mathematics Education from MSU Denver in the fall of 2022. He currently works as a math teacher at John F. Kennedy High School in Denver. Parker believes in creating student-centered classrooms that cultivate growth mindsets. In his free time, he enjoys rock climbing, playing basketball and spending time with his family.

Heather M. Falconer is Assistant Professor of Professional and Technical Writing and faculty member of the Maine Center for Research in STEM Education at the University of Maine, Orono. Falconer's research has appeared in journals such as *Written Communication*, *The WAC Journal*, and the *Journal of Hispanic Higher Education*, as well as multiple edited collections. Her book, *Masking inequality with good intentions,* is available through the Practices & Possibilities series/The WAC Clearinghouse.

Ann E. Fink (she/they) holds a Ph.D, an MSW, and an LSW. She is a neuroscientist, educator, ethicist, artist, social worker, and therapist. They received a doctorate in neuroscience from UCLA; their publications on the neurobiology of memory and emotion have appeared in the *Journal of Neuroscience, Journal of Neurophysiology, PNAS, AJOB Neuroscience,* and other journals. Ann's interdisciplinary work addresses the ethics of neuroscience in relation to identity, mental health, and social justice. She also uses comics in scholarly writing and teaching as part of the Graphic Medicine community. Among other appointments, Ann was previously Wittig Fellow in Feminist Biology at UW-Madison and Professor of Practice in Biological Sciences at Lehigh University. Ann currently adjuncts with the Lehigh University College of Education in addition to her position as Behavioral Health Consultant at Rutgers University's Counseling, Alcohol and other Drug Assistance, and Psychiatric Services (CAPS) within Rutgers Student Health Services.

Royce A. Francis is Associate Professor in the Department of Engineering Management and Systems Engineering at George Washington University. His overall research vision is to conduct research, teaching, and service that facilitates sustainable habitation of the built environment. This vision involves three thrusts: 1) infrastructure management, including resilience and risk analysis; 2) regulatory risk assessment and policy-focused research, especially for environmental contaminants and infrastructure systems; and, 3) engineering education research exploring the linkages between professional identity formation and engineering judgment. Dr. Francis received the Ph.D. from Engineering and Public Policy and Civil and Environmental Engineering at Carnegie Mellon University, M.S. in Civil and Environmental Engineering from Carnegie Mellon University, and the B.S. in Civil Engineering from Howard University.

Jessica Griffith is Assistant Professor of English (Professional Writing) at Jacksonville State University. Her interests include field wide research related to curriculum and program evaluation within technical and professional communication.

Carolyn Gubala teaches upper-division writing and serves as the assistant director of online writing instruction in the University Writing Program at the University of California, Davis. Her teaching and research interests include programmatic and pedagogical studies in technical professional communication.

Janelle M. Johnson is Professor of STEM Education in secondary teacher education at Metropolitan State University of Denver, a Hispanic-Serving Institution. She taught K–12 math and science with English Learners and now teaches multicultural education and science methods. Her research focuses on interdisciplinary STEM equity and inclusive approaches to teaching and learning. Most of her publications and presentations are co-authored with preservice and inservice teachers and she is the co-editor of "STEM21: Equity in Teaching and Learning to Meet Global Challenges of Standards, Engagement, and Transformation" (2018). Dr. Johnson is on the board of the Colorado Association of Science Teachers, the Director of the Colorado STEM Ecosystem, and served on the Council of State Science Supervisors ACESSE Network in the disrupting ableism affinity group.

Jennifer C. Mallette is Associate Professor in the Department of Writing Studies at Boise State University, where she collaborates with engineering faculty to support student writers. Her research builds on those collaborations, examining best practices for integrating writing into engineering curriculum; she also explores women's experiences in engineering settings through the context of writing. Her research has appeared in a number of journals and edited collections, most recently *The Journal of Writing Assessment, The Routledge Handbook for Scientific Communication*, and *Writing Beyond the University: Preparing Lifelong Learners for Lifewide Writing*.

LaKeisha McClary is Assistant Professor of Chemistry at The George Washington University where she teaches Introductory Quantitative Analysis Laboratory, a writing-in-the-disciplines course for chemistry majors and minors that fulfills a graduation requirement. Dr. McClary received the 2020 WID Award for Best Assignment Design for the design-it-yourself project that is the hallmark of the course. In addition to teaching the laboratory, she also teaches general chemistry, organic chemistry, and a writing course for chemistry majors engaged in semesters-long undergraduate research. Dr. McClary served as the Co-Chair of the Diversity, Equity, and Inclusion Committee of the Association for Writing Across the Curriculum and is a member of the American Chemical Society. Her research exploring chemistry students' thinking and reasoning has appeared in *International Journal of Science Education, Journal of Research in Science Teaching*, and *Journal of Chemical Education*.

Lisa Melonçon is Professor of Technical Communication at Clemson university. Her teaching and research focus on programmatic dimensions of technical and professional communication curricula, research methodologies, and the rhetoric of health and medicine.

Amanda Myers is a student at Metropolitan State University of Denver, pursuing a degree in English Secondary Education. She is passionate about promoting diversity, equity, and inclusion in education. She believes that Multicultural Education, Culturally Relevant Pedagogy, Funds of Knowledge, and cross-curricular work are essential aspects of teaching and research. Throughout her academic journey, Amanda has been actively pursuing various educational pursuits, including playing an active role in the Noyce STEM Scholarship program and serving as a Research Assistant to Dr. Janelle Johnson. She has also worked as a writing tutor in the STEM Learning Center on campus, demonstrating a commitment to helping fellow students succeed. Amanda is also a member of the Noyce Undergraduate Research group, where she collaborates with her peers and mentors to enhance students' learning experiences through research on broadening family engagement in STEM.

Jennifer Newell-Caito is Senior Lecturer of Biochemistry at the University of Maine, Orono in the Department of Molecular and Biomedical Sciences. Dr. Newell earned her Ph.D. in Biochemistry from the University of Rochester and completed her post-doctoral fellowship at Vanderbilt University. She currently teaches a one-year General, Organic, and Biochemistry course for non-majors and an Analytical Biochemistry Laboratory course for majors that is a requirement for graduation. She has also several ongoing pedagogical research projects focused on relationship-centered learning, Universal Design, student metacognition, and ungrading in first year and upper-level collegiate courses. Dr. Newell has co-created unique interactive digital modules for learning general and organic chemistry. In addition, her research laboratory focusing on the antioxidant effect of plant extracts in *Caenorhabditis elegans* (worms).

Madeline Onstott teaches middle school science in Littleton, Colorado. Her background in science education and passion for inclusion work well together when it comes to teaching students of different abilities and interests.

Ebtissam Oraby is Teaching Assistant Professor at George Washington University, where she teaches Arabic language and literature. She is a Teacher-Consultant for the Shenandoah Valley Writing Project. She is a scholar of curriculum and pedagogy specializing in multilingual education, and Arab and Muslim cultures. Her research interests lie at the intersection of philosophy, language, religion, and education. Her research investigates notions of alterity and affect in the multilingual classroom. Her most current research examines how elementary school students engage with a science curriculum rooted in Muslim ways of knowing.

Marie C. Paretti is Professor of Engineering Education at Virginia Tech, where she is Director of the Virginia Tech Engineering Communications Center (VTECC), Associate Director of the Virginia Tech Center for Coastal Studies, and Education Director of the interdisciplinary Disaster Resilience and Risk Management graduate program. She received a B.S. in chemical engineering and an M.A. in English from Virginia Tech, and a Ph.D. in English from the University of Wisconsin-Madison.

Her research focuses on communication and collaboration, design education, and identity (including race, gender, class, and other demographic identities) in engineering. She was awarded a CAREER grant from the National Science Foundation to study expert teaching in capstone design courses and is PI or co-PI on numerous NSF grants exploring communication, teamwork, design, identity, and inclusion in engineering. Drawing on theories of situated learning and identity development, her research explores the ways in which engineering education supports students' professional development in a range of contexts across multiple dimensions of identity.

Kylie E. Quave is Assistant Professor of Writing and of Anthropology at the George Washington University. She teaches first-year writing courses focused on the social production of knowledge in the sciences. Her research on equity and justice in teaching has recently appeared in *Advances in Archaeological Practice*, while her research on local responses to Inka and Spanish imperialism in the South American Andes has recently been published in the *Journal of Anthropological Archaeology* and *Journal of Archaeological Science: Reports*.

Rachel Riedner is Professor of Writing and of Women's, Gender, and Sexuality Studies at The George Washington University where she serves as Associate Dean of Undergraduate Studies. Her focus in her dean's role is on integrating a diversity perspective into all areas of liberal arts education. Dr. Riedner is the author of two books, multiple articles, and was the recipient of a Fulbright Specialist Grant to develop writing curriculum at the University of Tromsø, Norway. At GW, she has collaborated with STEM colleagues to integrate writing into course design and curriculum. This interest in writing in STEM has led to an NSF grant with Dr. Royce Francis and Dr. Marie Paretti that explores identity formation through writing in engineering education.

Joseph Schneiderwind is a full-time math teacher at Daniel C. Oakes High School in Castle Rock, Colorado. He currently teaches geometry, college algebra, and probability and statistics where he places particular emphasis on cross-curricular cooperation and interest- and application-based projects.

Sally B. Seraphin is Assistant Professor of Neuroscience and directs the Laboratory of Evolutionary Neuroscience at Trinity College, in Hartford, CT. She teaches courses on brain and behavior, human motivation and emotion, neuroscience methods, principles of neuroscience, social neuroscience, cultural neuroscience, and neurolaw. Her writing on STEM pedagogy has been featured in *Psychology Learning & Teaching* and the *Journal of Undergraduate Neuroscience Education (JUNE)* where she is also an Associate Editor. To support public dialog between academics, artists, and artisans, she founded and serves as Chief Editor for *The Thinking Republic* web magazine.

Riya Sharma graduated with a B.S. from The George Washington University, where she studied data science and political science (public policy focus) with a minor in journalism.

Holly Shelton is Assistant Professor of Composition, Rhetoric, and Linguistics at George Fox University in Newberg, Oregon. She has taught various configurations of STEM writing and often engages issues of linguistic diversity and genre pedagogy in her scholarship.

Katie Weaver completed her secondary education licensure in science as a post baccalaureate student after completing her B.A. in Chemistry from MSU Denver in 2018. She was a Noyce Scholar and is passionate about equity, diversity, and inclusion in education. In her free time Katie loves to swim, read, fish, do yoga, go hiking and hang out with friends.

Tanya Zarlengo is Associate Professor of Instruction serving as the Associate Director of Professional and Technical Communication at the University of South Florida. Her teaching and researching focus on the programmatic aspects of professional and technical communication, with an emphasis on the service course.

www.ingramcontent.com/pod-product-compliance
Lightning Source LLC
Chambersburg PA
CBHW060551080526
44585CB00013B/518